CALIFORNIA HSP Mat

3 in 1
Practice Book

Practice
Reteach
Spiral Review

Grade 3

Visit *The Learning Site!*
www.harcourtschool.com

Printed in the United States of America

ISBN 13: 978-0-15-383384-7
ISBN 10: 0-15-383384-X

3 4 5 6 7 8 9 10 0956 17 16 15 14 13 12 11 10 09

Contents

Unit 1: PLACE VALUE, ADDITION AND SUBTRACTION

Chapter 1: Understand Place Value

		RW	PW
1.1	Algebra: Patterns on a Hundred Chart	RW1	PW1
	Spiral Review Week 1	SR1	
1.2	Place Value to 1,000	RW2	PW2
1.3	Place Value to 10,000	RW3	PW3
1.4	Expanded Form	RW4	PW4
1.5	PS Workshop Strategy: Use Logical Reasoning	RW5	PW5
	Spiral Review Week 2	SR2	

Chapter 2: Compare, Order, and Round Numbers

		RW	PW
2.1	Compare Numbers	RW6	PW6
2.2	Order Numbers	RW7	PW7
2.3	PS Workshop Skill: Use a Number Model	RW8	PW8
2.4	Round to the Nearest Ten	RW9	PW9
	Spiral Review Week 3	SR3	
2.5	Round to the Nearest Hundred	RW10	PW10
2.6	Round to the Nearest Thousand	RW11	PW11

Chapter 3: Addition

		RW	PW
3.1	Algebra: Addition Properties	RW12	PW12
3.2	Algebra: Missing Addends	RW13	PW13
	Spiral Review Week 4	SR4	
3.3	Estimate Sums	RW14	PW14
3.4	Add with Regrouping	RW15	PW15
3.5	Model 3-Digit Addition	RW16	PW16
3.6	Add 3- and 4-Digit Numbers	RW17	PW17
	Spiral Review Week 5	SR5	
3.7	PS Workshop Strategy: Predict and Test	RW18	PW18

Chapter 4: Subtraction

		RW	PW
4.1	Estimate Differences	RW19	PW19
4.2	Subtract with Regrouping	RW20	PW20
4.3	Model 3-Digit Subtraction	RW21	PW21
	Spiral Review Week 6	SR6	
4.4	Subtract 3- and 4-Digit Numbers	RW22	PW22
4.5	Subtract Across Zeros	RW23	PW23
4.6	PS Workshop Skill: Choose the Operation	RW24	PW24
4.7	Algebra: Number Patterns	RW25	PW25
4.8	PS Workshop Skill: Estimate or Exact Answer	RW26	PW26
	Spiral Review Week 7	SR7	

Unit 2: MULTIPLICATION CONCEPTS AND FACTS

Chapter 5: Understand Multiplication

		RW	PW
5.1	Algebra: Relate Addition to Multiplication	RW27	PW27
5.2	Algebra: Model with Arrays	RW28	PW28
5.3	Multiply with 2	RW29	PW29
5.4	Multiply with 4	RW30	PW30
	Spiral Review Week 8	SR8	
5.5	Algebra: Multiply with 1 and 0	RW31	PW31
5.6	PS Workshop Strategy: Draw a Picture	RW32	PW32

Chapter 6: Multiplication Facts

		RW	PW
6.1	Multiply with 5 and 10	RW33	PW33
6.2	Multiply with 3	RW34	PW34
	Spiral Review Week 9	SR9	
6.3	Multiply with 6	RW35	PW35
6.4	Algebra: Practice the Facts	RW36	PW36
6.5	PS Workshop Strategy: Act it Out	RW37	PW37

Chapter 7: Facts and Strategies

		RW	PW
7.1	Multiply with 8	RW38	PW38
	Spiral Review Week 10	SR10	
7.2	Algebra: Patterns with 9	RW39	PW39
7.3	Multiply with 7	RW40	PW40

Key: PW Practice Workbook   RW Reteach Workbook   SR Spiral Review

Contents

7.4 Algebra: Practice the Facts....................... RW41 PW41
7.5 PS Workshop Strategy: Make A Table................ RW42 PW42

Spiral Review Week 11.......... SR11

Chapter 8: Facts and Properties

Spiral Review Week 11.......... SR11

8.1 Find a Rule.................. RW43 PW43
8.2 Missing Factors............ RW44 PW44
8.3 Multiply 3 Factors......... RW45 PW45
8.4 Multiplication Properties..................... RW46 PW46

Spiral Review Week 12.......... SR12

8.5 PS Workshop Skill: Mutlistep Problems........ RW47 PW47

Unit 3: DIVISION CONCEPTS AND FACTS

Chapter 9: Understand Division

9.1 Model Division.............. RW48 PW48
9.2 Relate Division and Subtraction............ RW49 PW49
9.3 Model with Arrays.......... RW50 PW50

Spiral Review Week 13.......... SR13

9.4 Algebra: Multiplication and Division.................. RW51 PW51
9.5 Algebra: Fact Families .. RW52 PW52
9.6 PS Workshop Strategy: Write a Number Sentence RW53 PW53

Chapter 10: Division Facts

10.1 Divide by 2 and 5 .... RW54 PW54

Spiral Review Week 14.......... SR14

10.2 Divide by 3 and 4 .... RW55 PW55
10.3 Division Rules for 1 and 0 .... RW56 PW56
10.4 Algebra: Practice the Facts....................... RW57 PW57
10.5 PS Workshop Skill: Choose the Operation ... RW58 PW58

Spiral Review Week 15.......... SR15

Chapter 11: Facts through 10

11.1 Divide by 6 .... RW59 PW59
11.2 Divide by 7 and 8 .... RW60 PW60
11.3 PS Workshop Strategy: Work Backward............ RW61 PW61
11.4 Divide by 9 and 10 .... RW62 PW62

Spiral Review Week 16.......... SR16

11.5 Algebra: Practice the Facts....................... RW63 PW63
11.6 Algebra: Find the Cost........................ RW64 PW64
11.7 Algebra: Expressions and Equations RW65 PW65

Unit 4: GEOMETRY AND MEASUREMENT

Chapter 12: Plane Figures

12.1 Line Segments and Angles RW66 PW66

Spiral Review Week 17.......... SR17

12.2 Types of Lines.............. RW67 PW67
12.3 Identify Plane Figures RW68 PW68
12.4 Triangles...................... RW69 PW69
12.5 Quadrilaterals.............. RW70 PW70

Spiral Review Week 18.......... SR18

12.6 Compare Plane Figures RW71 PW71
12.7 PS Workshop Strategy: Find a Pattern............... RW72 PW72

Chapter 13: Solid Figures

13.1 Identify Solid Figures..... RW73 PW73
13.2 Faces, Edges, and Vertices RW74 PW74

Spiral Review Week 19.......... SR19

13.3 Model Solid Figures RW75 PW75
13.4 Combine Solid Figures RW76 PW76
13.5 PS Workshop Skill: Identify Relationships RW77 PW77

Chapter 14: Perimeter, Area, and Volume

14.1 Perimeter....................... RW78 PW78

Spiral Review Week 20.......... SR20

Key: PW Practice Workbook RW Reteach Workbook SR Spiral Review

Contents

14.2 Estimate and Measure Perimeter....... RW79 PW79
14.3 Area of Plane Figures ... RW80 PW80
14.4 Area of Solid Figures..... RW81 PW81
14.5 Estimate and Find Volume RW82 PW82

Spiral Review Week 21........... SR21

14.6 PS Workshop Skill: Use a Model RW83 PW83

Unit 5: MULTIPLY AND DIVIDE BY 1-DIGIT

Chapter 15: Multiply by 1-Digit

15.1 Algebra: Multiples of 10 and 100 .... RW84 PW84
15.2 Arrays with Tens and Ones....................... RW85 PW85
15.3 Model 2-Digit Multiplication................. RW86 PW86

Spiral Review Week 22........... SR22

15.4 Estimate Products RW87 PW87
15.5 Multiply 2-Digit Numbers....................... RW88 PW88
15.6 PS Workshop Strategy: Solve A Simpler Problem............ RW89 PW89

Chapter 16: Multiply Greater Numbers

16.1 Algebra: Multiples of 10, 100, and 1,000 .... RW90 PW90

Spiral Review Week 23........... SR23

16.2 PS Workshop Strategy: Find a Pattern............... RW91 PW91
16.3 Multiply 3-Digit Numbers....................... RW92 PW92
16.4 Multiply 4-Digit Numbers....................... RW93 PW93
16.5 Multiply Money Amounts RW94 PW94

Spiral Review Week 24........... SR24

Chapter 17: Divide by 1-Digit

17.1 Model Division............... RW95 PW95
17.2 Algebra: Division Patterns RW96 PW96
17.3 Estimate Quotients........ RW97 PW97
17.4 Divide 2- and 3-Digit Numbers RW98 PW98

Spiral Review Week 25........... SR25

17.5 PS Workshop Strategy: Solve a Simpler Problem............ RW99 PW99
17.6 Divide Money Amounts RW100 PW100
17.7 Model Division with Remainders.......... RW101 PW101

Unit 6: FRACTIONS AND DECIMALS

Chapter 18: Understand Fractions

18.1 Model Part of a Whole RW102 PW102

Spiral Review Week 26........... SR26

18.2 Model Part of a Group RW103 PW103
18.3 Equivalent Fractions.... RW104 PW104
18.4 Compare and Order Fractions RW105 PW105
18.5 PS Workshop Strategy: Make A Model RW106 PW106

Spiral Review Week 27........... SR27

Chapter 19: Add and Subtract Like Fractions

19.1 Add Like Fractions RW107 PW107
19.2 Add Like Fractions RW108 PW108
19.3 Subtract Like Fractions...................... RW109 PW109
19.4 Subtract Like Fractions..................... RW110 PW110

Spiral Review Week 28........... SR28

19.5 PS Workshop Skill: Too Much/Too Little Information RW111 PW111

Chapter 20: Understand Decimals

20.1 Model Tenths.............. RW112 PW112
20.2 Model Hundredths....... RW113 PW113
20.3 Decimals Greater than One..................... RW114 PW114

Spiral Review Week 29........... SR29

Key: PW Practice Workbook RW Reteach Workbook SR Spiral Review

Contents

20.4 Relate Fractions, Decimals, and Money.. RW115 PW115
20.5 Add and Subtract Decimals and Money... RW116 PW116
20.6 PS Workshop Skill: Make a Model.............. RW117 PW117

Unit 7: DATA AND PROBABILITY, MONEY AND TIME

Chapter 21: Data and Probability

21.1 Record Outcomes RW118 PW118

Spiral Review Week 30........... SR30

21.2 PS Workshop Skill: Make a Graph RW119 PW119
21.3 Probability: Likelihood of Events RW120 PW120
21.4 Possible Outcomes RW121 PW121
21.5 Experiments RW122 PW122
21.6 Line Plots.................... RW123 PW123

Spiral Review Week 31........... SR31

21.7 Predict Future Events.. RW124 PW124
21.8 PS Workshop Strategy: Make an Organized List RW125 PW125

Chapter 22: Money and Time

22.1 Compare Money Amounts RW126 PW126
22.2 Model Making Change....................... RW127 PW127
22.3 PS Workshop Skill: Compare Strategies RW128 PW128

Spiral Review Week 32........... SR32

22.4 Add Money Amounts... RW129 PW129
22.5 Subtract Money Amounts RW130 PW130
22.6 Multiply and Divide Money Amounts RW131 PW131
22.7 Tell Time...................... RW132 PW132
22.8 A.M. and P.M. RW133 PW133

Spiral Review Week 33........... SR33

22.9 Model Elapsed Time ... RW134 PW134
22.10 Use a Calendar RW135 PW135
22.11 Sequence Events RW136 PW136

Unit 8: CUSTOMARY AND METRIC MEASUREMENT

Chapter 23: Customary Measurement

23.1 Length RW137 PW137
23.2 Estimate and Measure Inches........... RW138 PW138

Spiral Review Week 34........... SR34

23.3 Estimate and Measure Feet and Yards............ RW139 PW139
23.4 Capacity RW140 PW140
23.5 Weight RW141 PW141
23.6 Estimate or Measure... RW142 PW142
23.7 Algebra: Rules for Changing Units............ RW143 PW143

Spiral Review Week 35........... SR35

23.8 PS Workshop Skill: Choose a Unit RW144 PW144
23.9 Fahrenheit Temperature................ RW145 PW145

Chapter 24: Metric Measurement

24.1 Length RW146 PW146
24.2 Centimeters and Decimeters RW147 PW147
24.3 Meters and Kilometers RW148 PW148

Spiral Review Week 36........... SR36

24.4 Capacity RW149 PW149
24.5 Mass............................ RW150 PW150
24.6 Algebra: Rules for Changing Units............ RW151 PW151
24.7 PS Workshop Skill: Compare Strategies RW152 PW152

Key: PW Practice Workbook RW Reteach Workbook SR Spiral Review

Name________________________________

Algebra: Patterns on a Hundred Chart

You can use a hundred chart to find number patterns.

Use the hundred chart to find the next number in the pattern: 3, 6, 9, 12.

1	2	3	4	5	6	7	8	9	10
11	12	13	14	15	16	17	18	19	20
21	22	23	24	25	26	27	28	29	30
31	32	33	34	35	36	37	38	39	40
41	42	43	44	45	46	47	48	49	50
51	52	53	54	55	56	57	58	59	60
61	62	63	64	65	66	67	68	69	70
71	72	73	74	75	76	77	78	79	80
81	82	83	84	85	86	87	88	89	90
91	92	93	94	95	96	97	98	99	100

Shade in the numbers already provided for you in the pattern.

There are 2 numbers not shaded between each shaded number and the next, ending at 12.

Skip two numbers to follow the pattern and shade in 15.

So, 15 is the next number in the pattern.

Use the hundred chart . Find the next number in the pattern.

1. 2, 4, 6, 8, ☐
2. 40, 44, 48, 52, ☐
3. 22, 24, 26, 28, ☐
4. 8, 11, 14, 17, ☐
5. 63, 61, 59, 57, ☐
6. 82, 86, 90, 94, ☐
7. 38, 37, 36, 35, ☐

1	2	3	4	5	6	7	8	9	10
11	12	13	14	15	16	17	18	19	20
21	22	23	24	25	26	27	28	29	30
31	32	33	34	35	36	37	38	39	40
41	42	43	44	45	46	47	48	49	50
51	52	53	54	55	56	57	58	59	60
61	62	63	64	65	66	67	68	69	70
71	72	73	74	75	76	77	78	79	80
81	82	83	84	85	86	87	88	89	90
91	92	93	94	95	96	97	98	99	100

Name________________________________

Algebra: Patterns on a Hundred Chart

Use the hundred chart. Find the next number in the pattern.

1	2	3	4	5	6	7	8	9	10
11	12	13	14	15	16	17	18	19	20
21	22	23	24	25	26	27	28	29	30
31	32	33	34	35	36	37	38	39	40
41	42	43	44	45	46	47	48	49	50
51	52	53	54	55	56	57	58	59	60
61	62	63	64	65	66	67	68	69	70
71	72	73	74	75	76	77	78	79	80
81	82	83	84	85	86	87	88	89	90
91	92	93	94	95	96	97	98	99	100

1. 1, 3, 5, 7,

2. 6, 5, 4, 3,

3. 10, 15, 20, 25,

4. 15, 12, 9, 6,

5. 10, 20, 30, 40,

6. 65, 63, 61, 59,

Use the hundred chart. Tell whether each number is *odd* or *even*.

7. 7 ________

8. 36 ________

9. 50 ________

10. 77 ________

11. 98 ________

12. 90 ________

13. 8 ________

14. 24 ________

15. 21 ________

16. 33 ________

17. 9 ________

18. 85 ________

Name ______________________________ **Week 1**

Spiral Review

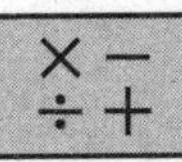

1. What is 4,601 in word form? ______________________________

____

2. What is 9,729 in word form?

____

____

3. What is seven hundred four in standard form? ________

4. What is one thousand, eight hundred in standard form? ________

5. Write the number in standard form.

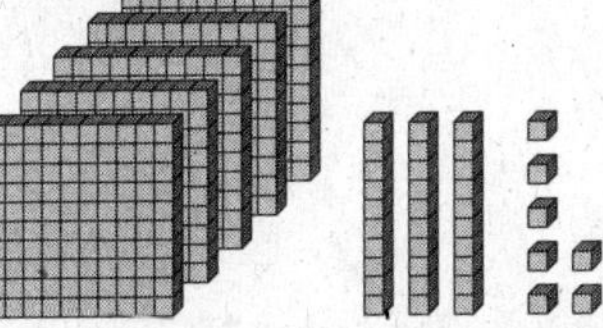

____

For 10–11, a class takes a survey about pets. Write the results as tally marks.

10. 3 students have dogs. ________

11. 6 students have fish. ________

12. Look at the table at the right. How many students were absent on Monday? ________

Absences	
Day	**Students Absent**
Monday	𝍸
Tuesday	|
Wednesday	|||

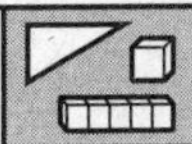

Name the plane figure

6. △ ________

7. ▭ ________

8. □ ________

9. ○ ________

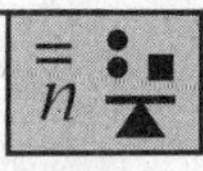

For 13–15, write the basic fact.

13. Taylor thinks of an addition fact. One of the addends is 7. The sum is also 7. What is a fact that Taylor could be thinking of?

____

14. Chris thinks of an addition fact. The sum is 11. One of the addends is 8. What is a fact that Chris could be thinking of?

____

15. Rosa thinks of an addition fact. The sum is 15. One of the addends is 10. What is a fact that Rosa could be thinking of?

____

Name________________________

Place Value to 1,000

Numbers are written using **digits**. Digits are the symbols 0, 1, 2, 3, 4, 5, 6, 7, 8, and 9.

The location of each digit in a number tells you about the value of the number.

Write the value of the underlined digit in: 624.

Use a place value chart to find the value.

	Hundreds	Tens	Ones	
4			4	single-digit number
24		2	4	two-digit number
624	6	2	4	three-digit number

The underlined digit is 6.

The digit 6 is in the hundreds column on the place value chart.

6 hundreds is the same as 600.

So, the value of the underlined digit is 600.

Write the value of the underlined digit. Fill in the place value chart to help you solve.

1. 784 ____________________

Hundreds	Tens	Ones

2. 784 ____________________

Hundreds	Tens	Ones

Write the value of the underlined digit.

3. 10 __________

4. 39 __________

5. 216 __________

6. 399 __________

Name____ **Lesson 1.2**

Place Value to 1,000

Write the value of the underlined digit.

1. $81\underline{8}$ ______ **2.** $1\underline{9}1$ ______ **3.** $\underline{8}17$ ______ **4.** $\underline{9}02$ ______

5. $25\underline{3}$ ______ **6.** $7\underline{0}4$ ______ **7.** $6\underline{4}0$ ______ **8.** $\underline{3}97$ ______

Write each number in standard form.

9. 300 + 40 + 2 ______ **10.** 500 + 60 + 1 ______ **11.** 200 + 10 + 9 ______

12. seven hundred three ______

13. four hundred ninety-nine ______

Problem Solving and Test Prep

14. Female elk can weigh up to six hundred pounds. In standard form, how many non-zero digits does this weight contain?

______

15. Male mountain lions usually weigh one hundred sixty pounds. In a place value chart of this weight, which digit would go in the hundreds place?

______

16. Which shows six hundred five written in standard form?

A 605
B 650
C 600 + 5
D 600 + 50

17. Which shows four hundred forty written in standard form?

A 400
B 440
C 444
D 400 + 40

Name____________________ Lesson 1.3

Place Value to 10,000

Numbers aren't always written with digits. Numbers can be written in other ways too.

These number forms all represent the same number.

Standard form: 1,409

Expanded form: 1,000 + 400 + 9

Word form: one thousand, four hundred nine

Write 1,000 + 300 + 8, in standard form.

Use a place value chart:

THOUSANDS	HUNDREDS	TENS	ONES
1	3	0	8
1,000 = *one thousand*, so put 1 in the thousands place.	300 = *three hundred*, so put 3 in the hundreds place.	There is no number written for the tens place, so the tens value is *zero*.	8 = *eight ones*, so put 8 in the ones place.

So, the standard form of 1,000 + 300 + 8 is 1,308.

Write two thousand, thirty-four in standard form.

THOUSANDS	HUNDREDS	TENS	ONES
2	0	3	4
Two thousand = 2,000, so put 2 in the thousands place.	There is no number written in the hundreds place, so the hundreds value is *zero*.	Thirty = 30, so put 3 in the tens place.	Four = 4, so put 4 in the ones place.

So, the standard form of two thousand, thirty-four is 2,034.

Write each number in standard form.

1. 6,000 + 200 + 40 + 1

____

2. eight thousand three hundred

____

3. two thousand six hundred

____

4. 5,000 + 500 + 9

____

Name____

Lesson 1.3

Place Value to 10,000

Write each number in standard form.

1. 9,000 + 8 ______

2. six thousand, one hundred twelve ______

3. four thousand, two hundred two ______

4. 2,000 + 700 + 30 + 4 ______

5. 3,000 + 700 + 20 + 4 ______

6. 5,000 + 200 + 9 ______

7. 6,000 + 9 ______

8. 9,000 + 600 + 30 + 8 ______

9. seven thousand four ______

10. four hundred seventy-seven ______

Write the value of the underlined digit.

11. $\underline{9},876$ ______

12. $7,\underline{2}19$ ______

13. $\underline{3},147$ ______

14. $4,2\underline{9}6$ ______

Problem Solving and Test Prep

15. Write a 4-digit number that contains the digits 0, 1, 2, and 3. What is the value of the first digit in your number?

______

16. Harry will have eaten 1,500 peanut butter and jelly sandwiches by the time he graduates from high school. How would you write 1,500 in word form?

______

17. Which number shows five thousand three hundred two?

A 532
B 5,032
C 5,302
D 5,320

18. Which is the value of the underlined digit in $\underline{7},318$?

A 7
B 70
C 700
D 7,000

Name________________________________ Lesson 1.4

Expanded Form

Write this number in expanded form: 3,890.

Use a place value chart to find the values in 4-digit numbers.

Here is 3,890 in a place value chart.

THOUSANDS	HUNDREDS	TENS	ONES
3	8	9	0

What you are actually doing is writing the sum of the value of each digit.
Do not include digits for which the value is 0.

Use the place value chart to find the value of each digit.
The value of 3 is 3 thousand ⟶ 3,000.
The value of 8 is 8 hundred ⟶ 800.
The value of 9 is 9 tens ⟶ 90.
The value of 0 is 0.

So, the expanded form of 3,890 is 3,000 + 800 + 90.

Write each number in expanded form.

1. 7,341 ____________________

2. 2,001 ____________________

3. 8,274 ____________________

4. 2,589 ____________________

5. 6,015 ____________________

6. 7,408 ____________________

Name____________________

Expanded Form

Complete the expanded form.

1. 2,333: 2,000 + 300 + ☐ + 3

2. 1,405: 1,000 + 400 + ☐

Write each number in expanded form.

3. 27,018 ____________________

4. 96,343 ____________________

5. 93,615 ____________________

6. 32,012 ____________________

7. 15,009 ____________________

8. 88,888 ____________________

Problem Solving and Test Prep

9. Fast Fact Lassen Peak in California is 10,457 feet high. What is the value of the digit 1 in 10,457?

____

10. During a single day, 37,465 people attended the California State Fair. How would you write 37,465 in word form?

____

11. Which shows twenty-four thousand, three written in expanded form?

A 24,003
B 24,300
C 20,000 + 4,000 + 3
D 20,000 + 200 + 3

12. Which is the value of the digit 9 in 987,654?

A 9
B 900
C 90,000
D 900,000

Name____ Lesson 1.5

Problem Solving Workshop Strategy: Use Logical Reasoning

Julie is thinking of an even number between 12 and 29. The sum of the digits is the same as the digit in the tens place. What is Julie's number?

Read to Understand

1. What information is given?

Plan

2. How can drawing a diagram help you solve the problem?

Solve

3. How can the table below help you solve the problem?

	12	13	14	15	16	17	18	19	20	21	22	23	24	25	26	27	28	29
Even	Y	N	Y	N	Y	N	Y	N	Y	N	Y	N	Y	N	Y	N	Y	N
Sum of Digits	3		5		7		9		2		4		6		8		10	
Digit in Tens Place	1		1		1		1		2		2		2		2		2	

____

____

4. What is Julie's number?

____

Check

5. Does your answer make sense?

____

____

Use logical reasoning to solve.

6. Twelve students take jazz or ballet lessons. Eight students take jazz, and seven students take ballet. How many students take both jazz and ballet?

____

7. Fifteen teachers supervise lunch or recess. Nine teachers supervise lunch, and ten teachers supervise recess. How many teachers supervise both lunch and recess? ________________

Name ______________________

Lesson 1.5

Problem Solving Workshop Strategy: Use Logical Reasoning

Problem Solving Strategy Practice

Use logical reasoning to solve.

1. Mario's locker number is between 80 and 99. The sum of the digits is 13. The tens digit is 3 more than the ones digit. What is Mario's locker number?

____

2. In a spelling bee, Cal, Dawn, and Amy were the top three finishers. Cal came in second. Dawn did not finish first. Who finished first?

____

3. Eight students tried out for the band or the chorus. Five students tried out for the band, the rest tried out for the chorus. How many students tried out for the chorus?

____

4. Earl answered 2 more questions correctly than Anna did. Anna answered 3 fewer questions correctly than Juanita did. Juanita answered 21 questions correctly. How many questions did Earl answer correctly?

____

Mixed Strategy Practice

5. Doug has 170 stamps in his collection. His first book of stamps has 30 more stamps in it than his second book. How many stamps are in Doug's first book?

____

6. Mr. Burns ran 14 miles last week. He only ran on Monday, Tuesday, and Wednesday. If Mr. Burns ran 3 miles on Tuesday and 5 miles on Wednesday, then how many miles did he run on Monday?

____

7. **USE DATA** Josie is 2 spans plus 3 ells tall. How many feet tall is Josie?

____

Unusual Measurements	
1 fathom	= 6 feet
2 spans	= 3 feet
3 ells	= 1 foot

Name ______________________ **Week 2**

Spiral Review

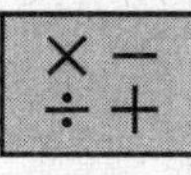

For 1–5, write the value of the underlined digit.

1. 9,420 ______

2. 1,609 ______

3. 2,093 ______

4. 3,826 ______

5. 7,824 ______

For 10–12, a class takes a survey about their favorite sports. Use 🯅 to replace the tally marks for each sport. Let 🯅 stand for 1 person.

Favorite Sport	
Sport	**Tally**
Swimming	𝍸
Karate	𝍸 𝍷
Soccer	𝍷𝍷

10. Swimming ______

11. Karate ______

12. Soccer ______

For 6–9, draw a line to match the object with the word that describes the shape.

6.  cone

7. 

cube

8. 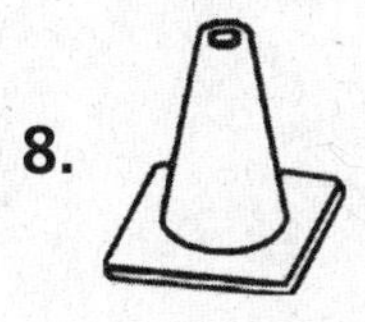 sphere

9. 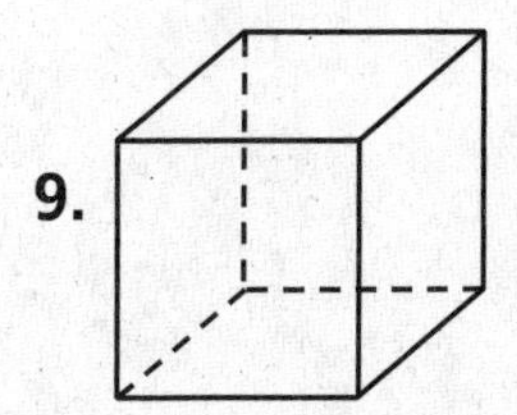 rectangular prism

For 13–15, write a number sentence to solve. Then solve.

13. There were 18 children at the library in the morning. There were 9 children at the library in the afternoon. How many more children were at the library in the morning than in the afternoon?

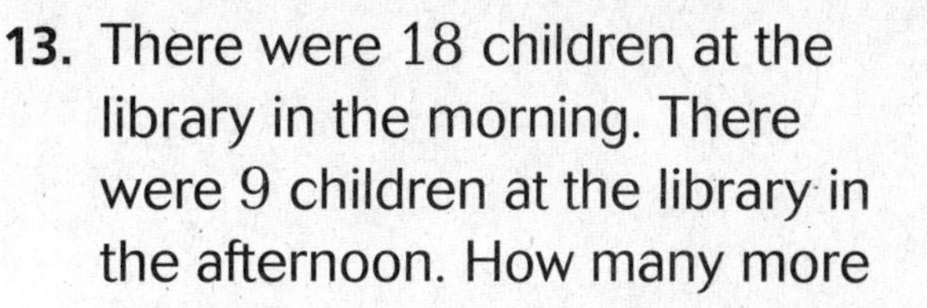

____

14. Lila has 13 markers. She finds 2 more markers. How many markers does Lila have in all?

____

15. Sid sees 14 robins. He also sees 6 blue jays. How many more robins does Sid see than blue jays?

____

Name____ Lesson 2.1

Compare Numbers

When you **compare** two numbers, you use symbols to show whether one is *greater than, less than,* or *equal to* the other.

greater than: > **less than: <** **equal to: =**

Compare the numbers. Write <, >, or = for the ○.

5,228 ○ 5,628.

Arrange the numbers, one above the other, so their place values align.

	thousands	hundreds	tens	ones
5,228	**5**	**2**	**2**	**8**
5,628	**5**	**6**	**2**	**8**

Compare the lined-up digits, beginning with the thousands place.

Use the symbols <, >, and = to compare.

Thousands: 5,228
5,628 5 = 5

Hundreds: 5,228
5,628 2 < 6

Since 2 < 6, then 5,228 < 5,628.

So, 5,228 (<) 5,628.

Compare the numbers. Write <, >, or = for each ○.

1. 378 ○ 387 **2.** 729 ○ 728 **3.** 4,318 ○ 4,318

4. 768 ○ 765 **5.** 1,034 ○ 1,043 **6.** 7,257 ○ 7,275

Name______________________________

Lesson 2.1

Compare Numbers

Compare the numbers. Write <, >, or = for each ◯.

1. 78 ◯ 87
2. 100 ◯ 99
3. 529 ◯ 592
4. 964 ◯ 946
5. 3,624 ◯ 3,624
6. 4,284 ◯ 284
7. 4,321 ◯ 4,312
8. 94 ◯ 940
9. 724 ◯ 724
10. 870 ◯ 87
11. 1,638 ◯ 1,863
12. 9,574 ◯ 9,745
13. 924 ◯ 944
14. 1,001 ◯ 1,101
15. 8,277 ◯ 8,177

Problem Solving and Test Prep

16. **Fast Fact** The tallest building in the United States is the Sears Tower in Illinois. It stands 1,450 feet tall. The tallest building in Canada is the CN Tower, which stands 1,815 feet tall. Compare the heights of these two buildings.

____

17. A 3rd grade has 384 students. A 4th grade has 348 students. Compare the number of students in each grade level.

____

18. Which number is less than 952 but greater than 924?

A 925
B 952
C 955
D 1,000

19. Which number is greater than 1,786 but less than 1,791?

A 1,678
B 1,768
C 1,786
D 1,790

Name____

Order Numbers

When you put more than two numbers in order, you compare the digits, starting with the place value farthest to the left. Set the numbers vertically to help you compare.

Write the numbers in order from greatest to least.

784, 699, 808

You can use base-ten blocks to visualize the numbers.

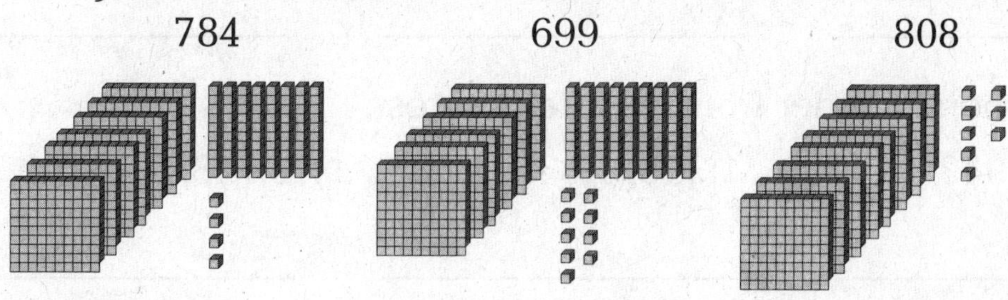

Step 1 Compare the hundreds of all three numbers.

7̲84
6̲99
8̲08

808 has the most flats, hundreds, so it has the greatest value.

Step 2 Compare the hundreds of the other two numbers.

7̲84
6̲99

784 has the most flats, hundreds, out of these two numbers, so it has the second greatest value.

So, the numbers from greatest to least are: 808, 784, 699.

Write the numbers in order from greatest to least.

1. 684, 799, 708 ______________

2. 369, 396, 693 ______________

3. 2,100; 2,010; 2,000 ______________

Write the numbers in order from least to greatest.

4. 437, 509, 402 ______________

5. 117, 171, 111 ______________

6. 1,004; 906; 960 ______________

Name____ Lesson 2.2

Order Numbers

Write the numbers in order from greatest to least.

1. 782, 780, 785 ______

2. 3,012; 3,644; 3,128 ______

3. 6,225; 6,237; 6,244 ______

4. 921, 929, 927 ______

5. 8,215; 8,152; 8,521 ______

6. 9,305; 9,350; 9,503 ______

Write the numbers in order from least to greatest.

7. 949, 941, 943 ______

8. 1,358; 1,835; 1,583 ______

9. 2,748; 2,751; 2,739 ______

10. 351, 355, 352 ______

11. 4,157; 4,175; 4,159 ______

12. 5,764; 5,674; 5,746 ______

Problem Solving and Test Prep

13. Fast Fact Dinosaurs lived long ago, and ranged in weight. The ankylosaurus weighed about 7,000 pounds, the stegosaurus weighed about 4,000 pounds, and the iguanodon weighed about 9,900 pounds. Which dinosaur weighed about the least?

______

14. Reasoning I am a number that is greater than 81 but less than 95. The sum of my digits is 15. What number am I?

______

15. Which number is greater than 872 but less than 902?

A 912

B 852

C 892

D 902

16. Which number is greater than 498 but less than 507?

A 497

B 499

C 507

D 510

Name____________________ **Lesson 2.3**

Problem Solving Workshop Skill: Use a Model

A bird show had 2,498 visitors in June, 2,675 visitors in July, and 2,189 visitors in August. Write the number of visitors at the bird show in order from greatest to least.

1. How can you organize the data?

____

2. Which two categories of data are given?

____

3. How can you use the number line below to order the data?

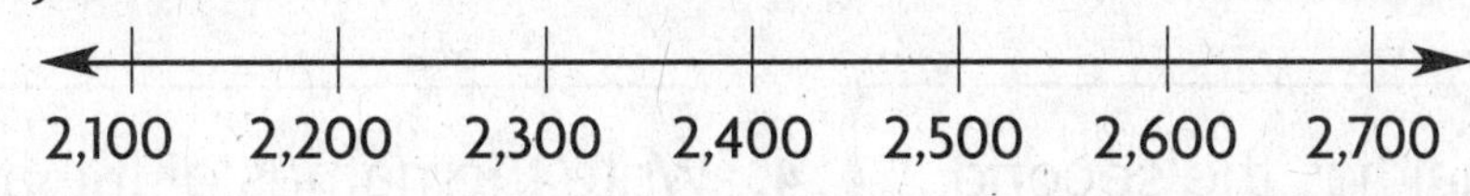

____

____

4. Where do the least number of visitors place on the number line; more to the left, or more to the right? _______________
5. Where do the greatest number of visitors place on the number line; more to the left, or more to the right? _______________
6. What is the order from greatest to least, of visitors to the bird show?

____

Use a model to solve.

7. On Friday 579 people attended a school play. On Saturday 724 people attended the play. On Sunday 595 people attended the play. What is the number of play attendees per night in order from least to greatest? Use the number line provided.

____

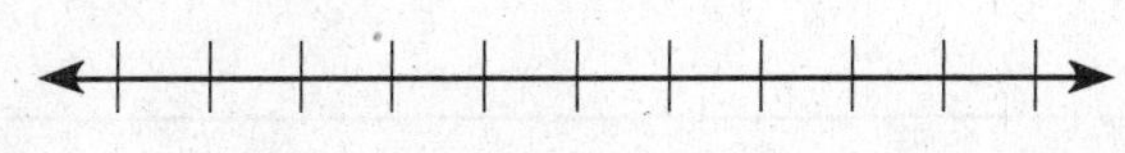

8. Zoo elephants eat 150 pounds of food a day, wild African elephants eat 770 pounds of food a day, and Asiatic elephants eat 650 pounds of food a day. What is the order of the types of elephants from greatest to least according to the amount of food each type eats in a day. Use the number line provided.

____

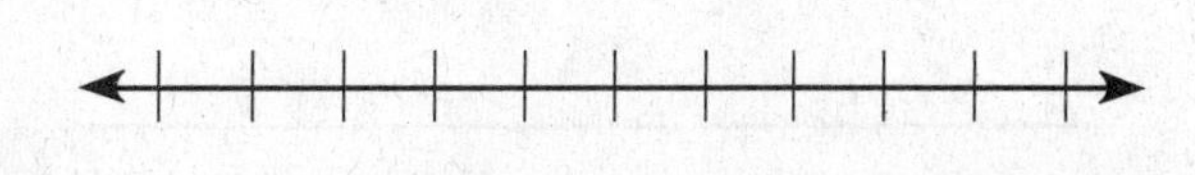

Name________________________________ **Lesson 2.3**

Problem Solving Workshop Skill: Use a Model

Problem Solving Skill Practice

USE DATA For 1–4, use the number line representing the weights of animals, below.

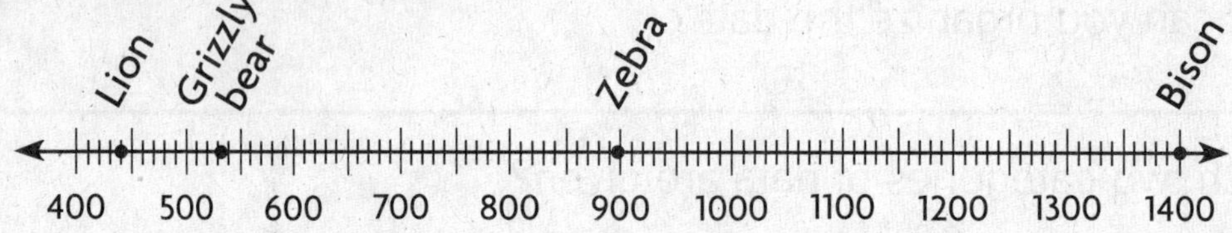

1. Which animal weighs the most?

____

2. Which animal weighs the least?

____

3. Which animal has the second greatest weight?

____

4. Write the names of the animals in order from greatest weight to least weight.

____

Mixed Applications

5. Charlie collected 5 marbles on Monday, 3 marbles on Wednesday, and 2 marbles on Tuesday. How could you put the numbers of marbles Charlie collected in order based on the days that they were collected from earliest to latest?

____

6. Patrick brought 4 pencils to school on Monday, 3 pencils to school on Tuesday, and a bagged lunch to school on Wednesday. How many pencils did Patrick bring to school on Monday and Tuesday combined?

____

7. Luke put three numbers in order from least to greatest. The total amount of digits in the numbers he ordered is 4. Are any of the three numbers Luke put into order made up of more than 2 digits?

____

8. Lana won two spelling bees last year. She told her mother that in the number 1,020 the number in the hundreds place has a value of 0. Is what Lana told her mother correct?

____

Name______________________ Lesson 2.4

Round to the Nearest Ten

To **estimate**, you must find a number that is close to an exact amount. You can find an estimate by **rounding**. When you round a number you find a number that tells you *about how much* or *about how many*.

Round the number below to the nearest ten.

981

Use the number line below.

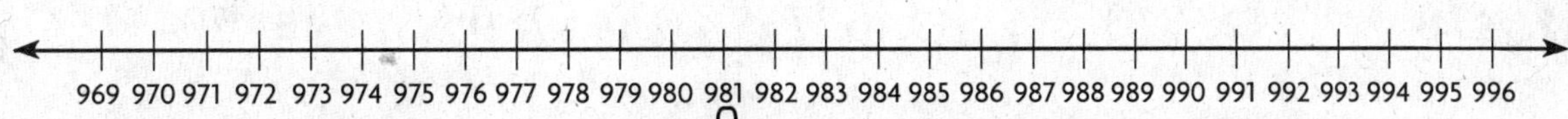

Place your finger on 981. Move left to 980.

You moved one number to the left to reach 980.

Place your finger on 981. Move right to 990.

You moved nine numbers to the right to reach 990.

981 is closer to 980 than to 990.

So, 981 rounded to the nearest ten is 980.

Use the number line. Round to the nearest ten.

1. 736

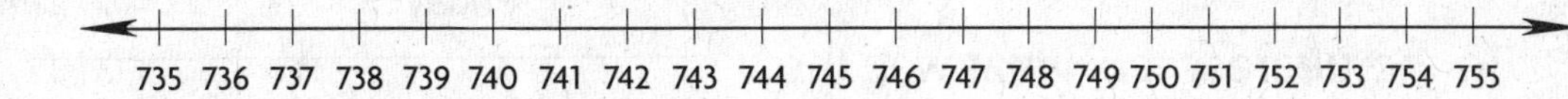

____

2. 418

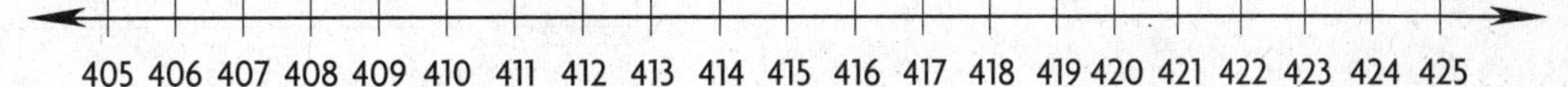

____

Round to the nearest ten.

3. 17 _______ **4.** 38 _______ **5.** 61 _______ **6.** 16 _______ **7.** 45 _______

8. 172 _______ **9.** 404 _______ **10.** 693 _______ **11.** 281 _______ **12.** 457 _______

Name________________________________

Round to the Nearest Ten

Round the number to the nearest ten.

1. 52 ______ **2.** 47 ______ **3.** 95 ______ **4.** 107 ______ **5.** 423 ______

6. 676 ______ **7.** 209 ______ **8.** 514 ______ **9.** 673 ______ **10.** 19 ______

11. 478 ______ **12.** 313 ______ **13.** 627 ______ **14.** 789 ______ **15.** 204 ______

Problem Solving and Test Prep

USE DATA For 16–17, use the table below.

16. To the nearest ten, what was the number of sea lions spotted on Friday?

____

17. To the nearest ten, what was the number of sea lions spotted from Friday to Sunday?

____

Sea Lions Spotted Off the Pier	
Day	**Number of Sea Lions Spotted**
Friday	48
Saturday	53
Sunday	65

18. The number of stamps in Krissy's collection, rounded to the nearest ten, is 670. How many stamps could Krissy have?

A 679
B 676
C 669
D 664

19. On a number line, the number labeled X is closer to 350 than it is to 360. Which number could X be?

A 354
B 356
C 361
D 365

Spiral Review

For 1–5, compare. Use <, >, or = for each ◯.

1. 546 ◯ 748

2. 208 ◯ 200

3. 969 ◯ 996

4. 6,399 ◯ 6,399

5. 3,000 ◯ 2,999

For 6–9, use the solid figure. Draw around the faces. Circle the face or faces of the solid figure.

6. 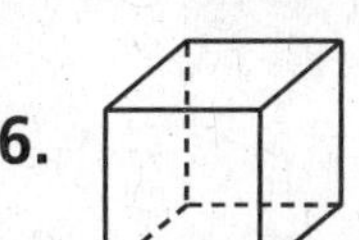  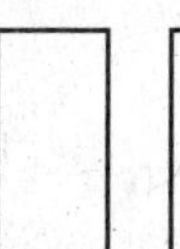 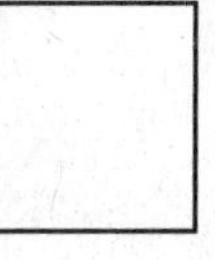

7. 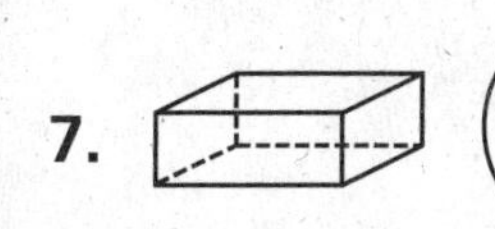  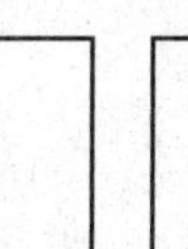 

8. 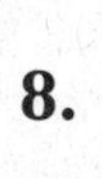 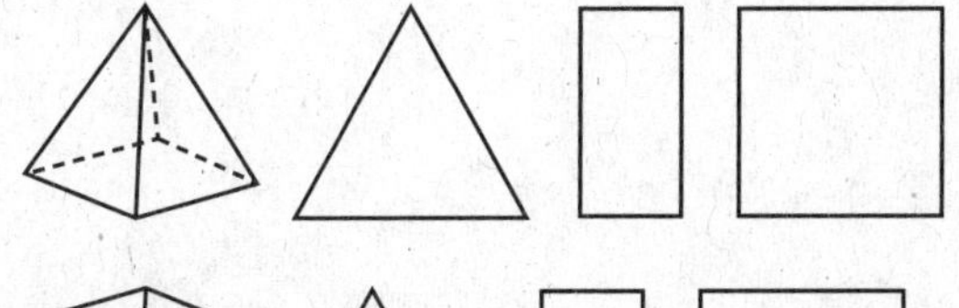

9. 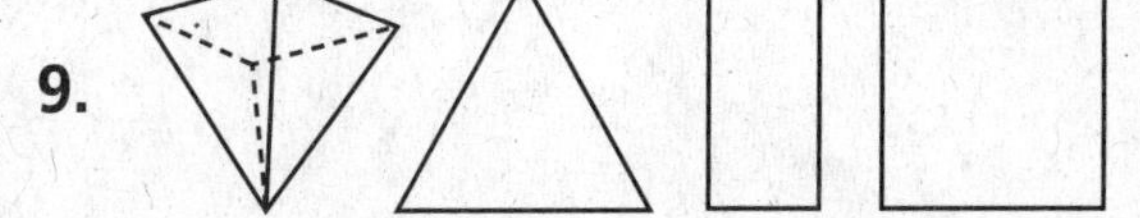

For 10–11, a class takes a survey about how they come to school. Write the results as tally marks.

10. 7 students ride bikes to school.

____

11. 11 students ride the bus.

____

12. Look at the table at the right. How many students went on vacation over Spring Break?

Spring Break	
Activity	**Students**
Stay Home	𝍸 |||
Visit Family	𝍸 |
Vacation	𝍸 𝍸

____

For 13–15, compare amounts to solve. Use <, > or =.

13. Mary Ann made 160 decorations for the school party. Nick made 158 decorations. Who made more decorations?

____

14. José ran 217 laps during the school year. Carla ran 220 laps during the school year. Who ran more laps?

____

15. Manny answered 118 addition problems correctly. Nora answered 118 addition problems correctly. Who answered more addition problems correctly?

____

Name________________________________

Round to the Nearest Hundred

To find an answer that is close to the exact amount, estimate.

Rounding is one way to estimate. Rounding tells you *about how much* or *about how many.*

Use *rounding rules* to round to the nearest hundred.

Ask the question:

Is the tens digit less than, greater than, or equal to 5?

Rounding Rule 1:
If the tens digit is less than 5, the hundreds digit stays the same and the tens digit and ones digit both become zero.

Rounding Rule 2:
If the tens digit is greater than or equal to 5, the hundreds digit increases by 1 and the tens digit and ones digit both become zero.

Round 3,366 to the nearest hundred.

Look at the digit in the tens place: 3,3(6)6

6 is greater than 5.

Since $6 > 5$, the hundreds digit increases by 1, and the tens digit and ones digit both become 0.

So, 3,366 rounded to the nearest hundred is 3,400.

Round to the nearest hundred.

1. 109 ________ **2.** 257 ________ **3.** 316 ________ **4.** 455 ________

5. 561 ________ **6.** 785 ________ **7.** 629 ________ **8.** 918 ________

9. 3,111 ________ **10.** 4,999 ________ **11.** 1,456 ________ **12.** 2,332 ________

NS 1.4 Round off numbers to 10,000 to the nearest ten, hundred, and thousand.

Name____ **Lesson 2.5**

Round to the Nearest Hundred

Round the number to the nearest hundred.

1. 349 ________ **2.** 251 ________ **3.** 765 ________ **4.** 3,218 ________ **5.** 6,552 ________

6. 4,848 ________ **7.** 5,298 ________ **8.** 6,342 ________ **9.** 7,112 ________ **10.** 412 ________

11. 901 ________ **12.** 5,451 ________ **13.** 2,982 ________ **14.** 9,216 ________ **15.** 1,543 ________

Problem Solving and Test Prep

USE DATA For 16–17, use the table below.

16. To the nearest hundred, how many feet below sea level is California's lowest point?

____

17. To the nearest hundred, how many square miles is California's water area?

____

California Geography	
Feature	**Size**
Coastline	840 miles
Lowest Point (below sea level)	282 feet
Water Area	7,734 square miles

18. Which number does NOT round to 500, when rounded to the nearest hundred?

A 451

B 499

C 533

D 552

19. On a number line, point P is closer to 300 than to 200. Which number could point P stand for?

A 219

B 247

C 273

D 202

Name____

Round to the Nearest Thousand

You can use a place value chart to help round to the nearest thousand.

Round 2,501 to the nearest thousand.

Thousands	Hundreds	Tens	Ones
2	5	0	1

First, look at the digit in the hundreds place: 2,501

5 is equal to 5.

Round up.

Add 1 to the digit in the thousands place.

Replace the hundreds, tens, and ones digits with zeros.

So, 2,501 rounded to the nearest thousand is 3,000.

Round 3,274 to the nearest thousand.

Thousands	Hundreds	Tens	Ones
3	2	7	4

First, look at the digit in the hundreds place: 3,274

2 is less than 5.

Round down.

The digit in the thousands place remains the same.

Replace the hundreds, tens, and ones digits with zeros.

So, 3,274 rounded to the nearest thousand is 3,000.

Round to the nearest thousand.

1. 1,239 ______ **2.** 9,339 ______ **3.** 9,612 ______ **4.** 5,432 ______ **5.** 7,777 ______

6. 5,602 ______ **7.** 1,345 ______ **8.** 6,002 ______ **9.** 1,806 ______ **10.** 7,980 ______

Name____________________

Lesson 2.6

Round to the Nearest Thousand

Round to the nearest thousand.

1. 8,732 **2.** 6,541 **3.** 3,498 **4.** 9,261 **5.** 2,674

________ ________ ________ ________ ________

Round to the nearest thousand, to the nearest hundred, and to the nearest ten.

6. 4,192 **7.** 5,647 **8.** 7,526 **9.** 2,796 **10.** 3,365

________ ________ ________ ________ ________

________ ________ ________ ________ ________

________ ________ ________ ________ ________

Problem Solving and Test Prep

USE DATA For 11–12, use the table below.

11. To the nearest thousand, how many people were involved with the largest lifesaving operation?

____

12. To the nearest thousand, how many kg were the largest twin battleship projectiles?

____

Guinness World Records	
Largest...	**Number**
Lifesaving Operation	2,735 people
Twin Battleship Projectiles	1,450 kg
Area of Sail	1,833 sq. meters

13. Mickey's class collected 1,521 soup labels. Which number is 1,521, rounded to the nearest thousand?

A 1,000
B 1,500
C 1,520
D 2,000

14. On a number line, the number labeled Z is closer to 6,000 than it is to 5,000. Which number could Z NOT be?

A 5,420
B 5,501
C 6,001
D 6,449

Name________________________________ **Lesson 3.1**

Algebra: Addition Properties

You can use addition properties to help you add.

Commutative Property of Addition (Order Property of Addition)
You can add numbers in any order and still get the same sum.

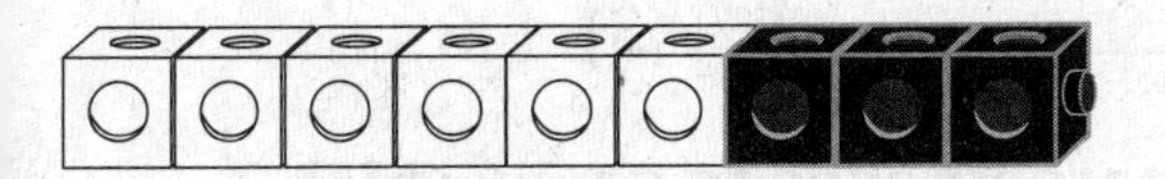

$6 + 3 = 9$

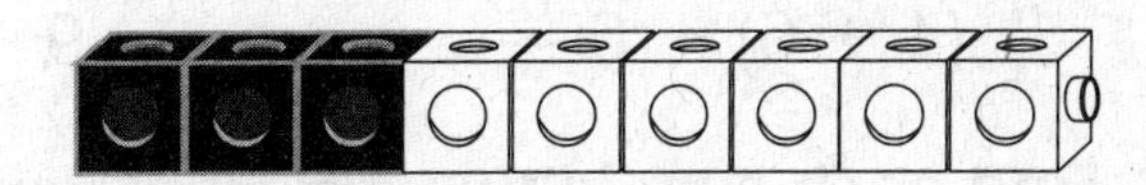

$3 + 6 = 9$

Associative Property of Addition (Grouping Property of Addition)
You can group addends in different ways and still get the same sum.

$(2 + 7) + 1 = 10$

$2 + (7 + 1) = 10$

Find each sum.

1. $3 + 5 =$ ___
$5 + 3 =$ ___

2. $2 + (1 + 4) =$ ___
$(2 + 1) + 4 =$ ___

3. $(5 + 5) + 2 =$ ___
$2 + (5 + 5) =$ ___

4. $6 + 4 =$ ___
$4 + 6 =$ ___

5. $8 + (2 + 4) =$ ___
$(8 + 2) + 4 =$ ___

6. $(7 + 8) + 3 =$ ___
$7 + (8 + 3) =$ ___

7. $0 + 2 =$ ___
$2 + 0 =$ ___

8. $6 + (2 + 5) =$ ___
$(6 + 2) + 5 =$ ___

9. $(4 + 7) + 9 =$ ___
$4 + (7 + 9) =$ ___

Name________________________

Lesson 3.1

Algebra: Addition Properties

Find each sum.

1. $4 + 7 =$ ____
   $7 + 4 =$ ____
2. $1 + (8 + 5) =$ ____
   $(1 + 8) + 5 =$ ____
3. $(3 + 9) + 4 =$ ____
   $3 + (9 + 4) =$ ____
4. $4 + (6 + 6) =$ ____
   $(4 + 6) + 6 =$ ____
5. $1 + 9 =$ ____
   $9 + 1 =$ ____
6. $5 + (3 + 3) =$ ____
   $(5 + 3) + 3 =$ ____

Find each sum in two different ways. Use parentheses to show which numbers you added first.

7. $7 + 3 + 5 =$ ____
8. $9 + 4 + 2 =$ ____
9. $62 + 18 + 5 =$ ____
10. $25 + 4 + 6 =$ ____
11. $42 + 1 + 9 =$ ____
12. $0 + 16 + 16 =$ ____
13. $9 + 7 + 9 =$ ____
14. $14 + 6 + 3 =$ ____
15. $50 + 6 + 30 =$ ____
16. $21 + 42 + 1 =$ ____

Problem Solving and Test Prep

17. On a nature walk, Sarah sees 3 squirrels, 5 chipmunks, and 8 birds. How many animals does Sarah see in all?

    ____

18. On Monday Ramon saw 4 squirrels and 8 birds, in the park. On Tuesday he saw 8 squirrels and 4 birds in the park. On Monday and Tuesday how many animals did Ramon see in all?

    ____

19. Which is the sum? $3 + 10 =$ ____

    A 0
    B 3
    C 13
    D 30

20. Which property is shown in the number sentence below?

    $8 + (9 + 4) = (8 + 9) + 4$

    A zero
    B commutative
    C identity
    D associative

Name________________________________

Algebra: Missing Addends

You can use related facts to help you find a missing addend.

What is the missing addend?

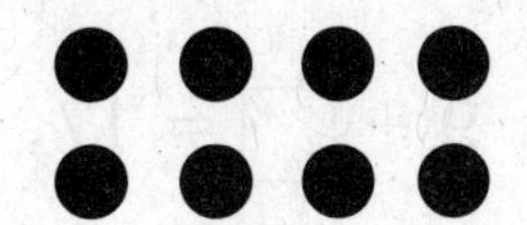

8 +  = 10

You can use a related subtraction fact.

10 − 8 = [2]

You can use an addition fact.

8 + [2] = 10

So, 8 + 2 = 10

Find the missing number. You can use counters to help solve.

1. 2 + ☐ = 8

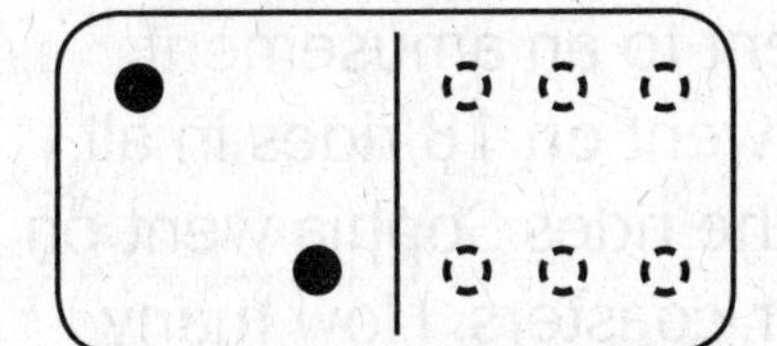

2. ☐ + 6 = 10

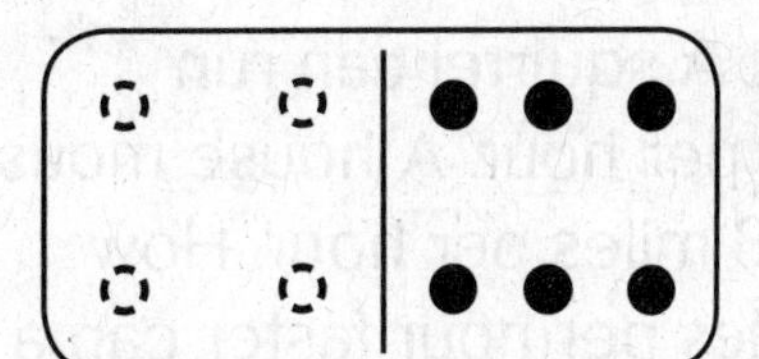

3. 8 + ☐ = 10

4. ☐ + 3 = 3

5. 1 + ☐ = 7

6. ☐ + 4 = 5

7. ☐ + 4 = 13

8. 2 + ☐ = 5

9. ☐ + 5 = 12

10. 9 + ☐ = 13

11. 2 + ☐ = 7

12. ☐ + 6 = 15

13. 2 + ☐ = 8

14. ☐ + 0 = 8

Name________________________________ Lesson 3.2

Algebra: Missing Addends

Find the missing addend. You may want to use counters.

1. $3 + \square = 10$
2. $\square + 9 = 14$
3. $\square + 6 = 11$
4. $\square + 2 = 5$
5. $\square + 7 = 13$
6. $2 + \square = 4$
7. $\square + 9 = 12$
8. $9 + \square = 17$
9. $6 + \square = 12$
10. $\square + 1 = 10$
11. $3 + \square = 8$
12. $\square + 4 = 4$

Find the missing number. You may want to use counters.

13. $9 + 9 = \_\_$
14. $3 + \square = 12$
15. $5 + 5 = \_\_$
16. $7 + 0 = \_\_$
17. $6 + 8 = \_\_$
18. $2 + \square = 10$
19. $\square + 5 = 12$
20. $\square + 0 = 3$
21. $8 + \square = 12$
22. $4 + 7 = \_\_$
23. $6 + \square = 11$
24. $2 + 7 = \_\_$

Problem Solving and TEST Prep

25. **Fast Fact** A squirrel can run 12 miles per hour. A house mouse can run 8 miles per hour. How many miles per hour faster can a squirrel run than a house mouse can run?

____

26. Sophia went to an amusement park. She went on 18 rides in all. Seven of the rides Sophia went on were roller coasters. How many rides that Sophia went on were not roller coasters?

____

27. Which is the sum?
$2 + 7 = \_\_$
A 5
B 6
C 8
D 9

28. Which is the missing addend for
$11 + \_\_ = 15$?
A 3
B 4
C 5
D 6

Spiral Review

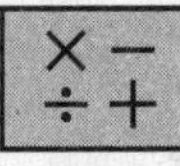

For 1–5, round each number to the nearest thousand.

1. 1,580 ________
2. 2,094 ________
3. 6,527 ________
4. 9,099 ________
5. 602 ________

For 6–9, look at the picture. Then circle the unit that would require a greater number if it were used instead.

The grasshopper is 3 cm long

6. If I measured in **(mm, m)**, I would have a greater number.

The book weighs 2000 g

7. If I measured in **(mg, kg)**, I would have a greater number.

The tub holds 3 L

8. If I measured in **(mL, kL)**, I would have a greater number.

The goal is 4 m high

9. If I measured in **(cm, km)**, I would have a greater number.

For 10–12, a class takes a survey about the insects they see on the playground. Let 🯅 stand for 1 person. Use 🯅 to replace the tally marks of eac insect seen below.

Favorite Sport	
Insect Seen	**Students**
Mosquito	𝍸
Grasshopper	|||
Butterfly	𝍸	

10. Mosquito ____________
11. Grasshopper ________
12. Butterfly ____________

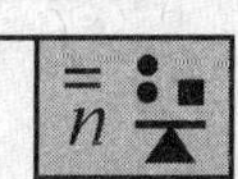

For 13–15, write a number sentence to solve.

13. Cory planted 17 vegetable seeds in the garden. He found 2 more seeds and planted them. How many seeds did Cory plant in all?

____

14. Finn picked 12 cucumbers from the garden. Then he picked 8 tomatoes. How many vegetables did Finn pick in all?

____

15. Sun Li counted 11 peppers on one plant and 5 peppers on another. How many peppers did Sun Li count in all?

____

Name____ Lesson 3.3

Estimate Sums

You can estimate to find out *about* how many.
An estimate is close to the exact answer.

A number line can help you estimate sums.

363 + 129 = ☐

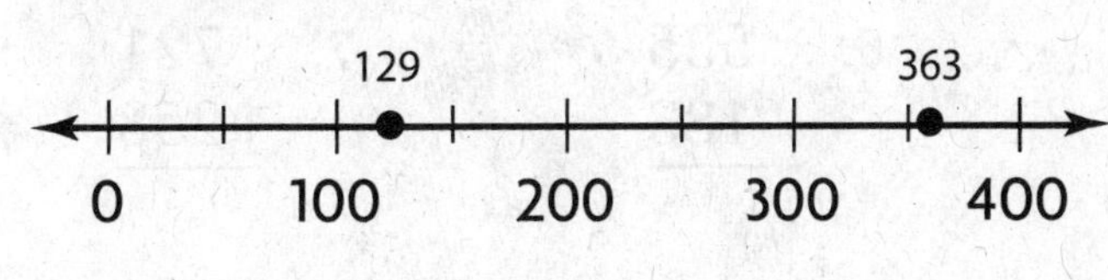

129 is closer to 100 than to 200.
363 is closer to 400 than to 300.

```
  363  →    400
+ 129  →  + 100
            500
```

2,617 + 4,251 = ☐

2,617 is closer to 3,000 than 2,000.
4,215 is closer to 4,000 than 5,000.

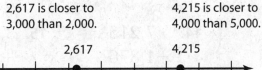

5,000

```
  2,617  →    3,000
+ 4,251  →  + 4,000
              7,000
```

Estimate each sum.

200

4,215
1,000
2,000

1. 46 + 68	**2.** 32 + 45	**3.** 219 + 538	**4.** 342 + 62
5. 391 + 259	**6.** 634 + 281	**7.** 3,213 + 4,628	**8.** 4,874 + 4,459
9. 4,817 + 228	**10.** 1,129 + 2,585	**11.** 7,089 + 482	**12.** 6,230 + 285

Name______________________________

Estimate Sums

Estimate each sum.

1. $64 + 29$

2. $45 + 21$

3. $14 + 37$

4. $423 + 17$

5. $271 + 349$

6. $535 + 183$

7. $721 + 258$

8. $183 + 134$

9. $661 + 32$

10. $387 + 97$

11. $3,294 + 2,523$

12. $1,622 + 4,097$

13. $5,206 + 3,851$

14. $7,215 + 1,376$

15. $3,609 + 897$

16. $6,832 + 625$

Problem Solving and Test Prep

17. One month, a group of fishermen caught 987 salmon. The next month, they caught 673 salmon. About how many salmon did the fishermen catch in all?

____

18. Julia went bird watching near Humboldt Bay. On Wednesday, she counted 38 birds. On Thursday, she counted 52 birds. About how many birds did Julia count in all?

____

19. Which is the estimated sum of 452 + 639?

A 800
B 900
C 1,000
D 1,100

20. Which is the estimated sum of 1,259 and 382?

A 400
B 700
C 1,400
D 1,700

Name______________________________

Add with Regrouping

You can use place-value models to add 2-digit numbers.

Find 24 + 39.

24 / 39	Step 1	Step 2	Step 3	Step 4
24 39	Add the ones. 4 ones + 9 ones 13 ones	Regroup as necessary. 13 ones is the same as 1 ten and 3 ones.	Add the tens. 1 ten + 2 tens + 3 tens 6 tens	Find the sum. 6 tens and 3 ones, or 63 So, 24 + 39 = 63.

Estimate. Then find each sum.

1. 19 + 64
2. 68 + 14 + 32
3. 53 + 36
4. 46 + 19
5. 53 + 20 + 68
6. 47 + 86
7. 25 + 32
8. 83 + 49
9. 67 + 28 + 12
10. 36 + 23 + 19

Name ______________________

Lesson 3.4

Add with Regrouping

Estimate. Then find each sum using place value or mental math.

1. $19 + 64$

2. $33 + 28 + 14$

3. $63 + 45$

4. $34 + 76$

5. $65 + 48 + 16$

6. $75 + 47$

7. $31 + 86$

8. $47 + 25$

9. $24 + 32 + 18$

10. $47 + 24 + 52$

11. $56 + 41 =$ _____

12. $83 + 15 =$ _____

13. $25 + 67 + 31 =$ _____

14. $29 + 67 =$ _____

15. $37 + 21 =$ _____

16. $49 + 34 + 61 =$ _____

Problem Solving and Test Prep

17. Kara bought 13 green apples and some red apples. She bought a total of 40 apples. How many red apples did Kara buy?

____

18. Manuel and his brother picked apples. Manuel picked 62 apples. His brother picked 39 apples. How many apples did Manuel and his brother pick in all?

____

19. Which is the sum?

$71 + 23 + 18 =$ _____

A 89
B 94
C 102
D 112

20. Which is the sum?

$65 + 28 =$ _____

A 83
B 92
C 93
D 98

Name____________________

Model 3-Digit Addition

You can use base-ten blocks to add.

Add 352 **and** 191.

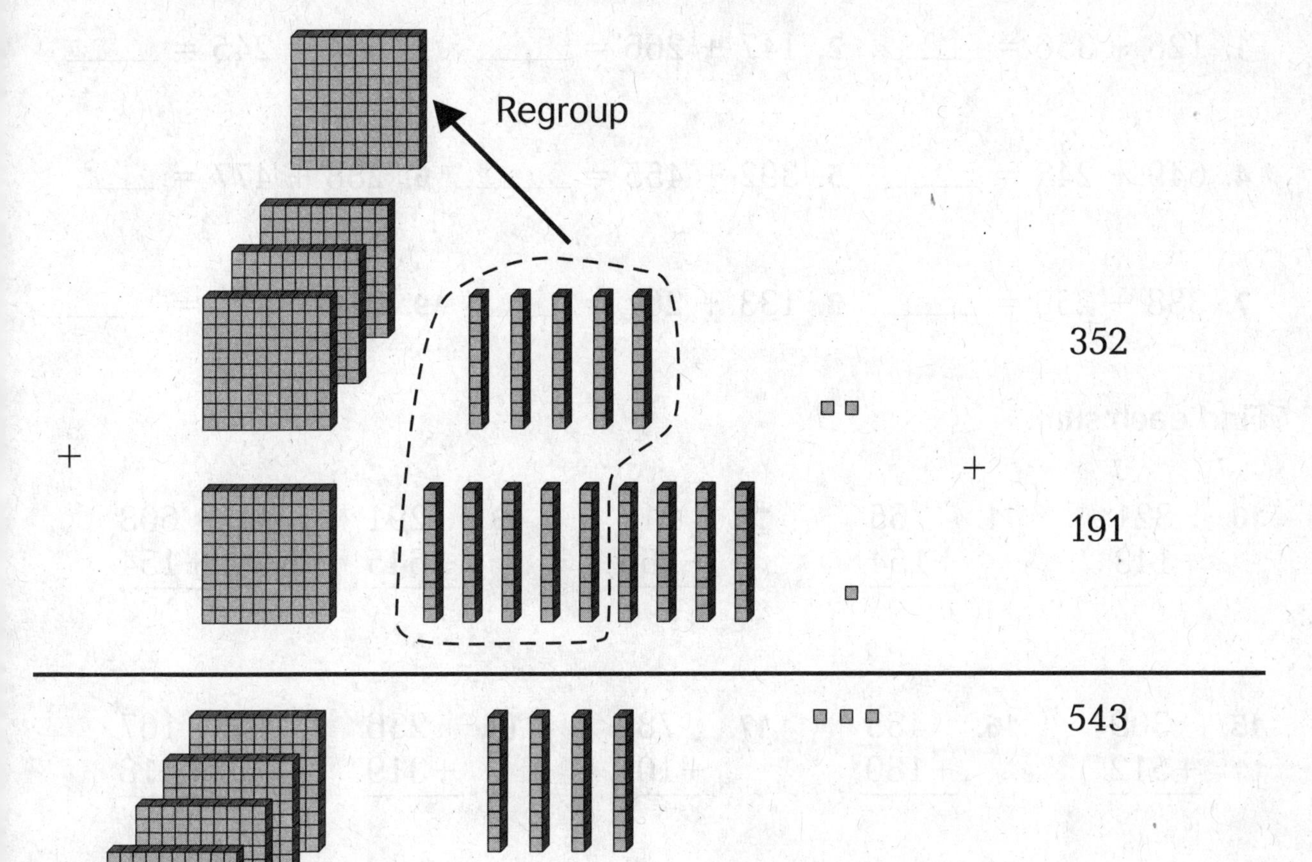

Use base-ten blocks to find each sum.

1. 233 + 474 = ☐ **2.** 193 + 324 = ☐ **3.** 175 + 228 = ☐

4. 541 + 286 = ☐ **5.** 445 + 346 = ☐ **6.** 119 + 385 = ☐

NS 2.1 Find the sum or difference of two whole numbers between 0 and 10,000.

Name______________________________________

Model 3-Digit Addition

Use base-ten blocks to find each sum.

1. $128 + 356 =$ _____ **2.** $147 + 266 =$ _____ **3.** $594 + 245 =$ _____

4. $649 + 248 =$ _____ **5.** $392 + 455 =$ _____ **6.** $288 + 477 =$ _____

7. $388 + 256 =$ _____ **8.** $133 + 267 =$ _____ **9.** $818 + 103 =$ _____

Find each sum.

10. $\begin{array}{r} 821 \\ +143 \\ \hline \end{array}$ **11.** $\begin{array}{r} 765 \\ +154 \\ \hline \end{array}$ **12.** $\begin{array}{r} 217 \\ +265 \\ \hline \end{array}$ **13.** $\begin{array}{r} 291 \\ +645 \\ \hline \end{array}$ **14.** $\begin{array}{r} 608 \\ +154 \\ \hline \end{array}$

15. $\begin{array}{r} 309 \\ +512 \\ \hline \end{array}$ **16.** $\begin{array}{r} 485 \\ +180 \\ \hline \end{array}$ **17.** $\begin{array}{r} 789 \\ +101 \\ \hline \end{array}$ **18.** $\begin{array}{r} 236 \\ +319 \\ \hline \end{array}$ **19.** $\begin{array}{r} 167 \\ +418 \\ \hline \end{array}$

20. $\begin{array}{r} 189 \\ +178 \\ \hline \end{array}$ **21.** $\begin{array}{r} 248 \\ +318 \\ \hline \end{array}$ **22.** $\begin{array}{r} 378 \\ +147 \\ \hline \end{array}$ **23.** $\begin{array}{r} 320 \\ +575 \\ \hline \end{array}$ **24.** $\begin{array}{r} 256 \\ +127 \\ \hline \end{array}$

25. $\begin{array}{r} 444 \\ +328 \\ \hline \end{array}$ **26.** $\begin{array}{r} 701 \\ +199 \\ \hline \end{array}$ **27.** $\begin{array}{r} 225 \\ +387 \\ \hline \end{array}$ **28.** $\begin{array}{r} 821 \\ +143 \\ \hline \end{array}$ **29.** $\begin{array}{r} 765 \\ +154 \\ \hline \end{array}$

30. $\begin{array}{r} 635 \\ +364 \\ \hline \end{array}$ **31.** $\begin{array}{r} 528 \\ +122 \\ \hline \end{array}$ **32.** $\begin{array}{r} 137 \\ +303 \\ \hline \end{array}$ **33.** $\begin{array}{r} 412 \\ +101 \\ \hline \end{array}$ **34.** $\begin{array}{r} 862 \\ +112 \\ \hline \end{array}$

Name ____________________

Add 3- and 4-Digit Numbers

You can use place-value charts to add.
Add 5,726 **and** 2,295.

Step 1

Add the ones.

	Th	H	T	O
			1	
	5,	7	2	6
+	2,	2	9	5
				1

11 ones = 1 ten, and 1 one

Step 2

Add the tens.

	Th	H	T	O
		1	1	
	5,	7	2	6
+	2,	2	9	5
			2	1

12 tens = 1 hundred, and 2 tens

Step 3

Add the hundreds.

	Th	H	T	O
	1	1	1	
	5,	7	2	6
+	2,	2	9	5
		0	2	1

10 hundreds = 1 thousand, and 0 hundreds

Step 4

Add the thousands.

	Th	H	T	O
	1	1	1	
	5,	7	2	6
+	2,	2	9	5
	8,	0	2	1

Find each sum.

1.

	Th	H	T	O
	2,	1	4	5
+		6	7	8

2.

	Th	H	T	O
	3,	2	8	7
+	2,	1	4	7

3.

	Th	H	T	O
	6,	5	2	6
+	2,	7	1	4

Estimate. Then find each sum.

4. 523 + 384

5. 667 + 328

6. 216 + 325

7. 1,375 + 4,417

8. 2,111 + 653

9. 6,272 + 3,412

10. 3,263 + 2,548

11. 4,323 + 674

12. 1,130 + 278

13. 5,973 + 1,238

Name____________________

Add 3- and 4-Digit Numbers

Estimate. Then find each sum.

1. 205 + 582 = ____

2. 725 + 237 = ____

3. 317 + 445 = ____

4. 377 + 429 = ____

5. 199 + 534 = ____

6. 2,627 + 4,312 = ____

7. 2,336 + 5,248 = ____

8. 7,743 + 1,185 = ____

9. 6,812 + 2,309 = ____

10. 3,476 + 358 = ____

11. 2,503 + 2,507 = ____

12. 7,883 + 1,374 = ____

13. 3,612 + 4,174 = ____

14. 1,975 + 585 = ____

15. 2,109 + 1,177 = ____

16. 832 + 415 = ____ **17.** 2,358 + 5,329 = ____ **18.** 4,210 + 688 = ____

Problem Solving and Test Prep

19. Margie flies 2,604 miles from Boston to Los Angeles for a vacation. She then flies the same distance to return home. How many miles does Margie fly in all?

____

20. Shawn has climbed 697 steps of the Eiffel Tower. He has 974 steps left to climb to reach the top. How many steps are on the Eiffel Tower?

____

21. Which is the sum of 2,485 and 821?

A 2,206
B 3,306
C 3,206
D 4,306

22. Which is the sum of 5,093 and 1,652?

A 3,441
B 6,645
C 5,745
D 6,745

Spiral Review

For 1–5, write each number in expanded form.

1. 49 ____________
2. 307 ____________
3. 622 ____________
4. 2,489 ____________
5. 6,002 ____________

For 8–10, answer the questions using the information in the line plot.

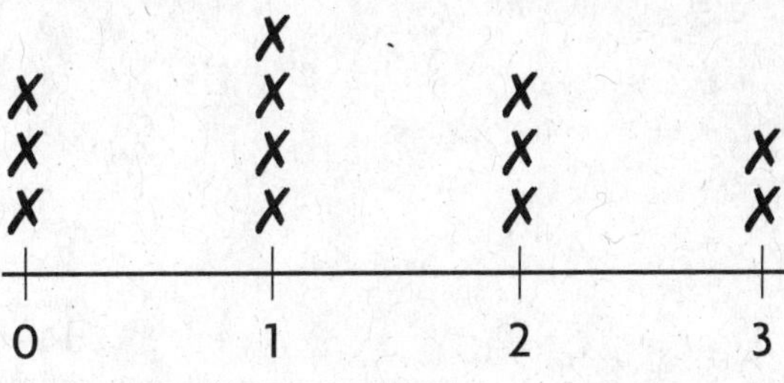

How Many Brothers or Sisters We Have

8. How many students have 1 brother or sister? ____________
9. What is the range of the data? ____________
10. What is the mode of the data? ____________

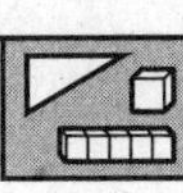

For 6–7, measure the length to the nearest inch.

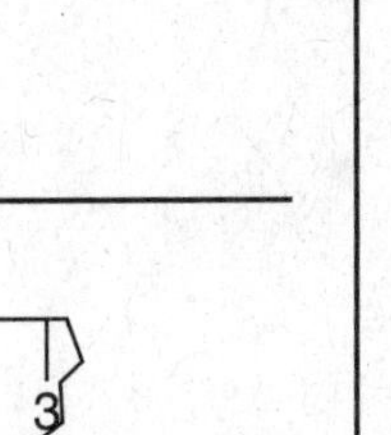

6. ____________

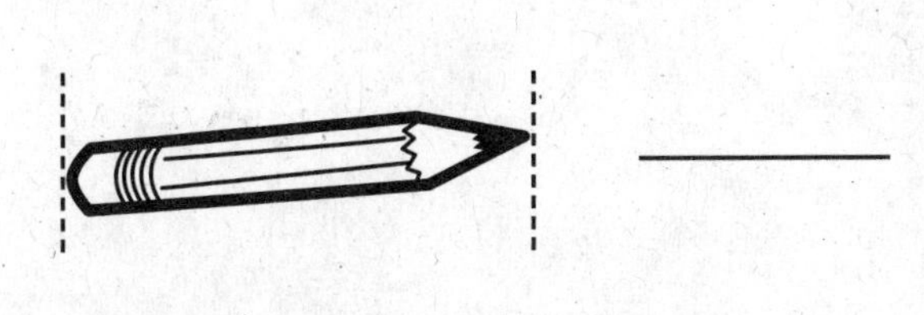

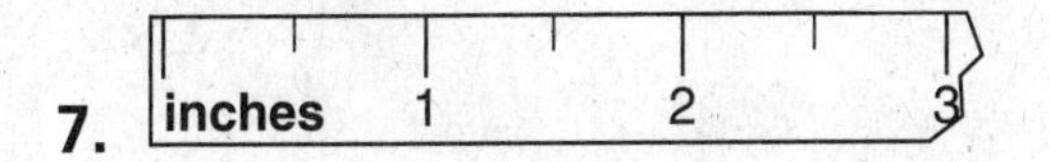

7. ____________

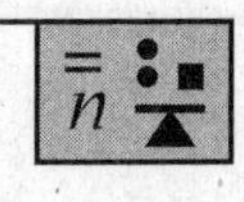

For 11–13, compare amounts to solve. Use <, >, or =.

11. Noni counted 1,618 cars on the long drive to her grandmother's house. Noni's brother counted 1,816 cars. Who counted fewer cars? ____________
12. Rosa's class collected 2,950 pennies to give to charity. Micah's class collected 2,950 pennies. Whose class collected fewer pennies? ____________
13. Bobby roasted 814 pumpkin seeds in one batch. In the second batch, he roasted 809 seeds. For which batch did Bobby roast more pumpkin seeds? ____________

Name____

Problem Solving Workshop Strategy: Predict and Test

In a survey, 100 students were asked to choose swimming or soccer as their favorite sport. Of those, 14 more students chose soccer than chose swimming. How many students chose soccer?

Read to Understand

1. Write the question as a fill-in-the-blank sentence.

____

Plan

2. How can you use the predict and test strategy to solve this problem?

____

Solve

3. Show how you solved the problem.

Predict		Test	
Swimming	**Soccer**	**Total**	**Notes**

4. Write your answer in a complete sentence.

____

Check

5. Is your answer reasonable? Explain.

____

____

Predict and test to solve.

6. There are 122 students signed up for soccer. Sixteen more girls than boys signed up. How many girls, and how many boys, signed up for soccer?

____

7. Max bought a baseball bat and a glove for $60. The baseball bat cost $14 more than the glove cost. How much did the glove cost?

____

Name____ Lesson 3.7

Problem Solving Workshop Strategy: Predict and Test

Problem Solving Strategy Practice

Predict and test to solve.

1. There were 300 people at the football game. There were 60 more students than adults at the game. How many students were at the football game?

____

2. The gym coach ordered 56 total, basketballs and soccer balls for next year. There were 10 fewer basketballs ordered than soccer balls ordered. How many of each type of ball were ordered?

____

Mixed Strategy Practice

USE DATA For 3–4, use the table.

3. Sami and Juan had the same number of baseball cards. Then Sami received some baseball cards for his birthday. How many cards did Sami receive for his birthday?

____

Baseball Cards Collected	
Name	**Number of Cards**
Sami	250
Pete	150
Juan	200

4. Pete has 50 baseball cards of players that are pitchers. He has 25 baseball cards of players that are catchers. The rest of his baseball cards are of players that are outfielders. How many cards are of players that are outfielders?

____

5. Lacy lost 35 barrettes. She then lost 15 more barrettes. At the end of the day she had 5 barrettes left. How many barrettes did lacy have to start?

____

6. Sarah, Jose, and Mike are sitting in a row. If you face them, Mike is not sitting on the left. Sarah is sitting to the right of Jose. Who is sitting in the middle?

____

Name______________________________

Estimate Differences

You can *estimate* to find a difference that is close to the exact answer.

A number line can help you estimate differences.

Estimate the difference. 435 – 268

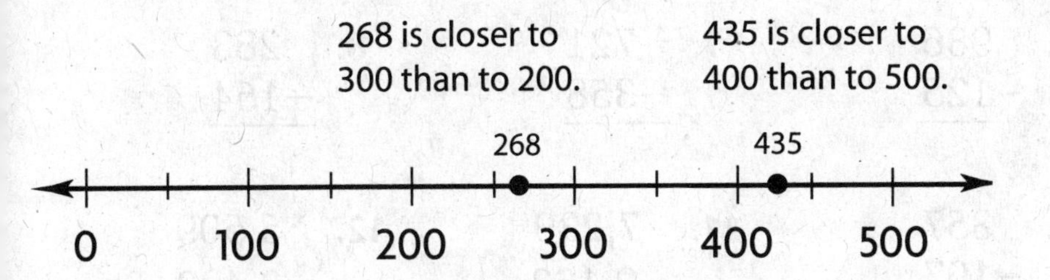

$$\begin{array}{rcr} 435 & \rightarrow & 400 \\ -\ 268 & \rightarrow & -\ 300 \\ \hline & & 100 \end{array}$$

Estimate the difference. 3,612 – 1,125

1,125 is closer to 1,000 than to 2,000.

3,612 is closer to 4,000 than to 3,000

1,125 3,612

0 1,000 2,000 3,000 4,000

$$\begin{array}{rcr} 3{,}612 & \rightarrow & 4{,}000 \\ -\ 1{,}125 & \rightarrow & -\ 1{,}000 \\ \hline & & 3{,}000 \end{array}$$

Estimate each difference.

1. 83 – 27

2. 65 – 17

3. 59 – 31

4. 94 – 45

5. 271 – 152

6. 794 – 258

7. 484 – 192

8. 751 – 258

9. 3,382 – 2,105

10. 4,609 – 1,285

11. 5,642 – 3,195

12. 6,823 – 1,298

Name ____________________

Estimate Differences

Estimate each difference.

1. 74 − 38 = ____
2. 512 − 26 = ____
3. 47 − 13 = ____
4. 65 − 32 = ____
5. 371 − 159 = ____
6. 986 − 125 = ____
7. 721 − 358 = ____
8. 283 − 154 = ____
9. 561 − 432 = ____
10. 357 − 197 = ____
11. 7,239 − 2,163 = ____
12. 5,509 − 3,492 = ____
13. 6,250 − 5,199 = ____
14. 7,215 − 1,522 = ____
15. 3,211 − 1,897 = ____
16. 9,132 − 2,625 = ____

Problem Solving and Test Prep

USE DATA For 17–18, use the table below.

Largest Saltwater Fish Caught	
Type of Fish	**Weight in Pounds**
Black Marlin	1,560
Pacific Blue Marlin	1,376
Striped Marlin	494
White Marlin	181

17. About how much more did the striped marlin weigh than the white marlin weighed?

____

18. About how much more did the black marlin weigh than the striped marlin weighed?

____

19. Lana estimated 923−452. She rounded each number to the nearest hundred and then subtracted. Which was Lana's estimate?

A 300
B 400
C 500
D 600

20. Which is the estimated difference?

5,659 − 2,382

A 3,000
B 4,000
C 5,000
D 6,000

Name____

Subtract With Regrouping

You can use base-ten blocks to subtract 2-digit numbers.

Subtract 53 – 28.

Subtract the ones.
Regroup as necessary.

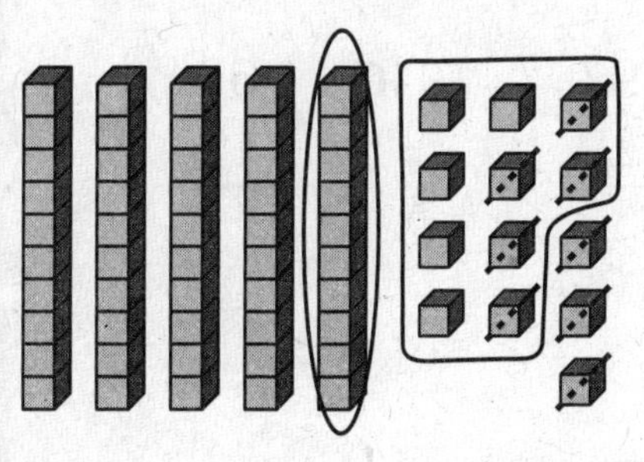

```
 4 13
  53
– 28
   5
```

Since 8 > 3, regroup 53 tens as 4 tens, 13 ones.

Subtract the tens

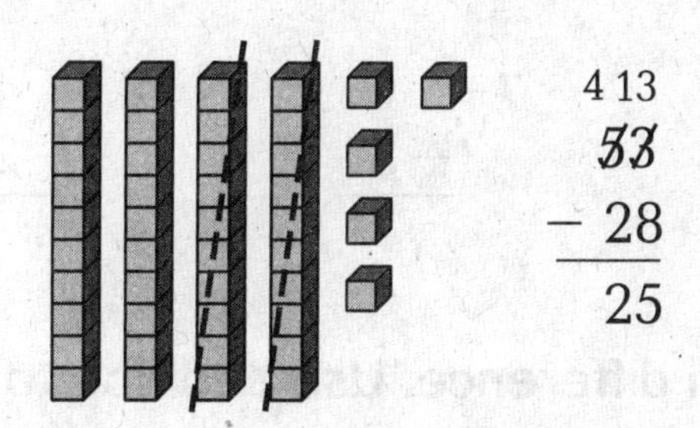

You can use addition to check:

25 + 28 = 53.

So, **53 – 28 =25.**

Find each difference. Use addition to check.

1. 65 – 43 = ☐ ________
2. 87 – 49 = ☐ ________
3. 93 – 28 = ☐ ________
4. 42 – 26 = ☐ ________
5. 73 – 52 = ☐ ________
6. 82 – 34 = ☐ ________
7. 33 – 14 = ☐ ________
8. 59 – 38 = ☐ ________
9. 96 – 27 = ☐ ________
10. 68 – 48 = ☐ ________
11. 75 – 63 = ☐ ________
12. 67 – 22 = ☐ ________

Name ______________________

Lesson 4.2

Subtract with Regrouping

Estimate. Then find each difference.

1. $\begin{array}{r} 79 \\ -53 \\ \hline \end{array}$ **2.** $\begin{array}{r} 35 \\ -14 \\ \hline \end{array}$ **3.** $\begin{array}{r} 63 \\ -45 \\ \hline \end{array}$ **4.** $\begin{array}{r} 76 \\ -58 \\ \hline \end{array}$ **5.** $\begin{array}{r} 55 \\ -16 \\ \hline \end{array}$

6. $\begin{array}{r} 82 \\ -47 \\ \hline \end{array}$ **7.** $\begin{array}{r} 68 \\ -31 \\ \hline \end{array}$ **8.** $\begin{array}{r} 47 \\ -25 \\ \hline \end{array}$ **9.** $\begin{array}{r} 97 \\ -19 \\ \hline \end{array}$ **10.** $\begin{array}{r} 63 \\ -17 \\ \hline \end{array}$

Find each difference. Use addition to check.

11. $56 - 41 =$ ______ **12.** $83 - 35 =$ ______

13. $67 - 31 =$ ______ **14.** $36 - 19 =$ ______

15. $66 - 15 =$ ______ **16.** $91 - 22 =$ ______

Problem Solving and Test Prep

17. A brown bear has an average height of 48 inches. An American black bear has an average height of 33 inches. What is the difference between these two bears' average heights?

______

18. An adult polar bear has a height of 63 inches. A polar bear cub has a height of 39 inches. What is the difference of these heights?

______

19. Which is the difference?

$72 - 48 =$ ______

A 24 **C** 34
B 26 **D** 36

20. At a fair, a drink stand sold 45 glasses of lemonade and 29 glasses of tea. How many more glasses of lemonade than glasses of tea were sold?

A 26 **C** 16
B 24 **D** 14

Name______________________________ Lesson 4.3

Model 3-Digit Subtraction

Use base-ten blocks to find 443 – 267.

Model 443.
Regroup 4 tens, 3 ones as 3 tens, 13 ones.
Subtract the ones.

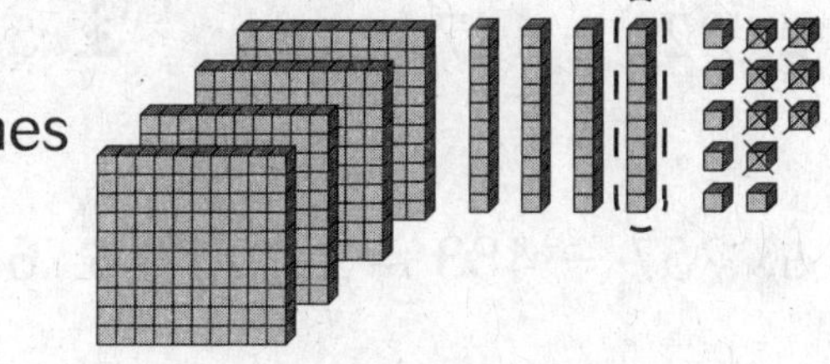

$$\begin{array}{r} \scriptstyle 3\ 13 \\ 44\not{3} \\ -\ 267 \\ \hline 6 \end{array}$$

Subtract the tens.
Regroup 4 hundreds, 3 tens as 3 hundreds, 13 tens.

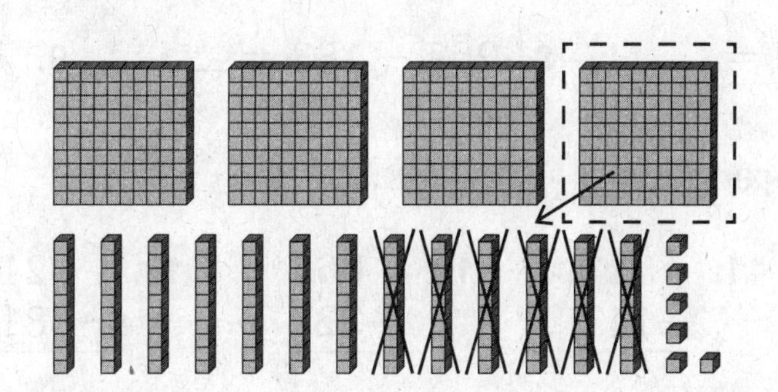

$$\begin{array}{r} \scriptstyle 3\ 13\ 13 \\ \not{4}\not{4}\not{3} \\ -\ 267 \\ \hline 76 \end{array}$$

Subtract the hundreds.

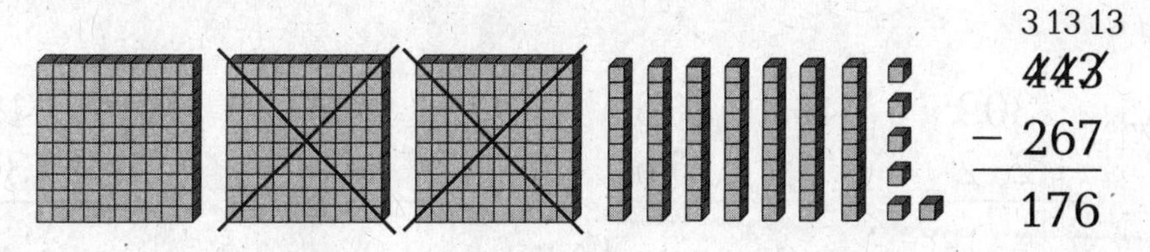

$$\begin{array}{r} \scriptstyle 3\ 13\ 13 \\ \not{4}\not{4}\not{3} \\ -\ 267 \\ \hline 176 \end{array}$$

So, 443 – 267 = 176.

Use base-ten blocks to find each difference.

1. $\begin{array}{r} 432 \\ -\ 218 \\ \hline \end{array}$ **2.** $\begin{array}{r} 633 \\ -\ 375 \\ \hline \end{array}$ **3.** $\begin{array}{r} 412 \\ -\ 108 \\ \hline \end{array}$

4. $\begin{array}{r} 746 \\ -\ 388 \\ \hline \end{array}$ **5.** $\begin{array}{r} 529 \\ -\ 483 \\ \hline \end{array}$ **6.** $\begin{array}{r} 385 \\ -\ 219 \\ \hline \end{array}$

Name ________________________

Model 3-Digit Subtraction

Use base-ten blocks to find each difference.

1. $494 - 271 =$ _____
2. $324 - 147 =$ _____
3. $549 - 255 =$ _____
4. $311 - 205 =$ _____
5. $757 - 483 =$ _____
6. $623 - 197 =$ _____
7. $388 - 265 =$ _____
8. $267 - 183 =$ _____
9. $706 - 258 =$ _____

Find each difference.

10. $765 - 154$
11. $821 - 143$
12. $665 - 327$
13. $821 - 581$
14. $387 - 198$
15. $309 - 212$
16. $485 - 276$
17. $784 - 359$
18. $319 - 236$
19. $418 - 276$
20. $189 - 178$
21. $548 - 318$
22. $707 - 629$
23. $845 - 563$
24. $956 - 127$
25. $752 - 382$
26. $607 - 199$
27. $387 - 225$
28. $900 - 459$
29. $765 - 150$
30. $777 - 444$
31. $228 - 116$
32. $939 - 540$
33. $442 - 378$
34. $808 - 102$

Name ______________________________ **Week 6**

Spiral Review

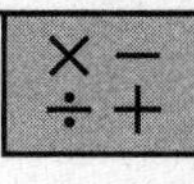

For 1–5, find each difference. Use addition to check.

1. $\begin{array}{r} 536 \\ -159 \\ \hline \end{array}$

2. $\begin{array}{r} 627 \\ -548 \\ \hline \end{array}$

3. $\begin{array}{r} 2{,}172 \\ -601 \\ \hline \end{array}$

4. $\begin{array}{r} 6{,}840 \\ -1{,}588 \\ \hline \end{array}$

5. $\begin{array}{r} 4{,}000 \\ -2{,}199 \\ \hline \end{array}$

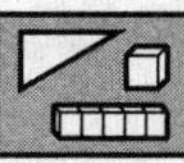

For 6–7, combine the given figures to make a new figure. Draw an outline of the new figure.

6. 

7.   

For 8–11, extend the patterns.

8. Skip count by twos:

   11, 13, ____, ____, ____, ____

9. Skip count by threes:

   20, 23, ____, ____, ____, ____

10. Skip count by fives:

    23, 28, ____, ____, ____, ____

11. Skip count by tens:

    7, 17, ____, ____, ____, ____

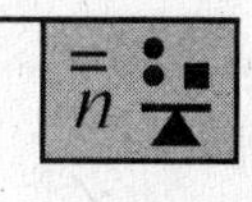

For 12–14, predict the next number in each pattern. Explain.

12. 29, 34, 39, 44, 49, ☐

    ____

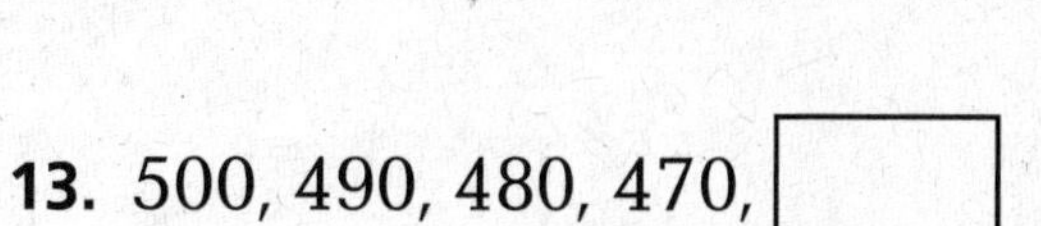

13. 500, 490, 480, 470, ☐

    ____

14. 130, 155, 180, 205, ☐

    ____

Name______________________________

Subtract 3- and 4-Digit Numbers

You can use place-value charts to subtract.

Find the difference. **5,166**
− 1,293

Estimate. 5,166 rounds to 5,000
1,293 rounds to 1,000
5,000 − 1,000 = 4,000

Step 1

Subtract the ones. Regroup if necessary.

	Th	H	T	O
	5,	1	6	6
−	1,	2	9	3
				3

Step 2

Subtract the tens. Regroup if necessary.

	Th	H	T	O
		0	16	
	5,	~~1~~	~~6~~	6
−	1,	2	9	3
			7	3

Step 3

Subtract the hundreds. Regroup if necessary.

	Th	H	T	O
	4	10	16	
	~~5,~~	~~1~~	~~6~~	6
−	1,	2	9	3
		8	7	3

Step 4

Subtract the thousands. Regroup if necessary.

	Th	H	T	O
	4	10	16	
	~~5,~~	~~1~~	~~6~~	6
−	1,	2	9	3
	3,	8	7	3

So, 5,166 − 1,293 = 3,873.

Since 3,873 is close to the estimate of 4,000, the answer is reasonable.

Estimate. Then find the difference.

1. 523 − 328

2. 767 − 238

3. 918 − 228

4. 4,327 − 1,784

5. 2,573 − 682

6. 6,272 − 3,460

7. 3,623 − 2,543

8. 5,239 − 638

9. 1,156 − 278

10. 5,973 − 1,483

Name______________________

Subtract 3- and 4-Digit Numbers

Estimate. Then find the difference.

1. 593 − 282 = ______
2. 377 − 188 = ______
3. 732 − 489 = ______
4. 654 − 386 = ______
5. 534 − 175 = ______

6. 4,657 − 2,132 = ______
7. 3,673 − 1,583 = ______
8. 7,526 − 5,649 = ______
9. 6,812 − 2,309 = ______
10. 3,476 − 967 = ______

11. 2,478 − 626 = ______
12. 7,388 − 6,374 = ______
13. 4,172 − 2,381 = ______
14. 5,672 − 825 = ______
15. 3,477 − 1,298 = ______

16. 784 − 547 = ______

17. 5,368 − 3,392 = ______

18. 5,265 − 389 = ______

Problem Solving and Test Prep

19. **Fast Fact** There are 1,785 Indo-Chinese tigers left in the wild and 500 Sumatran tigers left in the wild. What is the difference in the number of Indo-Chinese tigers and the number of Sumatran tigers, left in the wild?

____

20. There are 712 African mountain gorillas left in the wild. Ten years ago there were 581 of these gorillas left in the wild. How many more African mountain gorillas are in the wild today than there were 10 years ago?

____

21. Which is the difference between 3,945 and 2,194?

A 2,651 **C** 1,751
B 1,741 **D** 1,851

22. Which is the difference?

5,352 − 674

A 5,788 **C** 4,788
B 5,322 **D** 4,678

Name______________________________

Subtract Across Zeros

You can use place-value charts to subtract across zeros.
Find 5,000 – 3,159.

Step 1

In the ones column, 9 > 0. Where to begin regrouping? Thousands.

	Th	H	T	O
	4	~~10~~		
	~~5~~,	~~0~~	0	0
–	3,	1	5	9

Step 2

Regroup the hundreds.

	Th	H	T	O
	4	9 ~~10~~	10	
	~~5~~,	~~0~~	~~0~~	0
–	3,	1	5	9

Step 3

Regroup the tens.

	Th	H	T	O
	4	9 ~~10~~	9 ~~10~~	10
	~~5~~,	~~0~~	~~0~~	~~0~~
–	3,	1	5	9

Step 4

Subtract.

	Th	H	T	O
	4	9 ~~10~~	9 ~~10~~	10
	~~5~~,	~~0~~	~~0~~	~~0~~
–	3,	1	5	9
	1,	8	4	1

Find each difference.

1.

	Th	H	T	O
	7,	2	0	0
–	2,	1	5	9

2.

	Th	H	T	O
	3,	0	0	8
–	1,	3	4	7

3.

	Th	H	T	O
	6,	0	0	0
–	3,	4	2	6

Estimate. Then find the difference.

4. 206 – 129

5. 600 – 263

6. 980 – 538

7. 3,000 – 1,629

8. 3,400 – 2,188

9. 6,209 – 3,881

10. 4,019 – 1,462

11. 9,000 – 7,912

12. 8,100 – 2,253

13. 5,005 – 3,556

Name ____________________

Subtract Across Zeros

Estimate. Then find each difference.

1. $\begin{array}{r} 508 \\ -\ 175 \\ \hline \end{array}$

2. $\begin{array}{r} 400 \\ -\ 329 \\ \hline \end{array}$

3. $\begin{array}{r} 980 \\ -\ 246 \\ \hline \end{array}$

4. $\begin{array}{r} 5{,}600 \\ -\ 4{,}193 \\ \hline \end{array}$

5. $\begin{array}{r} 7{,}000 \\ -\ 5{,}823 \\ \hline \end{array}$

6. $\begin{array}{r} 6{,}088 \\ -\ 1{,}697 \\ \hline \end{array}$

7. $\begin{array}{r} 301 \\ -\ 213 \\ \hline \end{array}$

8. $\begin{array}{r} 7{,}508 \\ -\ 1{,}384 \\ \hline \end{array}$

9. $\begin{array}{r} 930 \\ -\ 429 \\ \hline \end{array}$

10. $\begin{array}{r} 5{,}000 \\ -\ 3{,}779 \\ \hline \end{array}$

Find each difference. Use addition to check.

11. 906 − 421 = ______

12. 4,000 − 3,724 = ______

13. 600 − 431 = ______

14. 5,002 − 3,687 = ______

15. 6,908 − 1,225 = ______

16. 830 − 315 = ______

Problem Solving and Test Prep

17. Juan wins tickets playing arcade games. He needs 400 tickets to purchase a beach ball. He already has 252 tickets. How many more tickets does Juan need?

____

18. Hannah already won 1,287 tickets playing arcade games. She needs 3,000 tickets to purchase a sweatshirt. How many more tickets does Hannah need?

____

19. Which is the difference?

4,092 − 1,628 = ______

A 3,464
B 2,564
C 2,474
D 2,464

20. Which is the difference?

$\begin{array}{r} 8{,}008 \\ -\ 2{,}369 \\ \hline \end{array}$

A 5,639
B 5,739
C 5,749
D 6,639

Name____________________

Problem Solving Workshop Skill: Choose the Operation

Brad has room for 100 baseball cards in his binder. He has 64 cards already. How many more cards does Brad need to fill his binder?

1. What are you asked to find? ____________________

____

2. Use the chart below to help you decide which operation you can use to solve this problem.

Operation	When to Use the Operation
Add	join groups to find how many in all, or the total
Subtract	take away or compare to find how many more, how many are left, or how many fewer

Which operation will you use to solve this problem? Explain your choice.

____

____

3. How many more cards does Brad need to fill his binder? Show your work.

____

Tell which operation you would use. Then solve the problem.

4. Samantha took pictures of her family's vacation. She took 173 pictures with a digital camera and 48 pictures with a single-use camera. How many more pictures did Samantha take with the digital camera than she did with the single-use camera?

____

____

5. Miguel has 123 trading cards in a binder. He has room for 97 more trading cards in the binder. How many trading cards can Miguel fit in his binder in all?

____

____

AF 1.1 Represent relationships of quantities in form of mathematical expressions, equations, or inequalities.

Name ____________________ Lesson 4.6

Problem Solving Workshop Skill: Choose the Operation

Problem Solving Skill Practice

Tell which operation you would use. Then solve the problem.

1. Julia read 128 pages in a book. She needs to read 175 more pages to finish the book. How many pages total, are in the book?

____

____

2. The library has 325 books about animals. Of these, 158 are checked out. How many books about animals are still in the library?

____

____

3. Kara plans to put together a puzzle. The puzzle contains 225 pieces. She has put together 137 pieces. How many more pieces does Kara need to put together to complete the puzzle?

____

____

4. Jeremy had 529 coins in his collection. He collected 217 more coins. How many coins are now in Jeremy's collection?

____

____

Mixed Applications

USE DATA For 5–6, use the table below.

5. How many glasses of lemonade were sold in all, from Monday to Friday? Will you need to use an estimate or an exact answer?

____

6. On Saturday, the lemonade stand sold 15 glasses of lemonade. How many more glasses were sold on Friday than were sold on Wednesday?

____

Glasses of Lemonade Sold

Day	Number of Glasses Sold
Monday	8
Tuesday	11
Wednesday	10
Thursday	7
Friday	15

7. The library contains 217 magazines, and 60 videos, that students can check out. Students have 109 magazines checked out. How many magazines are now available at the library?

____

Name____

Algebra: Number Patterns

A **pattern** is an ordered set of numbers or objects. You can use the relationships between the numbers or objects in a pattern to predict what comes next.

Find a pattern, then predict the next number in the pattern.

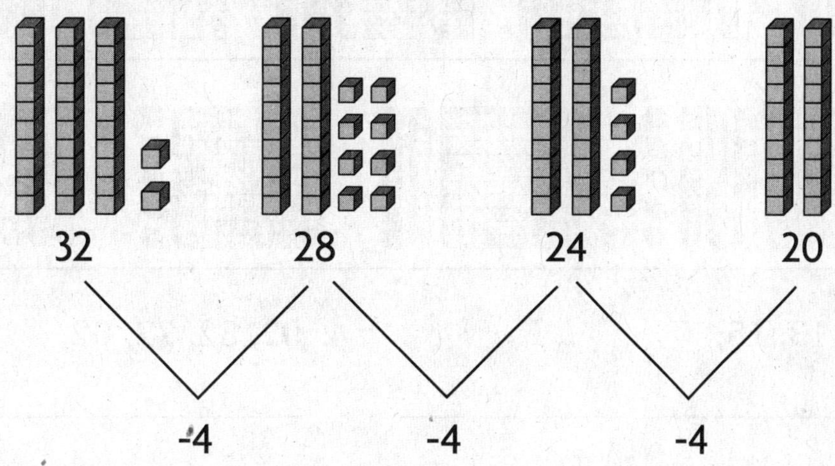

Each number is 4 less than the number before it.

This means you can predict the next number by subtracting 4.

$20 - 4 = 16$

So, 16 is the next number in the pattern.

Fill in the boxes to show the pattern.
Then predict the next number in the pattern.

1. 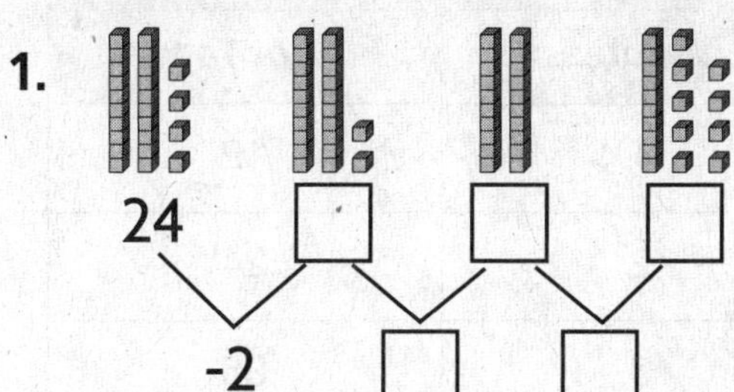

The next number in the pattern will be ____________.

Predict the next number in the pattern.

2. 20, 24, 28, 32, ☐ ___

3. 56, 46, 36, 26, ☐ ___

4. 54, 50, 46, 42, ☐ ___

5. 220, 200, 180, 160, ☐ ___

AF 2.2 Extend and recognize a linear pattern by its rules (e.g., the number of legs on a given number of horses may be calculated by counting by 4s or by multiplying the number of horses by 4).

Name____

Number Patterns

Predict the next number in each pattern. Explain.

1. 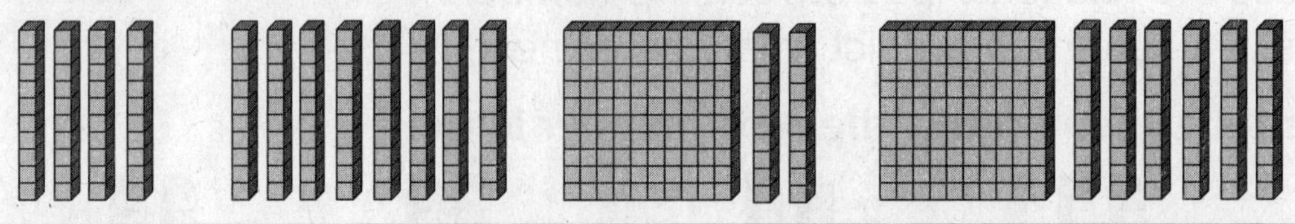

____

2. 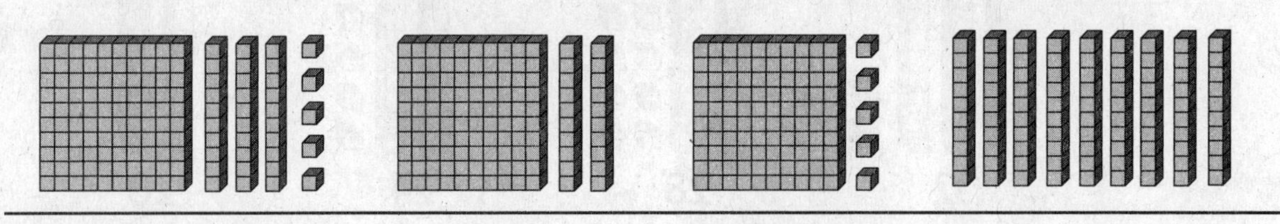

____

3. 7, 10, 13, 16, ______

4. 92, 82, 72, 62, ______

5. 40, 55, 70, 85, ______

6. 270, 280, 290, 300, ______

Problem Solving and Test Prep

USE DATA For 7–8, use the table below.

7. The table shows how many times George tapped his foot during 4 minutes. Describe the pattern.

____

8. **What If** George continued to tap his foot for a total of five minutes? How many foot taps would he have took part in then?

____

Foot Taps per Minute

Minutes	Foot Taps
1	12
2	24
3	36
4	48

9. Which will be the next number in the pattern?

27, 34, 41, 48,

A 57 **C** 54
B 55 **D** 51

10. Which will be the next number in the pattern?

320, 305, 290, 275,

A 270 **C** 260
B 265 **D** 255

Name ____________________

Lesson 4.8

Problem Solving Workshop Skill: Estimate or Exact Answer

A Boeing 777, that can carry up to 368, passengers is traveling from Dallas to Chicago. Can the airplane carry 1,080 people in 3 trips?

1. What are you asked to find? ____________________

2. Is an estimate enough to solve this problem? Explain. ____________________

3. Which information do you need to find an exact answer? Explain.

4. Show how to solve the problem in the space below.

5. Can the airplane carry 1,080 people in 3 trips? Explain. ____________________

Tell whether you need an exact answer or an estimate. Then solve.

6. On Saturday, 874 people visited a zoo. Of the visitors, 636 were children and the rest were adults. How many of the visitors to the zoo were adults?

7. The Mall of America, in Minnesota, contains 520 stores. Of those stores, 83 are restaurants. About how many stores in the Mall of America are not restaurants?

MR 2.5 Indicate the relative advantages of exact and appropriate solutions to problems and give answers to a specified degree of accuracy.

Name______________________________

Lesson 4.8

Problem Solving Workshop Skill: Estimate or Exact Answer

Problem Solving Skill Practice

Tell whether you need an exact answer or an estimate. Then solve.

1. A Boeing 777 can carry 368 passengers. A Boeing 747 can carry 416 passengers. About how many passengers can the airplanes carry in all?

____

2. Tony flies from Buffalo, NY to Springfield, IL. The first leg of his trip is 139 miles long. The second leg of his trip is 458 miles long. How many more miles is the second leg of his trip than the first leg of his trip?

____

Mixed Applications

USE DATA For 3–4, use the map below.

3. Tanya is flying from Flagstaff to San Francisco to visit friends. Her friends drive 16 miles to work each day. How many miles is Tanya's round trip flight?

____

4. How many more miles is it to fly from San Francisco to Flagstaff than from San Francisco to Fresno?

____

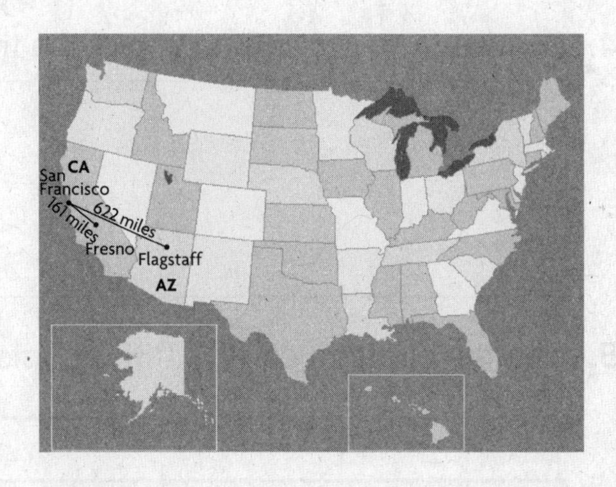

5. A Boeing 757 airplane can carry up to 208 passengers. Can the airplane carry 625 passengers if it takes 3 trips? Explain.

____

6. Luke has 3 coins in his pocket on Monday. On Tuesday he has 8 coins in his pocket. On Wednesday he has the same number of coins in his pocket as he had on Monday and Tuesday combined, minus 4 coins. On which day does Luke have the most coins in his pocket?

____

Spiral Review

For 1–5, write the value of the underlined digit.

1. $7{,}8\underline{1}6$ __________
2. $\underline{9}{,}217$ __________
3. $6{,}4\underline{2}2$ __________
4. $\underline{3}{,}405$ __________
5. $6{,}\underline{2}12$ __________

For 8–10, use the graph to answer the questions.

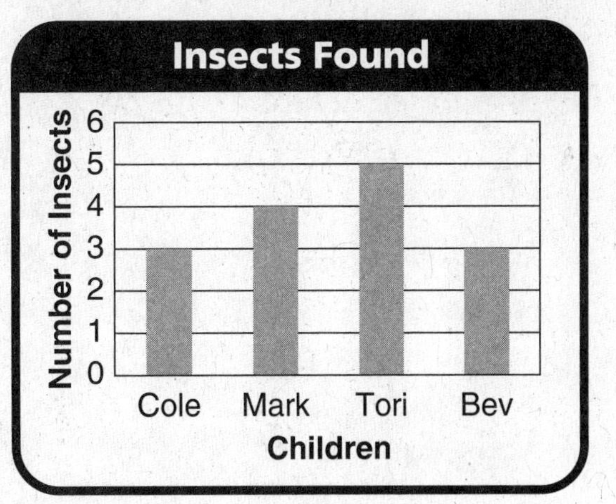

8. Who found the most insects?

____

9. Who found the same amount of insects? __________

10. How many insects did Mark find?

____

For 6–7, measure the length to the nearest centimeter.

6. 1 2 3 4 5 6 7 8 9 10 11 12 centimeters

7. 1 2 3 4 5 6 7 8 9 10 11 12 centimeters

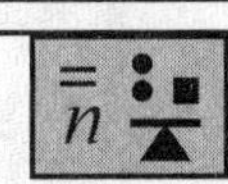

For 11–13, find each sum.

11. __________ $= 30 + 5$

12. $17 + 10 =$ __________

13. __________ $= 100 + 100$

Name____

Algebra: Relate Addition to Multiplication

Use counters to model 3 groups of 3. Then write an additional sentence and a multiplication sentence.

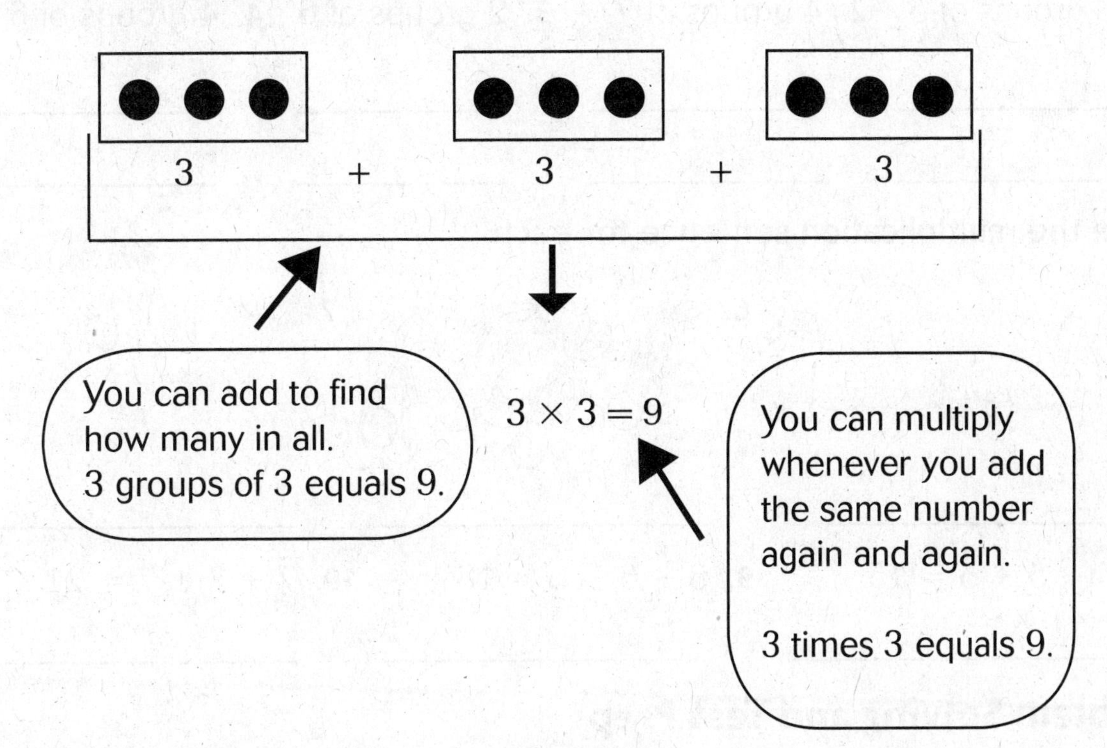

There are 9 counters.

Use counters to model. Then write an addition sentence and a multiplication sentence for each.

1. 4 cups of 3 blueberries

____

____

2. 2 cups of 3 blueberries

____

____

3. 5 cups of 2 blueberries

____

____

4. 3 cups of 5 blueberries

____

____

5. 7 cups of 3 blueberries

____

____

6. 4 cups of 2 blueberries

____

____

Name______________________________

Algebra: Relate Addition to Multiplication

Use counters to model. Then write an addition sentence and a multiplication sentence for each.

1. 3 groups of 5 **2.** 4 groups of 7 **3.** 2 groups of 6 **4.** 4 groups of 6

______ ______ ______ ______

______ ______ ______ ______

Write the multiplication sentence for each.

5.  **6.**  **7.** 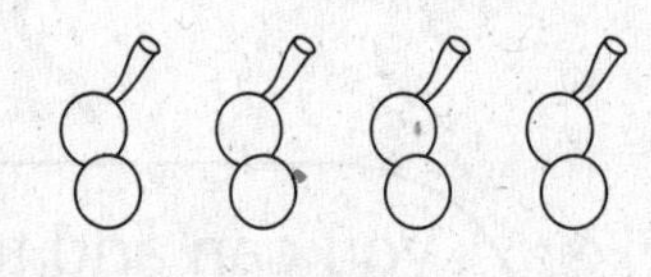

______ ______ ______

8. $5 + 5 + 5 = 15$ **9.** $6 + 6 + 6 = 18$ **10.** $7 + 7 + 7 = 21$

______ ______ ______

Problem Solving and Test Prep

11. Mike bakes apple bread. He uses 2 apples for every loaf of bread. He makes 4 loaves of bread. How many apples does Mike use in all?

______

12. Cynthia makes small pizzas. She puts 4 mushrooms on each pizza. How many mushrooms does Cynthia need to make 3 pizzas?

______

13. Which is another way to show $3 + 3 + 3 + 3$?

A 4×3

B 4×4

C 3×12

D 3×3

14. Which is another way to show $6 + 6 + 6$?

A 6×4

B 3×3

C 3×6

D 6×6

Name____ **Lesson 5.2**

Algebra: Model with Arrays

An **array** is a group of objects in rows and columns.

Write multiplication sentences for the arrays below.

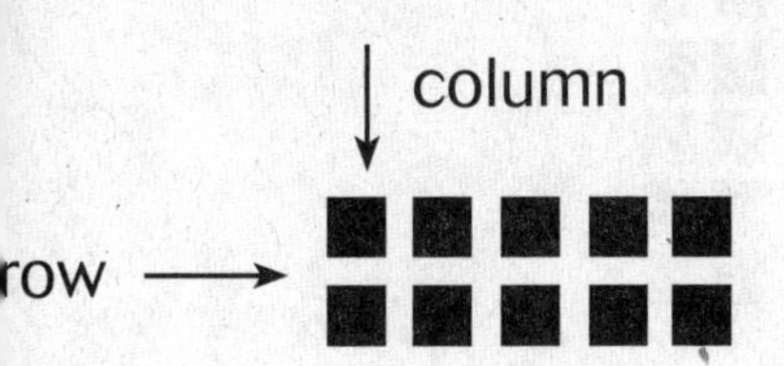

There are 2 rows and 5 columns.

Count. There are total of 10 tiles.
2 rows of 5 equals 10.
So, the multiplication sentence is $2 \times 5 = 10$.

There are 5 rows and 2 columns.

Count. there are total of 10 tiles.
5 rows of 2 equals 10. So, the multiplication sentence is $5 \times 2 = 10$.

Write a multiplication sentence for each array.

1.

____

____

2. 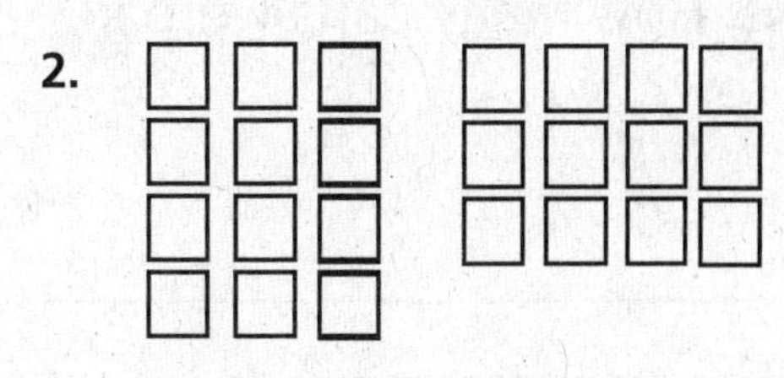

____

____

3. 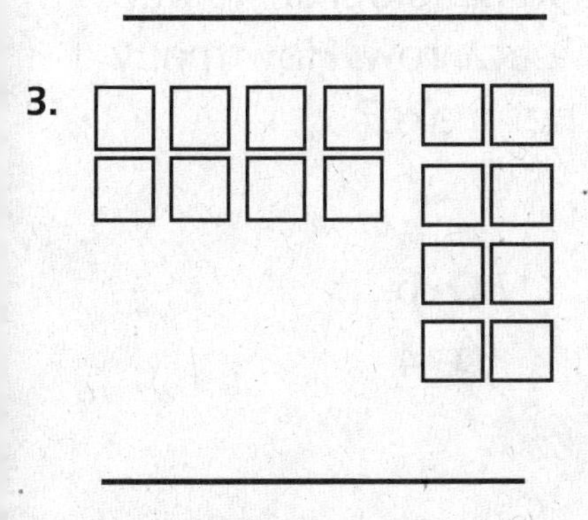

____

____

4. 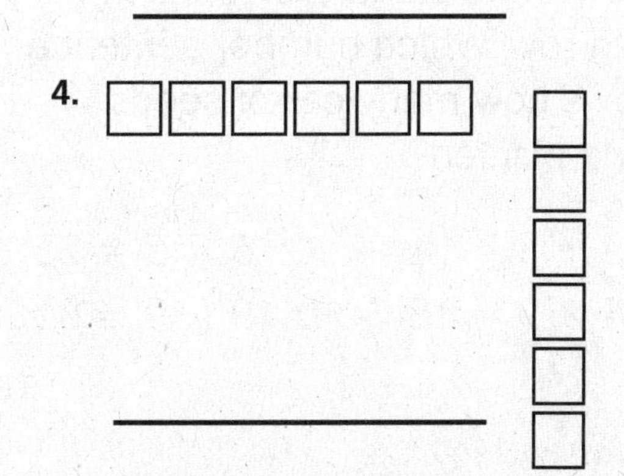

____

____

Name ____

Model with Arrays

Write a multiplication sentence for each array.

1. 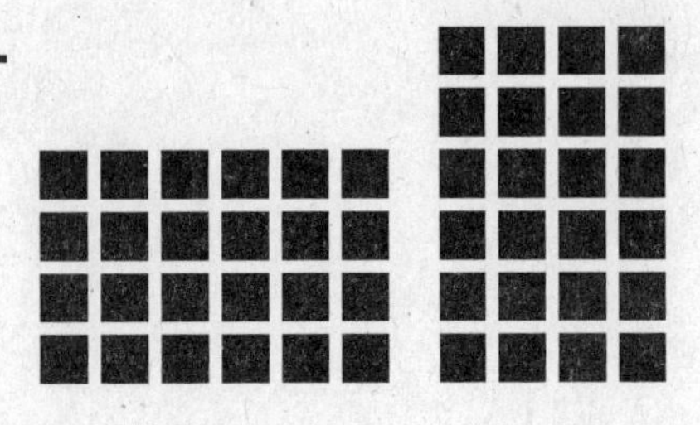

____

2. 

____

Write the multiplication sentence for each array. Then draw the array that shows the Commutative Property.

3. 

4. 

5. 

______________________ ______________________ ______________________

Problem Solving and Test Prep

6. Jerry put 30 cans of tomatoes in 6 rows. How many cans were in each row?

____

7. Maya pulled 6 carrots each, from 2 rows in her garden. She used 4 carrots to make soup. How many carrots does Maya have left?

____

8. Kayla planted carrot seeds in 5 rows. She planted 9 seeds in each row. Which number sentence shows how many carrot seeds Kayla planted?

A $9 + 5 = 14$
B $5 \times 9 = 45$
C $5 \times 5 = 25$
D $9 \times 9 = 81$

9. Chet stacked blocks to make a wall. He used 32 blocks. He put 8 blocks in each row. How many rows did Chet make?

A 8
B 5
C 6
D 4

Name____

Multiply with 2

Write a multiplication sentence for the counters below.

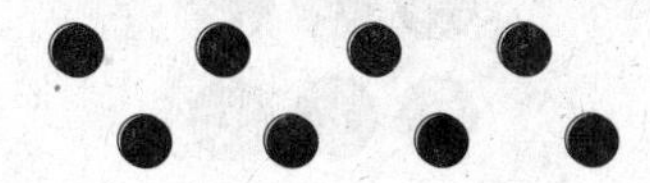

There are 2 groups of 4 counters.
Add the two groups together.
$4 + 4 = 8$
Multiply, since 4 occurred in the addition equation twice.
So the multiplication sentence for these counters is :
$2 \times 4 = 8$

Write the multiplicaion sentence for each.

1. 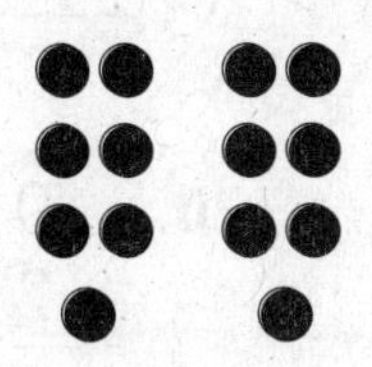

____

2. 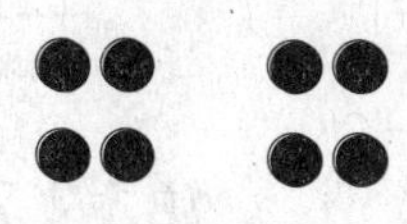

____

3. 

____

4. 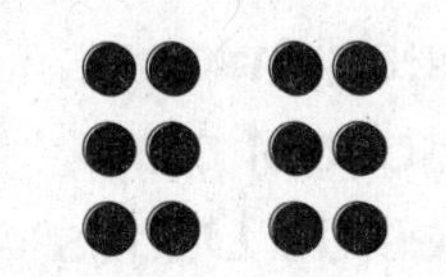

____

5. 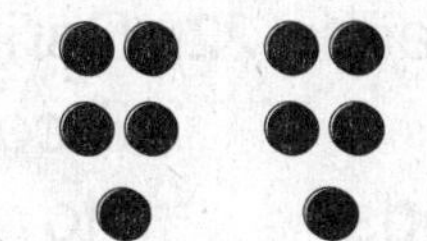

____

6. 

____

7. 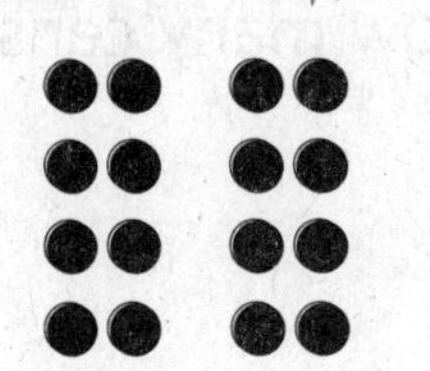

____

8. 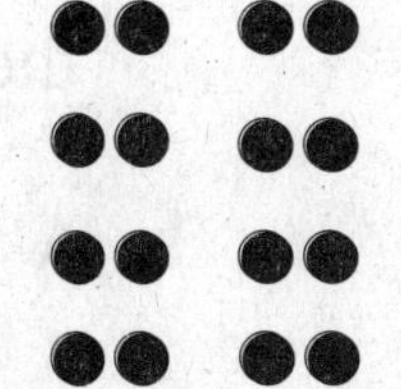

____

9. 

____

Name

Multiply with 2

Write a multiplication sentence for each.

1. ●● ●● ●● ●● ●●

2. ●●●●● ●●●●●

3. ●●● ●●●

4. ●● ●● ●● ●●

Find the product.

5. $2 \times 7 =$ ____
6. $5 \times 2 =$ ____
7. $2 \times 4 =$ ____
8. $3 \times 2 =$ ____

9. 2×3
10. 5×2
11. 2×8
12. 2×6
13. 3×2
14. 2×4

15. 6×2
16. 2×7
17. 2×2
18. 7×2
19. 4×2
20. 9×2

Problem Solving and Test Prep

21. Seven friends go for a swim. Each pays $2 to use the town pool. How much money do the friends spend at the pool?

22. Darius and Marvin each wear 3 costumes in the school play. How many costumes do Darius and Marvin wear?

23. Savannah and George each wore 4 costumes in the school play. Which number sentence shows Savannah and George's total number of costumes worn?

   A $2 \times 4 = 8$
   B $3 \times 2 = 6$
   C $5 + 2 = 7$
   D $4 \times 2 = 6$

24. There are 2 rows with 9 cans in each row. Which number sentence shows how many cans there are in all?

   A $9 + 2 = 11$
   B $9 \times 3 = 21$
   C $2 + 9 = 11$
   D $2 \times 9 = 18$

Name____ **Lesson 5.4**

Multiply with 4

Find the product: $5 \times 4 = \square$

Use counters to model 5 equal groups of 4. There are 5 groups, each containing 4 counters.

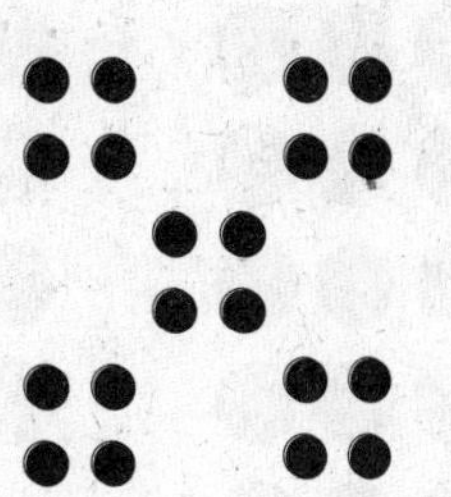

Count. There are a total of 20 counters. 5 groups of 4 equals 20.

So, $5 \times 4 = 20$

Find the product.

1.

2. 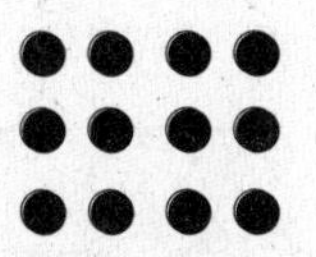

3. 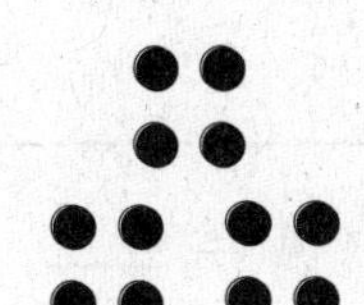

4. 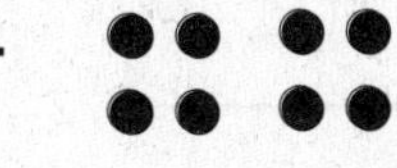

5. $4 \times 4 =$ ____

6. $4 \times 9 =$ ____

7. $5 \times 4 =$ ____

8. $6 \times 4 =$ ____

9. $\begin{array}{r} 4 \\ \times\ 8 \\ \hline \end{array}$

10. $\begin{array}{r} 3 \\ \times\ 4 \\ \hline \end{array}$

11. $\begin{array}{r} 4 \\ \times\ 3 \\ \hline \end{array}$

12. $\begin{array}{r} 4 \\ \times\ 8 \\ \hline \end{array}$

Name________________________________ Lesson 5.4

Multiply with 4

Find the product.

1. ●●●●
●●●●
●●●●
●●●●
●●●●
●●●●
●●●●
●●●●

2. 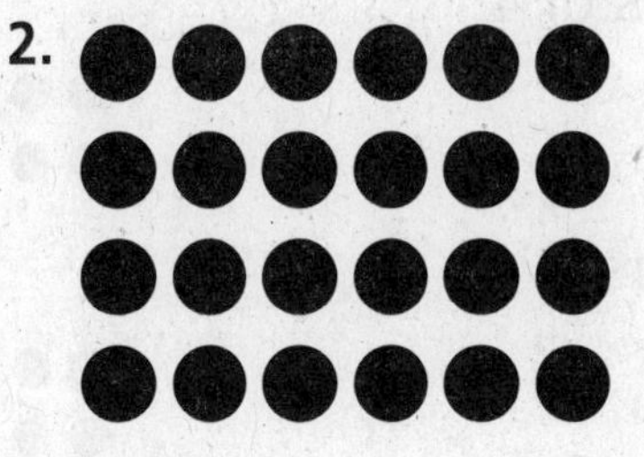

3. 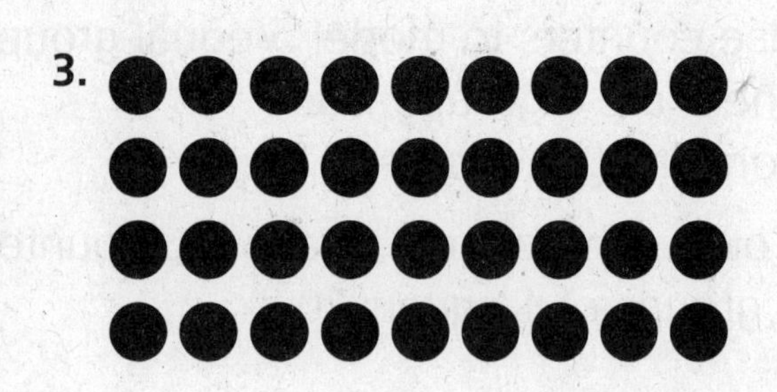

4. $4 \times 5 =$ ____

5. $4 \times 4 =$ ____

6. $2 \times 4 =$ ____

7. $3 \times 4 =$ ____

8. 4×3

9. 5×4

10. 4×6

11. 4×9

12. 7×4

13. 8×4

Problem Solving and Test Prep

14. **Reasoning** Mary's brother gave her some toy cars. These toy cars included 36 wheels in all. How many toy cars did Mary receive, if each toy car included 4 wheels?

15. Eli has 3 toy cars. Andy has 2 toy cars. Amanda has 4 toy cars. If each toy car includes 4 wheels, how many wheels do their cars include in all?

16. Sasha has 7 toy cars. How many wheels do Sasha's cars include?

A 11
B 21
C 24
D 28

17. There are 4 rows of 8 toy cars on a shelf. Which number sentence shows how many toy cars there are on the shelf?

A $8 + 4 = 12$
B $8 \times 4 = 36$
C $4 \times 8 = 32$
D $4 \times 7 = 28$

Name ______________________________ Week 8

Spiral Review

For 1–5, write the numbers in order from least to greatest.

1. 707, 139, 610, 601

____

2. 475, 919, 199, 105

____

3. 1,978; 2,559; 1,879; 1,421

____

4. 2,228; 3,366; 3,334; 2,316

____

5. 7,845; 7,942; 7,930; 7,854

____

For 10–12, use the pattern in the table to answer the questions.

number of spiders	1	2	3	4		6
number of legs	8	16	24		40	

10. How many legs are on 4 spiders? ______________

11. How many spiders have 40 legs in all? ______________

12. How many legs are on 6 spiders? ______________

For 6–9, draw a line to match the shape with the correct number of sides.

6. [square] 5

7. [circle] 0

8. [triangle] 4

9. [pentagon] 3

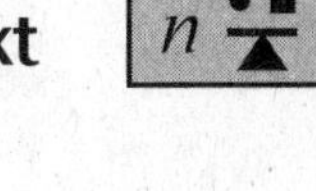

For 13–15, predict the next number in each pattern. Explain.

13. 117, 97, 77, 57, 37, ☐

____

14. 374, 474, 574, 674, ☐

____

15. 46, 52, 58, 64, ☐

____

Name___________________________

Multiply with 1 and 0

Multiply with 0

$4 \times 0 =$ ______

Draw a pair of pants with four pockets, each containing 0 marbles.

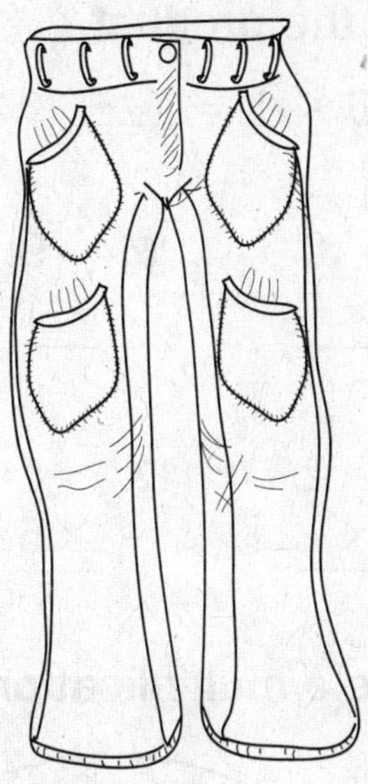

4 pockets × 0 marbles in each pocket equals 0 marbles.

Zero Property of Multiplication
Zero multiplied by any number is 0:

So, $4 \times 0 = 0$, and $0 \times 4 = 0$.

Multiply with 1

$1 \times 6 =$ ______

Draw a pair of pants with one pocket containing 6 marbles.

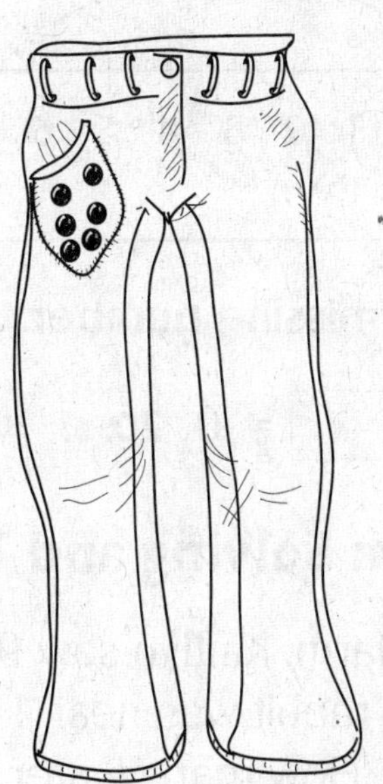

1 pocket × 6 marbles equals 6 marbles in all.

Identity Property of Multiplication

Any number multiplied by 1 is that number.

So, $1 \times 6 = 6$, and $6 \times 1 = 6$.

Find the product.

1. $4 \times 0 =$ ______
2. $1 \times 5 =$ ______
3. $0 \times 8 =$ ______
4. $6 \times 1 =$ ______

5. 1×8
6. 3×0
7. 1×0
8. 0×5
9. 2×1

NS 2.6 Understand the special properties of 0 and 1 in multiplication and division.

Name ______________________

Lesson 5.5

Multiply with 1 and 0

Find the product.

1. $6 \times 1 =$ ____ **2.** $0 \times 9 =$ ____ **3.** $1 \times 4 =$ ____ **4.** $8 \times 0 =$ ____

5. $\begin{array}{r} 0 \\ \times 6 \\ \hline \end{array}$ **6.** $\begin{array}{r} 9 \\ \times 1 \\ \hline \end{array}$ **7.** $\begin{array}{r} 4 \\ \times 0 \\ \hline \end{array}$ **8.** $\begin{array}{r} 5 \\ \times 1 \\ \hline \end{array}$ **9.** $\begin{array}{r} 3 \\ \times 0 \\ \hline \end{array}$ **10.** $\begin{array}{r} 1 \\ \times 8 \\ \hline \end{array}$

11. $\begin{array}{r} 2 \\ \times 1 \\ \hline \end{array}$ **12.** $\begin{array}{r} 1 \\ \times 6 \\ \hline \end{array}$ **13.** $\begin{array}{r} 1 \\ \times 4 \\ \hline \end{array}$ **14.** $\begin{array}{r} 0 \\ \times 1 \\ \hline \end{array}$ **15.** $\begin{array}{r} 3 \\ \times 1 \\ \hline \end{array}$ **16.** $\begin{array}{r} 1 \\ \times 0 \\ \hline \end{array}$

Write a multiplication sentence shown on each number line.

17.

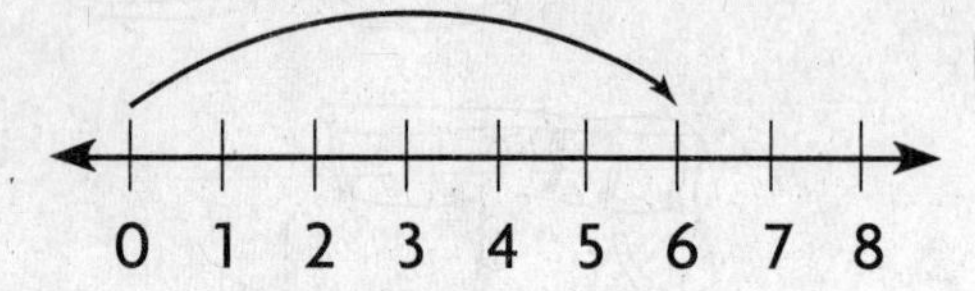

____

18.

____

Find the missing number.

19. $5 \times \square = 0$ **20.** $1 \times \square = 9$ **21.** $7 \times \square = 7$ **22.** $0 \times 52 =$ ____

Problem Solving and Test Prep

23. At a farm, Kaitlyn saw 9 rabbits. Each rabbit was near 1 water bowl. How many water bowls did Kaitlyn see at the farm?

____

24. Cody saw 8 calves on his visit to a farm. None of the calves had horns. How many horns did Cody see at the farm?

____

25. Chloe has 6 pockets. Each pocket contains 1 coin. Which number sentence shows how many coins Chloe has total?

A $1 + 6 = 7$ **C** $6 \times 1 = 6$

B $0 \times 6 = 0$ **D** $6 \times 0 = 6$

26. Len has 7 pockets. He has 0 coins in each pocket. Which number sentence shows how many coins Len has total?

A $7 \times 0 = 7$ **C** $7 \times 1 = 7$

B $0 \times 7 = 0$ **D** $1 + 7 = 8$

Name

Problem Solving Workshop Strategy: Draw a Picture

There are 7 rows of flute players in a marching band. There are 5 flute players in each row. How many flute players are in the marching band?

Read to Understand

1. Write the question as a fill-in-the-blank sentence.

Plan

2. How can drawing a picture help you solve the problem?

Solve

3. Draw a picture to find how many flute players are in the marching band.

Check

4. Is there another strategy you could use to solve this problem? Explain.

For 5–6, draw a picture to solve.

5. There are 4 rows of horn players in a marching band. Each row has 6 people. How many horn players are in this marching band?

6. There are 2 rows of trumpet players in a marching band. Each row has 4 people, each wearing a pair of gloves. How many trumpet players are in the marching band?

Problem Solving Workshop Strategy: Draw a Picture

Problem Solving Strategy Practice

Draw a picture to solve.

1. Mr. Jardin has 8 tomato plants. On each tomato plant there are 7 ripe tomatoes. How many ripe tomatoes does Mr. Jardin have?

____

2. In a marching band, there are 4 rows of horn players. Each row is made up of 9 horn players. How many horn players are in the marching band?

____

3. Four students have apple slices in their lunches. If each student receives 6 slices, how many apple slices do the 4 students have?

____

Mixed Strategy Practice

4. There are 8 drummers in a marching band. Each drummer has 2 drum sticks. How many drum sticks are there?

____

5. Matthew is making a large pizza for his party. There are 8 people at the party. Each person will eat 1 slice. How many slices should Matthew cut the pizza into?

____

6. At Adam's lunch table, 7 students have peas and no one has spinach. How many servings of peas and spinach are at Adam's lunch table?

____

7. **Pose a Problem** If two times as many students had eaten peas in exercise 6, then how many servings of peas and spinach would be at Adam's lunch table?

____

Name____ **Lesson 6.1**

Multiply with 5 and 10

Numbers that are multiplied are called factors.

The result of multiplying factors together is called the product.

Find the product: ____ = 5 × 5 .

You can use a number line to multiply by 5.

The first factor, 5, tells you to make 5 jumps.
The second factor, 5, tells you that each jump is 5 spaces.
So, 5 × 5 means 5 jumps of 5 spaces each.
Start at 0.

Stop at 25.

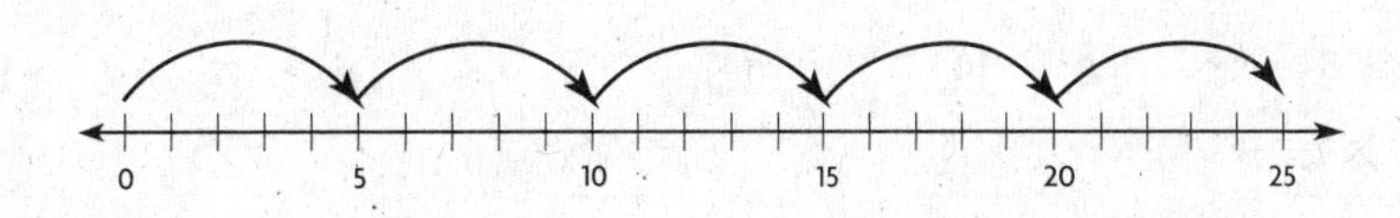

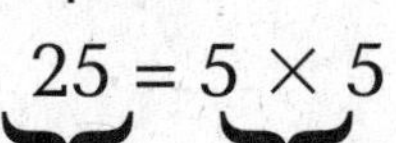

25 = 5 × 5

product factors

Find the product: 10 × 7 = ____ .

You can also use a number line to find products of 10, or you can *use zeros*.

You can use the 1s product to find the 10s product.

First, multiply with 1 instead of 10. 1 × 7 = 7.

Then, write zero after the 1s product. 70

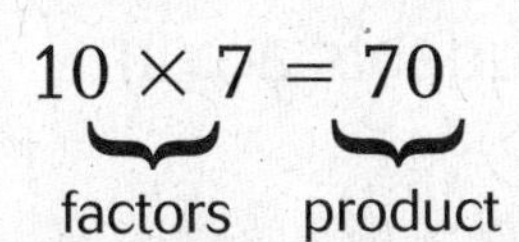

10 × 7 = 70

factors product

Use the number line to find the product.

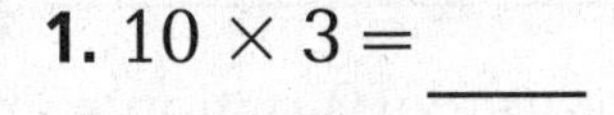

1. 10 × 3 = ____

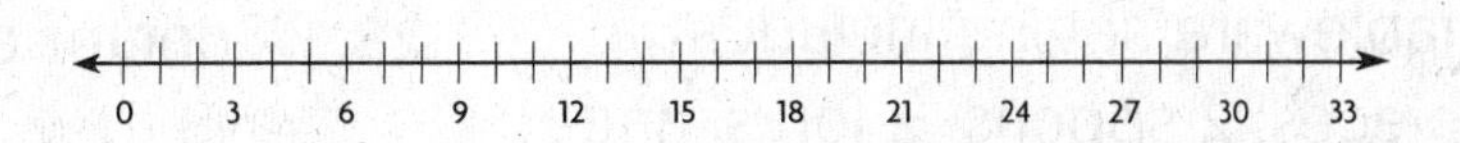

2. 6 × 5 = ____

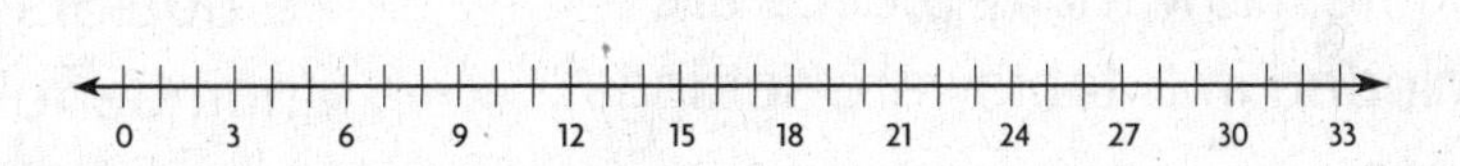

Find the product.

3. 5 × 4 = ____ **4.** 10 × 9 = ____ **5.** 0 × 5 = ____ **6.** ____ = 10 × 4

7. 8 × 5

8. 10 × 5

9. 5 × 3

10. 9 × 5

11. 2 × 10

12. 10 × 8

Name____

Multiply with 5 and 10

Find the product.

1. $10 \times 7 =$ ____ 2. ____ $= 5 \times 4$ 3. $8 \times 10 =$ ____ 4. ____ $= 5 \times 7$

5. $0 \times 10 =$ ____ 6. ____ $= 10 \times 4$ 7. $5 \times 1 =$ ____ 8. ____ $= 10 \times 3$

9. $2 \times 5 =$ ____ 10. $0 \times 10 =$ ____ 11. $10 \times 8 =$ ____ 12. ____ $= 5 \times 3$

13. 3×5 14. 10×1 15. 5×5 16. 4×5 17. 5×10 18. 10×7

19. 10×4 20. 9×5 21. 7×5 22. 5×1 23. 5×6 24. 10×9

Problem Solving and Test Prep

25. A car can carry up to 5 people. There are 6 cars. What is the maximum number of people who can ride in these cars at one time?

____

26. The entire school choir is standing in 6 rows with 10 students in each row. How many students are in the school choir?

____

27. A tableware setting includes 5 pieces: 2 spoons, 2 forks, and 1 knife. How many pieces are included in 8 tableware settings?

A 13

B 20

C 40

D 80

28. A doctor can help up to 10 patients each day. If an office employs 5 doctors, what is the maximum number of patients they can help each day?

A 15

B 50

C 100

D 150

Name________________________________

Multiply with 3

You can use a number line to multiply with 3.

Find the product: _____ = 7 × 3.

The factor 7 tells you to make 7 jumps. The factor 3 tells you that each jump should be 3 spaces.

So, 7 × 3 means 7 jumps of 3 spaces each.

Start at 0.

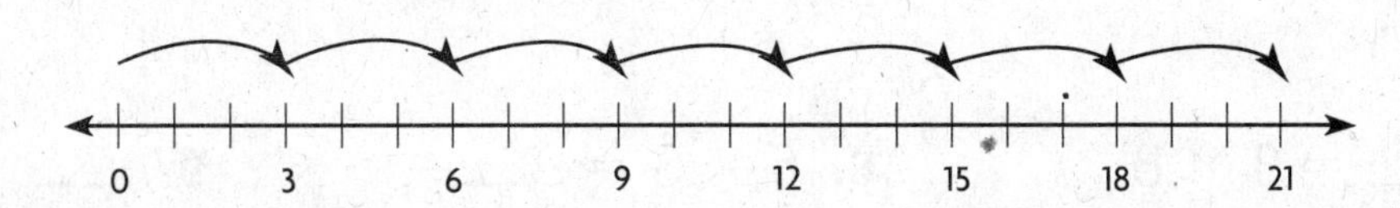

Stop at 21.
So, 21 = 7 × 3.

Use the number line to find the product.

1. 4 × 3 = _____

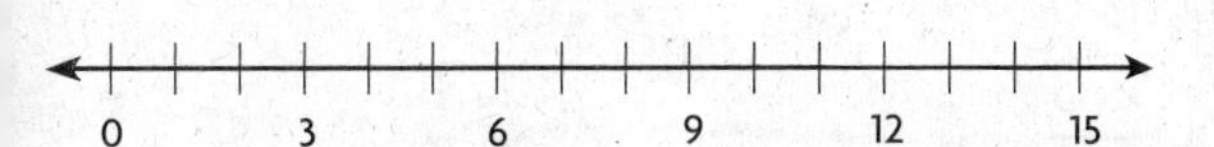

2. _____ = 5 × 3

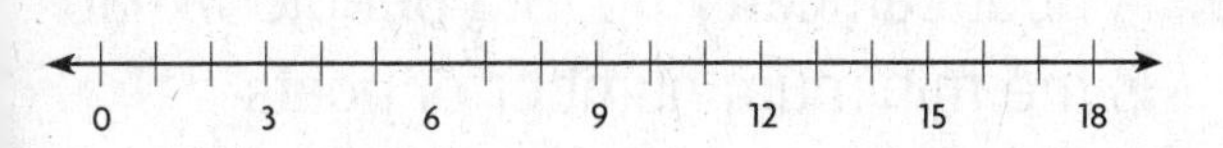

3. 9 × 3 = _____

0 3 6 9 12 15 18 21 24 27 30

Find the product.

4. 1 × 3 = _____ **5.** 3 × 8 = _____ **6.** _____ = 3 × 0 **7.** 3 × 5 = _____

8. 9 × 3 = ___ **9.** 3 × 7 = ___ **10.** 10 × 3 = ___ **11.** 3 × 3 = ___ **12.** 3 × 6 = ___ **13.** 2 × 3 = ___

NS 2.2 Memorize to automaticity the multiplication table for numbers between 1 and 10.

Name______________________________

Multiply with 3

Find the product.

1. $4 \times 3 =$ ____ **2.** $7 \times 3 =$ ____ **3.** ____ $= 3 \times 9$

4. ____ $= 5 \times 3$ **5.** ____ $= 3 \times 3$ **6.** $5 \times 3 =$ ____

7. ____ $= 3 \times 8$ **8.** $6 \times 3 =$ ____ **9.** ____ $= 3 \times 0$

10. $\begin{array}{r} 6 \\ \times 3 \\ \hline \end{array}$ **11.** $\begin{array}{r} 3 \\ \times 1 \\ \hline \end{array}$ **12.** $\begin{array}{r} 4 \\ \times 3 \\ \hline \end{array}$ **13.** $\begin{array}{r} 8 \\ \times 3 \\ \hline \end{array}$

14. $\begin{array}{r} 7 \\ \times 3 \\ \hline \end{array}$ **15.** $\begin{array}{r} 9 \\ \times 3 \\ \hline \end{array}$ **16.** $\begin{array}{r} 0 \\ \times 3 \\ \hline \end{array}$ **17.** $\begin{array}{r} 3 \\ \times 3 \\ \hline \end{array}$

Problem Solving and Test Prep

18. A design contains 5 triangles. How many sides do 5 triangles include?

____

19. A boat can carry up to 3 people. What is the minimum number of boats needed to carry 24 people? Explain.

____

20. There are 8 buns in each bag of hamburger buns. If you have 3 bags of hamburger buns, how many hamburger buns do you have in all?

A 8

B 11

C 16

D 24

21. A pint of ice cream serves 3 people. How many people are served by 5 pints of ice cream?

A 3

B 5

C 15

D 30

Name ______________________________ **Week 9**

Spiral Review

For 1–5, find each sum. Use subtraction to check.

1. $\begin{array}{r} 567 \\ +207 \\ \hline \end{array}$

2. $\begin{array}{r} 789 \\ +516 \\ \hline \end{array}$

3. $\begin{array}{r} 2{,}207 \\ +918 \\ \hline \end{array}$

4. $\begin{array}{r} 2{,}836 \\ +1{,}855 \\ \hline \end{array}$

5. $\begin{array}{r} 4{,}207 \\ +2{,}793 \\ \hline \end{array}$

For 6–9, draw a line to match the number of vertices to the correct shape.

6. 4 

7. 3 

8. 5 

9. 6 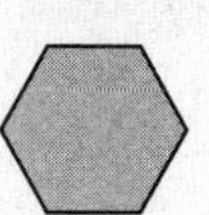

For 10–12, answer the questions using the information in the line plot.

6	7	8	9
X X	X X X X X	X X X	X X

How Many Rooms in Your House

10. How many more houses have 7 rooms than have 6 rooms?

____

11. What is the range of the data?

______

12. What is the mode of the data?

______

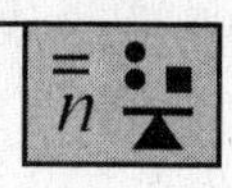

For 13–14, write a number sentence to solve.

13. Donna picked 29 lemons from her tree. She gives 14 to Scott. How many lemons does Donna have left?

____

14. Amber made 30 paper flowers for her party. She gave 10 to her mother to put on the porch. How many flowers does Amber have left?

____

Name________________________________

Lesson 6.3

Multiply with 6

You can multiply with 6 by using a number line.

Find the product: _____ = 6 × 4.

The factor 6 tells you to make 6 jumps. The factor 4 tells you that each jump should be 4 spaces.

So, 6 × 4 means 6 jumps of 4 spaces each.

Start at 0. Make 6 jumps of 4 spaces each.

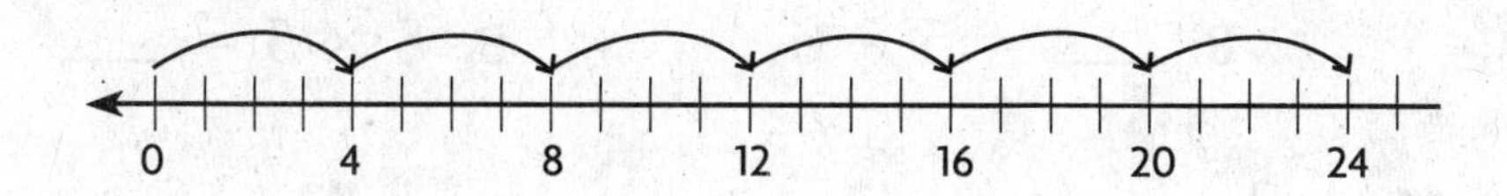

Stop at 24.

So, 24 = 6 × 4.

Use the number line to find the product.

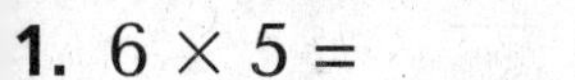

1. 6 × 5 = _____

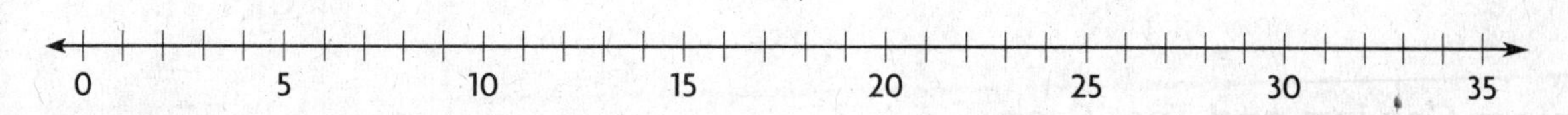

2. 6 × 6 = _____

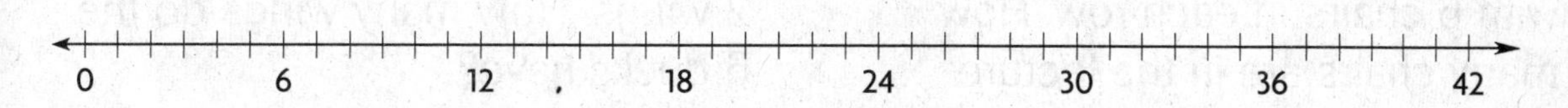

3. 6 × 8 = _____

Find the product.

4. 6 × 2 = _____ **5.** 6 × 0 = _____ **6.** 10 × 6 _____ **7.** 7 × 6 = _____

8.	**9.**	**10.**	**11.**	**12.**	**13.**
4 ×6	6 ×5	9 ×6	5 ×4	0 ×6	6 ×1

Name________________________

Multiply with 6

Find the product.

1. $9 \times 6 =$ ____
2. ____ $= 6 \times 8$
3. $4 \times 6 =$ ____
4. ____ $= 6 \times 7$
5. $6 \times 1 =$ ____
6. ____ $= 6 \times 6$
7. $6 \times 0 =$ ____
8. ____ $= 5 \times 6$
9. $5 \times 5 =$ ____
10. 4×6
11. 9×6
12. 6×8
13. 6×1
14. 6×9
15. 6×7
16. 2×6
17. 6×6

Problem Solving and Test Prep

18. A lecture room contains 9 rows, with 6 chairs in each row. How many chairs are in the lecture room?

    ____

19. Lila saw 6 ducks. Each duck has 2 wings. How many wings do the 6 ducks have?

    ____

20. Ken has 6 pages of stickers. Each page contains 8 stickers. How many stickers does Ken have?

    A 40
    B 46
    C 48
    D 60

21. Heavy-duty pickup trucks hold 6 tires. How many tires do 5 heavy-duty pickup trucks hold?

    A 30
    B 36
    C 55
    D 60

Name____ Lesson 6.4

Algebra: Practice the Facts

You can use a number line to practice multiplication facts.

Find the product: _____ = 9 × 4.

The factor 9 tells you to make 9 jumps. The factor 4 tells you that each jump should be 4 spaces.

So, 9 × 4 tells you to make 9 jumps of 4 spaces each.

Start at 0.

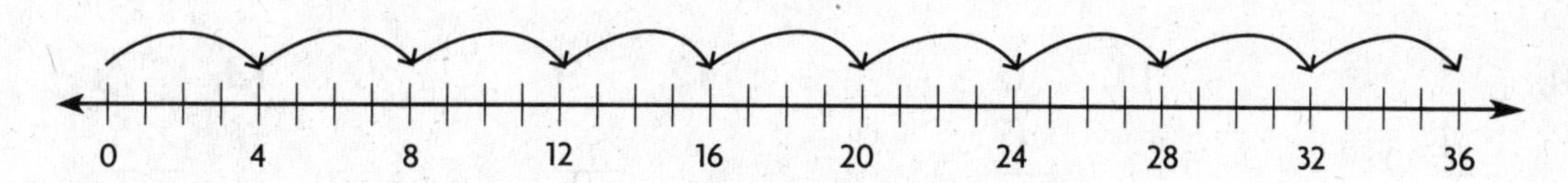

Stop at 36.

So, 36 = 9 × 4.

Use each number line to find the product.

1. 6 × 8 = _____

2. _____ = 5 × 7

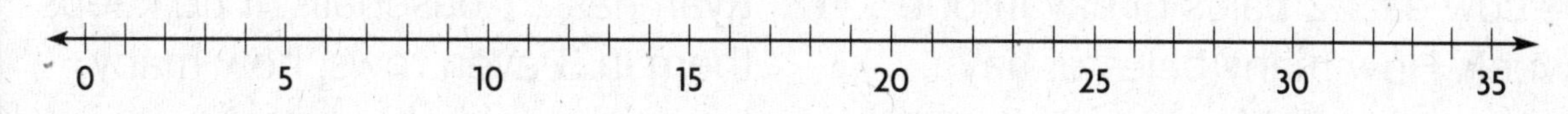

3. 6 × 6 = _____

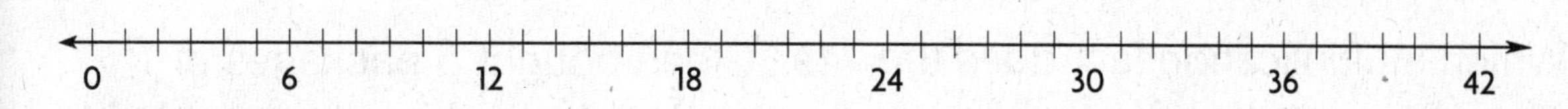

Find the product.

4. 5 × 10 = _____ 5. 6 × 0 = _____ 6. _____ = 4 × 3 7. 8 × 5 = _____

8. 5 × 7

9. 3 × 6

10. 3 × 3

11. 9 × 5

12. 4 × 5

13. 8 × 3

Name________________________________

Practice the Facts

Find the product.

1. $10 \times 8 =$ ____

2. $3 \times 0 =$ ____

3. ____ $= 4 \times 6$

4. ____ $= 9 \times 3$

5. $6 \times 5 =$ ____

6. ____ $= 2 \times 8$

7. ____ $= 1 \times 5$

8. $6 \times 10 =$ ____

9. $5 \times 3 =$ ____

10. $\begin{array}{r} 3 \\ \times 4 \\ \hline \end{array}$

11. $\begin{array}{r} 6 \\ \times 6 \\ \hline \end{array}$

12. $\begin{array}{r} 9 \\ \times 1 \\ \hline \end{array}$

13. $\begin{array}{r} 7 \\ \times 5 \\ \hline \end{array}$

Show two different ways to find each product.

14. $3 \times 7 =$ ____

15. ____ $= 5 \times 2$

Problem Solving and Test Prep

16. A cow eats 2 bales of hay in one week. How many bales of hay does a cow eat in 6 weeks?

____

17. Ryan has 21 baseballs. If he keeps them in 3 even rows, how many baseballs are in each row?

____

18. Which multiplication fact does the picture below show?

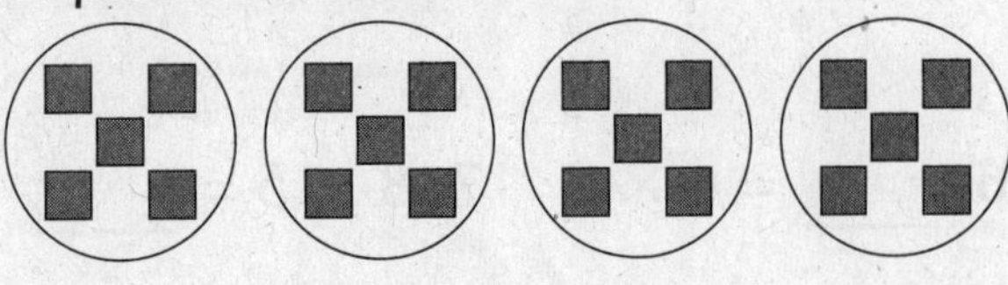

A $5 \times 3 = 15$ **C** $5 \times 5 = 25$

B $4 \times 5 = 20$ **D** $6 \times 5 = 30$

19. Glenn bought 5 packages of postcards. Each package included 10 postcards. How many postcards did Glenn buy? Explain.

____

____

____

Name____

Problem Solving Workshop Strategy: Act It Out

At Kim's party, the children line up to play a game. Lisa is in front of Tim. Matt is behind Avery but ahead of Lisa. Kim is in front of Brett. Brett is in front of Avery. In what order are the children lined up?

Read to Understand

1. What information is given? ____

____

Plan

2. Which strategy can you use to solve the problem? ____________________

Solve

3. How can you use the lettered circles to act out the problem?

____

____

____

L T M
A K B

4. In what order are the children lined up? ____________________

____

Check

5. What is another way you can act out the problem? Explain.

____

____

Act out the problem to solve.

6. Ms. Campbell has 4 bedrooms. In order to put 5 roses in each of her bedrooms, how many roses does Ms. Campbell need?

____

7. Four cars are in a line. The red car is ahead of the blue car. The green car is behind the blue car. The red car is behind the white car. Which color car is last in the line?

____

NS 2.2: Memorize to automaticity the multipication table for numbers between 1 and 10.

Name____ Lesson 6.5

Problem Solving Workshop Strategy: Act It Out

Problem Solving Strategy Practice

Act out the problem to solve.

1. Luis puts ice cubes into glasses for his friends' drinks. He puts 3 ice cubes into each glass. How many ice cubes does Luis need if he has 9 friends?

____

2. Rebecca hands out coupons. She gives 4 coupons to each customer. How many coupons does Rebecca hand out if she has 6 customers?

____

3. Four men are in a line. Fred is in front of Rex. Ken is behind Rex. William is in front of Fred. Who is first in line?

____

4. Vic is handing out colored pencils for drawing. Each student receives 5 colors. How many colored pencils does Vic hand out if there are 9 students?

____

Mixed Strategy Practice

5. Donald rolls sushi. It takes him 5 minutes to make each roll. How many minutes would it take Donald to make 7 rolls?

____

6. Tina has 4 dimes, 5 nickels, and 4 pennies. How many coins does Tina have in all?

____

USE DATA for 7–8, use the table below.

7. Jenny bought 3 packages of T-shirts. How many T-shirts did she buy in all?

____

8. Which contains more items, 3 packages of socks or 3 packages of headbands?

____

Clothing Packages

Item	Number in Package
Socks	6
T-shirts	2
Headbands	4

Name____

Lesson 7.1

Multiply with 8

You can use the Commutative Property and multiplication facts you already know to help you multiply.

Find the product of 8 × 3.

Use the facts you know.

3 rows of 8, or 3×8

$3 \times 8 = 24$

Both arrays represent the same product: **24**.

So, $8 \times 3 = 24$.

8 rows of 3, or 8×3

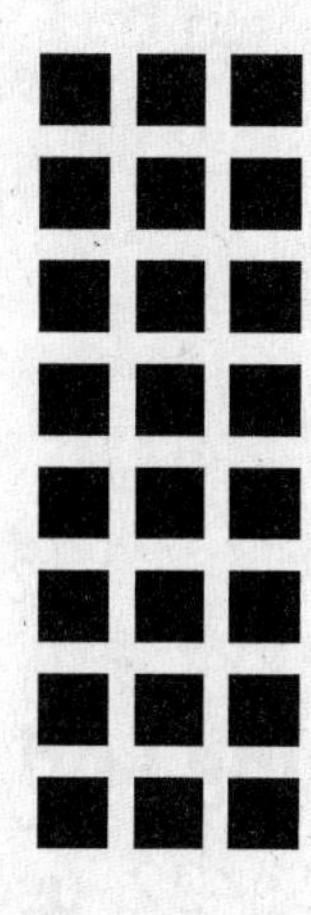

$8 \times 3 = 24$.

Complete each blank. Use the Commutative Property of Multiplication.

1. $8 \times 6 =$ ______
2. $8 \times 10 =$ ______
3. $8 \times 2 =$ ______
4. $4 \times 8 =$ ______
5. $8 \times 5 =$ ______
6. $8 \times 1 =$ ______

Use the Commutative Property of Multiplication to find each product.

7. 7×8
8. 8×6
9. 10×8
10. 8×4
11. 3×8
12. 8×1

Name ____________________

Multiply with 8

Find the product.

1. $8 \times 3 =$ ____ 2. $10 \times 8 =$ ____ 3. $1 \times 8 =$ ____ 4. $7 \times 5 =$ ____

5. $7 \times 9 =$ ____ 6. $8 \times 4 =$ ____ 7. $8 \times 9 =$ ____ 8. $4 \times 4 =$ ____

9. 8×7 10. 1×8 11. 3×7 12. 3×8 13. 6×3 14. 9×8

15. 6×8 16. 4×8 17. 2×9 18. 8×2 19. 8×8 20. 5×8

Problem Solving and Test Prep

USE DATA For 21–22, use the table.

Growth of Beanstalks in 1 Week	
Student	**Height of Beanstalk**
Kaylie	8 inches
Bret	6 inches
Amy	4 inches

21. If Kaylie's beanstalk grows the same amount every week, how tall will it be after 6 weeks?

____

22. If the beanstalks grow the same amount each week, how much taller than Amy's beanstalk will Kaylie's beanstalk be, after 8 weeks?

____

23. At the dog park, there are 8 dogs. Each dog is given 3 bones. How many bones are given out at the dog park?

A 21 **C** 23

B 24 **D** 28

24. There are 6 pieces of fruit in each bag. Sandra buys 8 bags. How many pieces of fruit does Sandra buy?

A 42 **C** 45

B 48 **D** 14

Spiral Review

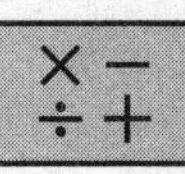

For 1–5, write the number in standard form.

1. $1,000 + 500 + 60 + 2$ ________

2. $9,000 + 80 + 3$ ________

3. $6,000 + 2$ ________

4. $3,000 + 400 + 70 + 1$ ________

5. $900 + 4$ ________

For 6–9, draw a line on the figure to make the given shapes.

6. 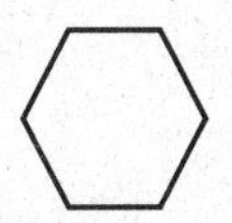 Make two trapezoids.

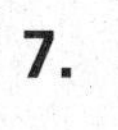

7. 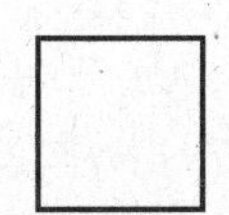 Make two triangles.

8. 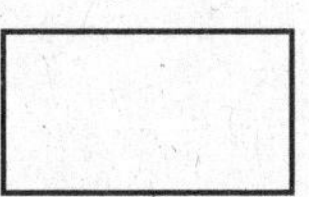 Make two squares.

9. 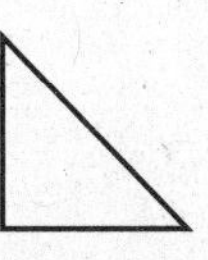 Make two triangles.

For 10–12, use the graph to answer the questions.

Beverage Served

Number

Apple juice, Lemonade, Orange juice, Milk

Beverage

Key: 1 Glass = 2 Cups.

10. What was the most popular beverage served? ________

11. How many cups of apple juice were served? ________

12. How many more cups of lemonade were served than cups of milk were served? ________

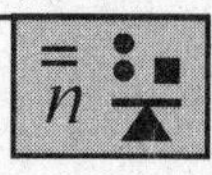

For 13-15, find each difference.

13. ______ $= 20 - 20$

14. $48 - 40 =$ ______

15. ______ $= 600 - 200$

Name____ Lesson 7.2

Algebra: Patterns with 9

You can use paterns to find facts of 9.

- The tens digit is one less than the number being multiplied by 9.

  $8 \times 9 = 72$

- The sum of the digits in the product is 9. $7 + 2 = 9$

Find the product.

$\square = \mathbf{9 \times 6}$

You know that the tens digit in the product will be one less than 6. $\square = \mathbf{9 \times 6}$

$$\begin{array}{r} \mathbf{6} \\ \mathbf{-\ 1} \\ \hline \mathbf{5} \end{array}$$

You know that the sum of the digits in the product will be 9.
What number plus 5 equals 9? $(4) + 5 = 9$

So, $54 = 9 \times 6$.

Use the pattern of 9 to find each product.

1. $4 \times 9 =$ ____ **2.** $6 \times 9 =$ ____ **3.** $3 \times 9 =$ ____

4. $7 \times 9 =$ ____ **5.** $2 \times 9 =$ ____ **6.** $8 \times 9 =$ ____

Name______________________________ **Lesson 7.2**

Algebra: Patterns with 9

Find each product.

1. ____ $= 9 \times 3$ **2.** $9 \times 4 =$ ____ **3.** ____ $= 9 \times 8$ **4.** $9 \times 5 =$ ____

5. $7 \times 9 =$ ____ **6.** ____ $= 3 \times 4$ **7.** $9 \times 9 =$ ____ **8.** ____ $= 5 \times 4$

9. 9×1 **10.** 9×2 **11.** 6×3 **12.** 9×6 **13.** 9×7 **14.** 9×8

Compare. Write <, >, or = for each ◯.

15. 5×8 ◯ 6×7 **16.** 9×3 ◯ 4×7 **17.** 3×6 ◯ 2×8

18. 4×3 ◯ 2×6 **19.** 9×4 ◯ 6×6 **20.** 9×5 ◯ 8×4

Problem Solving and Test Prep

21. A model of the solar system includes 9 planets. How many planets are in 8 models?

____

22. Bob has 4 plants. Ron has 9 times as many plants as Bob has. How many plants does Ron have?

____

23. A package of colored pencils contains 9 pencils. How many colored pencils are in 3 packages?

A 6
B 9
C 18
D 27

24. Ms. Lee took 9 children to the zoo. Each child brought 4 snacks. How many snacks did the 9 children bring to the zoo?

A 4
B 9
C 13
D 36

Name____

Lesson 7.3

Multiply with 7

You can multiply two numbers in either order.
The product is the same.

7×3

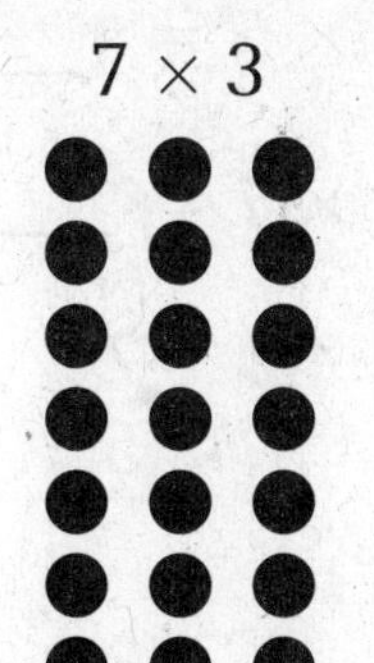

7 rows of 3

$7 \times 3 = 21$

3×7

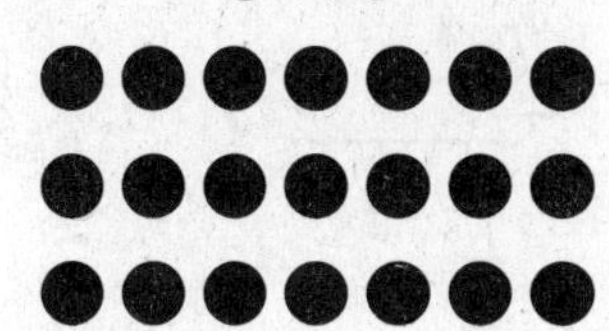

3 rows of 7

$3 \times 7 = 21$

You can also break apart 3×7 into two arrays to help you find the product.

Make an array that shows 3 rows of 7:

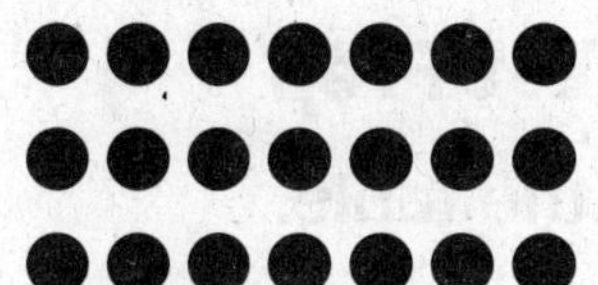

Break the array into smaller arrays:

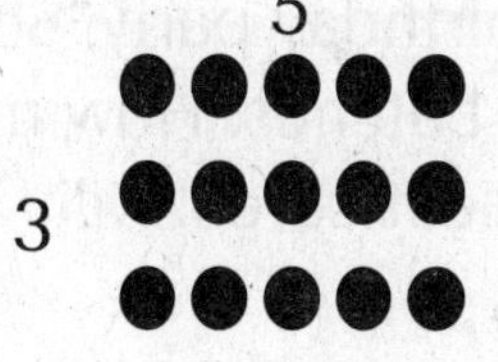

$3 \times 2 = 6$ $3 \times 5 = 15$

Add the products. $6 + 15 = 21$

So, $3 \times 7 = 21$.

Find the product.

1. $7 \times 3 =$ ________ **2.** $7 \times 5 =$ ________ **3.** $7 \times 4 =$ ________

4. $7 \times 2 =$ ________ **5.** $7 \times 1 =$ ________ **6.** $7 \times 9 =$ ________

7. $7 \times 6 =$ ________ **8.** $7 \times 0 =$ ________ **9.** $7 \times 10 =$ ________

10. $7 \times 2 =$ ________ **11.** $7 \times 7 =$ ________ **12.** $7 \times 5 =$ ________

Name________________________________ **Lesson 7.3**

Multiply with 7

Find the product.

1. $7 \times 3 =$ _____ **2.** $9 \times 7 =$ _____ **3.** $7 \times 8 =$ _____ **4.** $6 \times 5 =$ _____

5. $7 \times 1 =$ _____ **6.** $4 \times 7 =$ _____ **7.** $6 \times 8 =$ _____ **8.** $5 \times 7 =$ _____

9. 8×5 **10.** 2×7 **11.** 6×7 **12.** 7×7 **13.** 9×7 **14.** 7×5

15. 4×6 **16.** 7×4 **17.** 8×7 **18.** 9×3 **19.** 7×1 **20.** 7×6

Problem Solving and Test Prep

USE DATA For 21–22, use the table.

21. Molly is going to make snack mix for Ben's birthday party. She wants to make 7 batches. How many cups of wheat cereal will Molly need?

____

Snack Mix Recipe for 1 Batch	
Snack	**Number of Cups**
Wheat Cereal	4
Rice Crisps	2
Sesame Toasts	1

22. If Molly makes 7 batches of snack mix, how many cups of snacks will she need in all? _______________

23. Adriana is making muffins with a mold that holds 7 muffins. How many muffins can Adriana make with 4 molds?

A 14
B 21
C 28
D 35

24. A box holds 7 dog biscuits. Dan has 3 boxes of biscuits. How many biscuits does Dan have?

A 14
B 21
C 28
D 35

Name______________________________

Algebra: Practice the Facts

You can use a picture or an array to help you multiply.

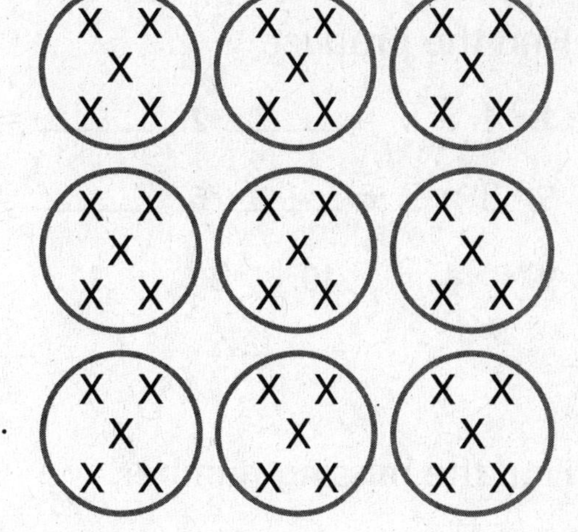

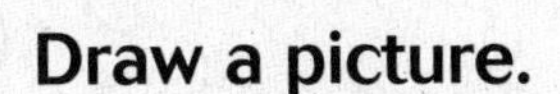

Draw a picture.

Draw a picture to find $9 \times 5 = \square$.

Draw 9 groups of 5.

Count by 5s; 5, 10, 15, 20, 25, 30, 35, 40, 45.

9 groups of 5 is 45.

So, $9 \times 5 = 45$.

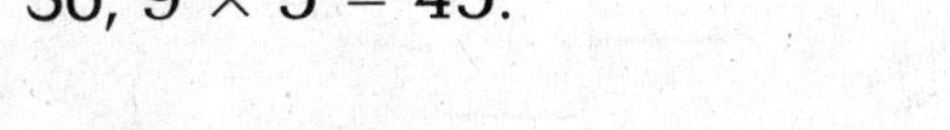

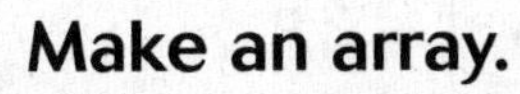

Make an array.

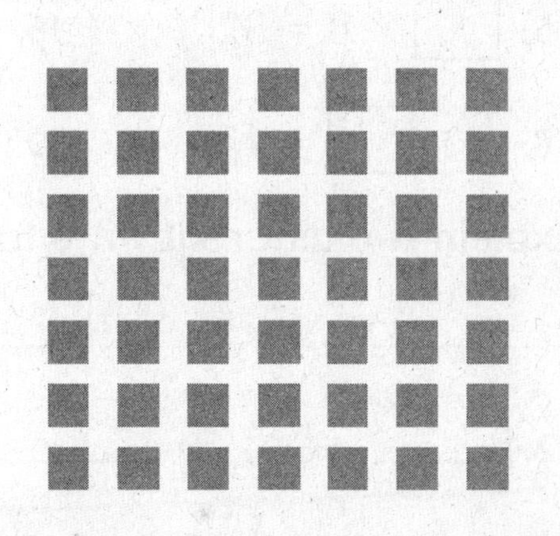

You can make an array to solve $7 \times 7 = \square$.

Think: There are 7 rows in the array.
There are 7 columns in the array.
Find the total number of tiles.

So, $7 \times 7 = 49$.

Draw a picture, or make an array, to find each product.

1. $5 \times 7 =$ ______

2. $9 \times 4 =$ ______

3. $6 \times 3 =$ ______

4. $4 \times 8 =$ ______

5. $5 \times 5 =$ ______

6. $7 \times 8 =$ ______

Name ______________________________ **Lesson 7.4**

Algebra: Practice the Facts

Find the product.

1. $4 \times 5 =$ ______ **2.** ______ $= 8 \times 9$ **3.** $7 \times 5 =$ ______ **4.** ______ $= 6 \times 6$

5. $3 \times 2 =$ ______ **6.** ______ $= 6 \times 7$ **7.** ______ $= 9 \times 4$ **8.** $5 \times 8 =$ ______

9. $\begin{array}{r} 4 \\ \times 9 \\ \hline \end{array}$ **10.** $\begin{array}{r} 2 \\ \times 7 \\ \hline \end{array}$ **11.** $\begin{array}{r} 8 \\ \times 8 \\ \hline \end{array}$ **12.** $\begin{array}{r} 9 \\ \times 3 \\ \hline \end{array}$ **13.** $\begin{array}{r} 5 \\ \times 8 \\ \hline \end{array}$ **14.** $\begin{array}{r} 7 \\ \times 6 \\ \hline \end{array}$

Find the missing number.

15. $\square \times 8 = 32$ **16.** $7 \times 8 = \square$ **17.** $\square \times 6 = 24$

18. $5 \times \square = 45$ **19.** $\square \times 9 = 27$ **20.** $6 \times \square = 48$

Explain two different ways to find the product.

21. ______ $= 9 \times 9$ ______________________________

22. ______ $= 8 \times 10$ ______________________________

Compare. Write $<$, $>$, or $=$ for each $\bigcirc$.

23. $3 \times 8 \bigcirc 4 \times 6$ **24.** $9 \times 5 \bigcirc 6 \times 8$ **25.** $4 \times 7 \bigcirc 9 \times 3$

Problem Solving and Test Prep

26. Each basketball team has 8 players. How many players are on 7 basketball teams?

____

27. Each tennis team has 9 players. How many players are on 3 tennis teams?

____

28. Which multiplication factor does this array show?

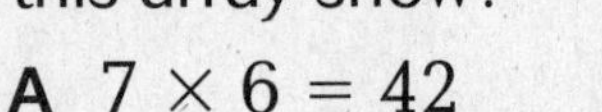

A $7 \times 6 = 42$

B $6 \times 8 = 48$

C $6 \times 7 = 48$

D $8 \times 6 = 42$

29. Which is greater than 9×4?

A 3×9

B 5×7

C 8×5

D 5×6

Name____________________

Problem Solving Workshop Strategy: Make a Table

At Angel Island State Park, there is a 5-mile trail. Mike hiked this trail 3 times. How many miles did Mike hike in all?

Read to Understand

1. Write the question as a fill-in-the-blank sentence.

____

Plan

2. What strategy can you use to solve?

____

Solve

3. Complete the table.

Mike's Hikes at Angel Island State Park			
Times	1	2	3
Miles			

4. How many miles did Mike hike in all?

____

Check

5. How can you check your answer? ____________________

____

Make a table to solve.

6. Belle swims 4 miles a day, 5 days a week. How many miles does Belle swim in 5 days? ____________

7. The store packages 6 fruit bars per bag. Eileen buys 6 bags. How many fruit bars does Eileen buy from the store? ____________

AF 2.1 Solve simple problems involving a functional relationship between two quantities (e.g., find the total cost of multiple items given the cost per unit).

Name ____________________ Lesson 7.5

Problem Solving Workshop Strategy: Make a Table

Problem Solving Strategy Practice

Make a table to solve.

1. Matt rode his bike 6 miles every day for 5 days. How many miles did Matt ride his bike during 5 days?

____

2. Jake sold 7 bags of oranges every Friday for 4 weeks. How many bags of oranges did Jake sell during 4 weeks?

____

Mixed Strategy Practice

3. Brad read 3 pages in his book each day. How many pages did Brad read in 8 days? Draw an array to solve.

____

4. Ken has a white shirt and a red shirt. He has a yellow, a blue, and a green tie. How many different one shirt – one tie combinations are there?

____

5. USE DATA How many tickets are needed to ride all the rides, one time? Show your work.

____

____

Tickets For Each Ride	
Ride	**Number of Tickets**
Pony Ride	6
Obstacle Course	2
Water Slide	9
Basketball Shoot	4

Name ______________________ **Week 11**

Spiral Review

1. What is 5,211 in word form? ______________

____

2. What is 8,009 in word form?

____

3. What is four hundred nineteen in standard form? ________

4. What is three thousand, six hundred forty-two in standard form? ________

5. Write the number in standard form.

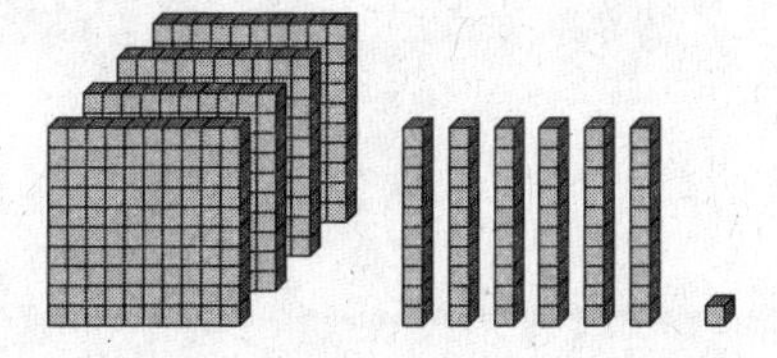

______

For 10–11, a class takes a survey about recess. Write the results as tally marks.

10. 8 students play basketball.

______

11. 4 students catch bugs.

______

12. Look at the table at the right. How many students brought sack lunches on Wednesday?

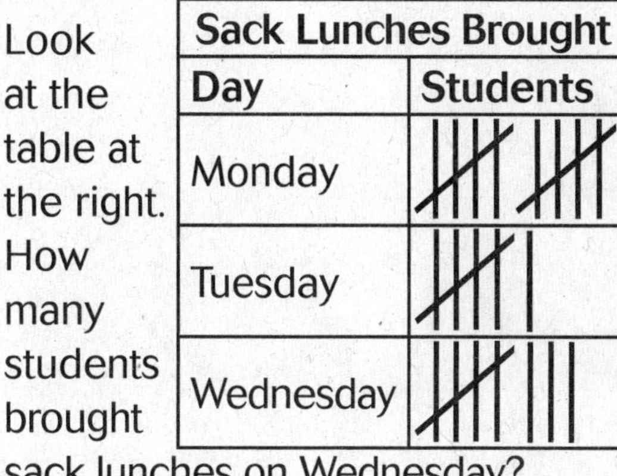

Sack Lunches Brought	
Day	**Students**
Monday	卌 卌
Tuesday	卌 |
Wednesday	卌 |||

____

For 6–9, label the figure as a *square, rhombus, trapezoid,* or *rectangle*. Use each label only once.

6. ______________

7. ______________

8. ______________

9. ______________

For 13–14, write a number sentence to solve.

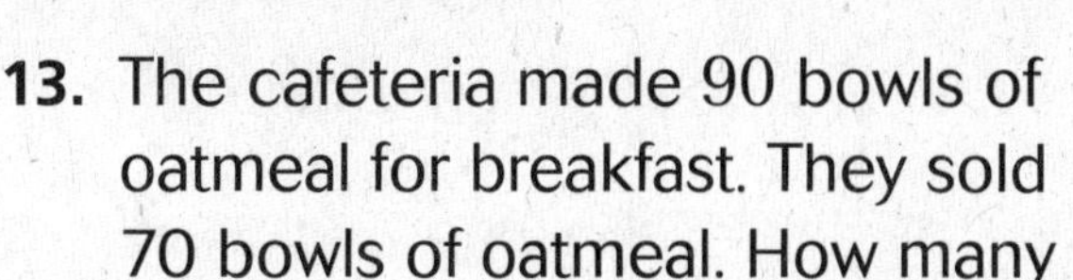

13. The cafeteria made 90 bowls of oatmeal for breakfast. They sold 70 bowls of oatmeal. How many bowls of oatmeal were left?

____

14. The Flower Fairy sold 25 roses on Grandparent's Day. The Plant Planet sold 20 roses. How many roses did the two stores sell in all?

____

Name ______________________________ **Lesson 8.1**

Algebra: Find a Rule

A pattern can be described by a rule such as *multiply by 5.*

Write a rule and complete the table below.

Campers	1	2	3	4	5
Flashlights	1	2	3	?	?

- Look at the first column of numbers.

Think: what can you do to 1 to make 1?

- You could multiply by 1.
- Look at the next column of numbers.

Think: What can be done to 2 to make 2?

- You could multiply by 1.
- Try the rule: multiply by 1.

Check the rule with the next column.

Does $3 \times 1 = 3$? Yes.

- Use the rule to complete the table.

So the rule is multiply the number of campers by 1.
The missing numbers are 4 and 5.

Write a rule for each table. Then complete the table.

1. rule: ______________________________

Boats	2	3	4	5
Life Jackets	10	15		

2. rule: ______________________________

Tickets	2	3	4	5
Cost	$6	$9		

Name ________________________________

Algebra: Find a Rule

Write a rule for each table. Then complete the table.

1. ______________________

Children	1	2	3	4	5
Number of Backpacks	5	10			

2. ______________________

Games	2	3	4	5	6
Players	6	9			

3. ______________________

Maps	1	2	3	4	5
Cost	$4	$8			

4. ______________________

Maps	3	4	5	6	7
Campers	6	8			

Problem solving and Test Prep

USE DATA For 5–6, use the table below.

Canoes	1	2	3	4
Campers	3	6	9	

5. Write a rule for the information in this table.

____

6. How many campers can fit into 4 canoes? ______________________

7. One rowboat holds 6 people. How many people can fit into 5 rowboats?

A 15 **C** 30
B 16 **D** 36

8. Each camper needs 2 graham crackers to make s'mores. How many graham crackers do 5 campers need to make s'mores?

A 10 **C** 25
B 20 **D** 50

Name______________________________ Lesson 8.2

Algebra: Missing Factors

A **variable** is a letter or a symbol that stands for an unknown number, or missing factor.

$2 \times \square = 10$

$2 \times b = 10$

variable

You can use counters to help find the variable.

Arrange the counters into equal groups and count how many are in each group.

Find the missing factor: $\square \times 2 = 14$.

Arrange 14 counters into equal groups of 2.

There are 7 groups of 2 counters.

So, $7 \times 2 = 14$. 7 is the missing factor, or variable.

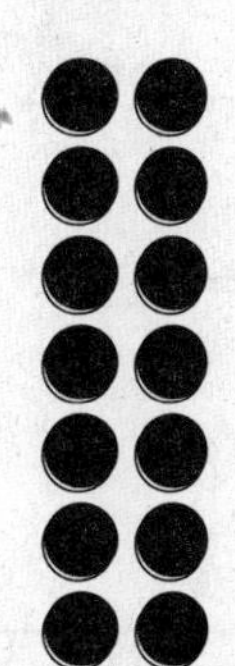

Find the missing factor. Use the array to help.

1. $\square \times 4 = 24$

2. $3 \times \square = 21$

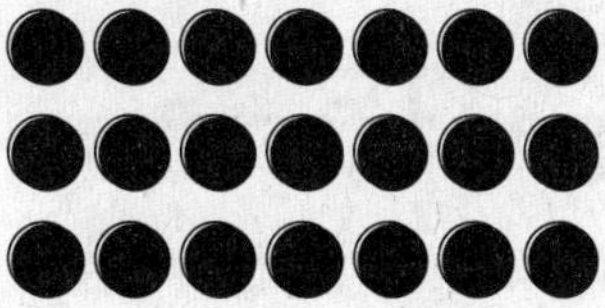

3. $\square \times 5 = 20$

4. $4 \times \square = 16$

Find the missing factor.

5. $\square \times 5 = 25$ **6.** $\square \times 4 = 24$ **7.** $3 \times \square = 12$ **8.** $\square \times 8 = 56$

Name________________________________

Algebra: Missing Factors

Find the missing factor.

1. $\square \times 5 = 30$

2. $\square \times 7 = 28$

3. $4 \times \square = 16$

4. $\square \times 9 = 27$

5. $9 \times \square = 36$

6. $\square \times 8 = 56$

7. $5 \times \square = 40$

8. $6 \times \square = 48$

9. $\square \times 3 = 18$

10. $n \times 7 = 56$ ______________

11. $5 \times k = 45$ ______________

12. $3 \times g = 12$ ______________

13. $d \times 5 = 10 + 5$ ______________

14. $4 \times t = 8 \times 3$ ______________

15. $a \times 7 = 30 - 2$ ______________

Problem Solving and Test Prep

16. Chloe goes camping. She brings enough food for 18 meals. She plans to eat 3 meals each day. How many days' worth of food did Chloe bring?

____

17. Lisa is having a cookout. She wants to make 18 hot dogs. The buns she is buying come in packs of 6. How many packs of buns does Lisa need to buy?

____

18. Which is the missing factor?

$\square \times 4 = 36$

A 6
B 7
C 8
D 9

19. Todd wants to bring juice to a picnic. There will be 24 people at the picnic. The juice comes in packages of 6. How many packages will Todd need to bring, so that each person receives one juice?

A 3
B 4
C 6
D 8

Name______________________________

Lesson 8.3

Algebra: Multiply 3 Factors

You can use the **Associative Property of Multiplication** to multiply with 3 factors. This property is called the grouping property because it allows you to change the grouping of factors while keeping the same product.

Find the product: $3 \times \mathbf{(3 \times 7)}$

Step 1: Solve inside the parentheses.
$3 \times 7 = \mathbf{21}$

Step 2: Multiply the product from inside the parentheses with the remaining factor.
$3 \times \mathbf{21} = 63$

So, $3 \times (3 \times 7) = 63$

Find the product: $\mathbf{(3 \times 3)} \times 7$

Step 1: Solve inside the parentheses.
$\mathbf{(3 \times 3)} = 9$

Step 2: Multiply the product from inside the parentheses with the remaining factor.
$\mathbf{9} \times 7 = 63$

Find the product.

1. $(2 \times 3) \times 5$ ______
2. $2 \times (4 \times 2)$ ______
3. $(5 \times 2) \times 3$ ______
4. $4 \times (3 \times 3)$ ______
5. $5 \times (7 \times 1)$ ______
6. $8 \times (4 \times 1)$ ______
7. $(3 \times 3) \times 9$ ______
8. $(2 \times 2) \times 6$ ______

Name ____________________

Lesson 8.3

Algebra: Multiply 3 Factors

Find the product. Write another way to group the factors.

1. $(4 \times 2) \times 3$ ________

2. $7 \times (2 \times 4)$ ________

3. $(5 \times 1) \times 9$ ________

4. $(3 \times 3) \times 2$ ________

5. $6 \times (2 \times 2)$ ________

6. $(4 \times 1) \times 4$ ________

7. $(2 \times 3) \times 6$ ________

8. $7 \times (2 \times 2)$ ________

Use parentheses. Find the product.

9. $2 \times 3 \times 5$ ________

10. $1 \times 7 \times 6$ ________

11. $3 \times 2 \times 6$ ________

12. $4 \times 2 \times 7$ ________

13. $3 \times 3 \times 9$ ________

14. $6 \times 4 \times 2$ ________

15. $7 \times 8 \times 1$ ________

16. $9 \times 3 \times 2$ ________

Find the missing factor.

17. $(3 \times \square) \times 5 = 30$

18. $7 \times (\square \times 2) = 42$

19. $(\square \times 4) \times 6 = 48$

Problem Solving and Test Prep

20. A roller coaster has 2 trains. Each train has 10 rows of seats. Each row has 2 seats. How many seats are on the roller coaster?

____

21. A roller coaster has 5 cars. Each car has 2 rows of seats. Each row has 2 seats. How many seats are on the roller coaster?

____

22. Which is the product?

$4 \times 5 \times 2 =$ ____.

A 18
B 20
C 40
D 50

23. A subway train has 2 cabs. Each cab has 5 rows. Each row has 5 seats. How many seats are on the subway train?

A 40
B 50
C 60
D 70

Name____ Lesson 8.4

Algebra: Multiplication Properties

The properties of multiplication can help you multiply two or more numbers.

Zero Property

The product of zero and any number equals 0.

Examples: $\mathbf{0} \times 5 = \mathbf{0}$ $8 \times \mathbf{0} = \mathbf{0}$

Identity Property

The product of 1 and any number equals that number.

Examples: $\mathbf{1} \times 5 = 5$ $8 \times \mathbf{1} = 8$

Associative Property

This is the grouping property. When you have at least 3 numbers, you can group numbers in any order using parentheses. The product stays the same.

Example:

$$2 \times (3 \times 4) = (2 \times 3) \times 4$$
$$2 \times 12 = 6 \times 4$$
$$24 = 24$$

Commutative Property

You can multiply two factors in any order and the product is the same.

Examples:

$$2 \times 8 = 8 \times 2 \qquad 8 \times 10 = 10 \times 8$$
$$16 = 16 \qquad 80 = 80$$

Find the product. Tell which property you used.

1. $8 \times 0 =$ _____ ______________________________
2. $(3 \times 4) \times 8 =$ _____ ______________________________
3. $1 \times 7 =$ _____ ______________________________
4. $(3 \times 3) \times 3 =$ _____ ______________________________
5. $10 \times 1 =$ _____ ______________________________
6. $6 \times 7 =$ _____ ______________________________
7. $0 \times 9 =$ _____ ______________________________
8. $6 \times (3 \times 2) =$ _____ ______________________________
9. $1 \times 4 =$ _____ ______________________________
10. $5 \times (2 \times 4) =$ _____ ______________________________

AF 1.5 Recognize and use the commutative and associative properties of multiplication.

Name____

Algebra: Multiplication Properties

Find the product. Tell which property you used.

1. 4×3 ______________

2. 1×9 ______________

3. 7×0 ______________

4. $(2 \times 3) \times 6$ ______________

5. 4×9 ______________

6. $2 \times (3 \times 3)$ ______________

7. 8×1 ______________

8. 7×3 ______________

9. 0×5 ______________

10. 6×7 ______________

11. $4 \times (5 \times 1)$ ______________

12. 6×3 ______________

Find the missing factor.

13. $6 \times \square = 8 \times 6$

14. $7 \times 0 = \square \times 7$

15. $(2 \times \square) \times 7 = 2 \times (2 \times 7)$

Problem Solving and Test Prep

16. Holly bought 4 balls of yarn. Each ball of yarn cost $7. How much money did Holly spend?

____

17. Alice wants to knit 3 hats. She needs 2 balls of yarn for each hat. How many balls of yarn does Alice need?

____

18. Which is an example of the Zero Property?

A $2 \times 1 = 2$

B $2 \times 7 = 7 \times 2$

C $2 \times 0 = 0$

D. $2 \times (2 \times 4) = (2 \times 2) \times 4$

19. Which is an example of the Associative Property?

A $4 \times 6 = 6 \times 4$

B $(2 \times 2) \times 5 = 2 \times (2 \times 5)$

C $0 \times 7 = 0$

D $8 \times 1 = 8$

Name ______________________________ **Week 12**

Spiral Review

For 1–5, find the product.

1. $4 \times 3 =$ ______
2. $7 \times 5 =$ ______
3. $8 \times 3 =$ ______
4. $10 \times 6 =$ ______
5. $2 \times 9 =$ ______

For 8–10, a class takes a survey about the music they listen to. Let 🯅 stand for 1 pictograph/person. Use 🯅 to replace the tally marks for each category of music below.

What Music We Listen To	
Music	**Students**
Classical	𝍸 |||
Pop	||||
Country	𝍸 |

8. Classical ______________
9. Pop ______________
10. Country ______________

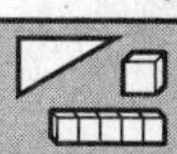
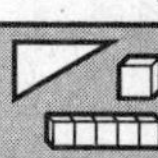

For 6–7, measure the length to the nearest inch.

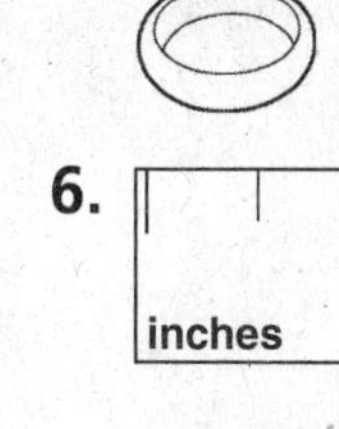

6. (ruler: 1, 2, 3, 4 inches)

7. (ruler: 1, 2, 3, 4 inches)

For 11–12, fill in the blank.

11. Bart thinks of a multiplication fact. One of the factors is 9. The product is also 9. What is a fact that Bart could be thinking of?

____

12. Cho thinks of a multiplication fact. The product is 24. One of the factors is 8. What is a fact that Cho could be thinking of?

____

Name________________________________ Lesson 8.5

Problem Solving Workshop Skill: Multistep Problems

The trainers give the horses apples as rewards. They are working with 6 horses. The trainers reward each horse with 3 apples. If the trainers started with 30 apples, how many apples are left?

1. What are you asked to find?

____

2. What information are you given?

____

____

3. Make a picture to better understand the problem.

4. Make a plan to solve the problem. What will you do first? What will you do next? Explain. _______________

____

5. If the trainers started with 30 apples, how many apples are left?

____

6. How can you check your answer? _______________

____

7. Toni has 3 backpacks. Each backpack includes 5 pockets. Lynn has 1 backpack with 6 pockets. How many pickets in all do Toni and Lynn have on their backpacks?

____

8. Kate goes to a fruit stand. A pint of blueberries costs $3. A bag of peaches costs $8. Kate buys 4 pints of blueberries and 2 bags of peaches. How much does Kate spend in all?

____

Name ______________________

Problem Solving Workshop Skill: Multistep Problems

Problem Solving Skill Practice

1. Paula scored 5 goals total, in two soccer games. One goal was later taken away. She then scored 3 goals in both of her next two games. How many soccer goals did Paula score in these 4 games?

____

2. A summer camp rented 2 canoes and 3 paddleboats. Each canoe holds 3 people and each paddleboat holds 4 people. How many people can go out on the canoes and paddleboats this summer camp rented?

____

3. Stan is at the circus. He buys 4 drinks and 2 sandwiches. The drinks cost $3 each, and sandwiches cost $4 each. How much money does Stan spend in all?

____

Mixed Applications

4. **Use Data** Jane went shopping for school supplies. She bought 2 packages of pens and 3 erasers. How much money did Jane spend in all?

____

School Supplies	
Item	**Cost**
Pens	$3 per package
Markers	$6 per package
Erasers	$1 each
Folders	50¢ each

5. David received a bicycle for his birthday. He rode his bicycle 7 miles the first week he had it, and 10 miles the second week he had it. In the third week after he had first received his bicycle, David rode his bicycle twice as many miles as he had ridden it the first two weeks combined. During which week did David ride his bicycle the farthest distance?

____

Name________________________________ Lesson 9.1

Model Division

When you **divide**, you separate objects, or individuals, into equal groups.

Find: 24 ÷ 6

Use 24 counters.

Place 4 counters in each cup, until you have used all 24 counters.

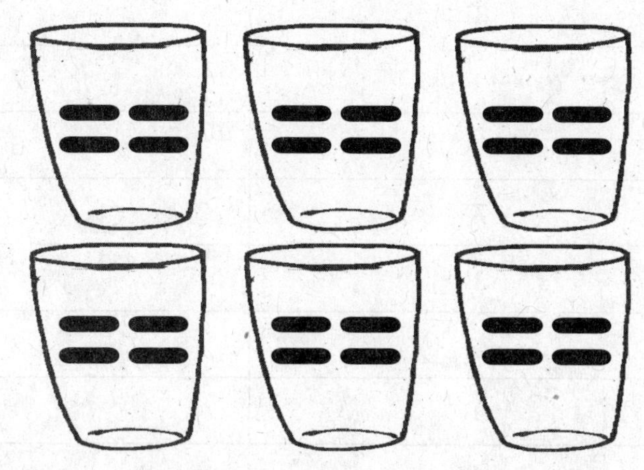

There are 6 equal groups, with 4 counters in each group.

So, 24 ÷ 6 = 4.

Draw counters in each cup. Then complete the table. Use coins and cups to help.

	Counters	Number of Equal Groups	Number in Each Group
1.	12		
2.	10		
3.	16		
4.	25		

Name ______________________

Model Division

Complete the table. Use counters to help.

	Counters	Number of Equal Groups	Number in Each Group
1.	24		4
2.	32	8	
3.	35		7
4.	18		6
5.	21	7	
6.	24	3	
7.	18		2
8.	36	9	
9.	36	6	
10.	12	3	
11.	20		4

Problem Solving and Test Prep

12. Gary has 45 stickers. He wants to put the same number of stickers on each of 9 pages in his sticker book. How many stickers will be on each page?

____

13. Alice has 18 shells. She wants to put the same number of shells in each of 3 jars. How many shells will be in each jar?

____

14. Which is the missing factor?

$7 \times \square = 21$

A 2

B 3

C 4

D 5

15. Al has 16 coins. He puts 4 coins in each of his boxes. How many boxes does Al have?

A 4

B 3

C 6

D 8

Name____

Relate Division and Subtraction

Use a number line or repeated subtraction to solve.

$8\overline{)32}$

Use a number line.

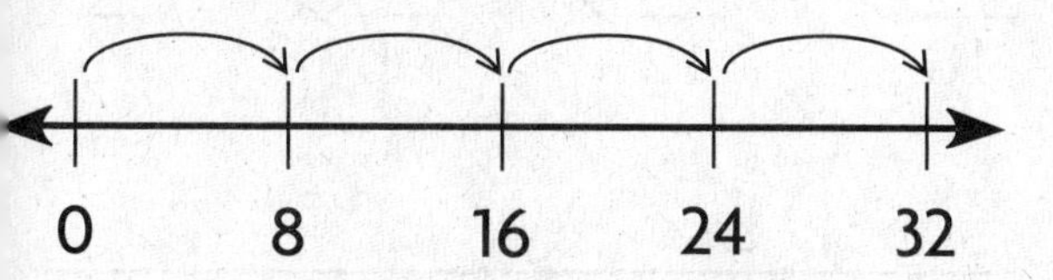

You made 4 jumps of 8.

So, $32 \div 8 = 4$.

Use repeated subtraction.

$32 - 8 = 24$ → $24 - 8 = 16$ → $16 - 8 = 8$ → $8 - 8 = 0$

You subtracted 8 four times

So, $32 \div 8 = 4$.

Use a number line or repeated subtraction to solve.

1. $16 \div 8 =$ ______ **2.** $20 \div 5 =$ ______ **3.** $21 \div 3 =$ ______

4. $18 \div 6 =$ ______ **5.** $10 \div 2 =$ ______ **6.** $9 \div 3 =$ ______

7. $36 \div 6 =$ ______ **8.** $42 \div 7 =$ ______ **9.** $45 \div 5 =$ ______

Name ______________________________ **Lesson 9.2**

Relate Division and Subtraction

Write a division sentence for each.

1.

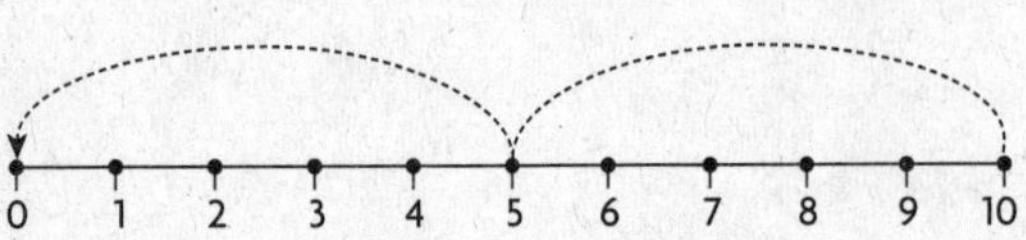

____

2. $\begin{array}{r} 24 \\ -6 \\ \hline 18 \end{array} \nearrow \begin{array}{r} 18 \\ -6 \\ \hline 12 \end{array} \nearrow \begin{array}{r} 12 \\ -6 \\ \hline 6 \end{array} \nearrow \begin{array}{r} 6 \\ -6 \\ \hline 0 \end{array}$

____

Use a number line or repeated subtraction to solve.

3. $12 \div 3 =$ ____ **4.** $20 \div 4 =$ ____ **5.** $21 \div 3 =$ ____ **6.** $15 \div 5 =$ ____

7. $27 \div 9 =$ ____ **8.** $32 \div 4 =$ ____ **9.** $9\overline{)36} =$ ____ **10.** $2\overline{)14} =$ ____

11. $3\overline{)18} =$ ____ **12.** $6\overline{)30} =$ ____ **13.** $4\overline{)28} =$ ____ **14.** $7\overline{)42} =$ ____

15. $12 \div 3 =$ ____ **16.** $5\overline{)50} =$ ____ **17.** $6 \div 2 =$ ____ **18.** $8\overline{)40} =$ ____

Problem Solving and Test Prep

15. Olivia went apple picking. She picked 48 apples. She put 6 apples in each of her baskets. How many baskets did Olivia use?

____

16. Randy has 72 photographs. He puts his photographs into 8 equal piles. How many photographs are in each pile?

____

17. Terri sets the table for 8 guests. She uses 16 plates. How many plates will each guest have?

A 2

B 24

C 3

D 8

18. Hal has 24 flowers in a bunch. He puts 4 flowers in each of his vases. How many vases does Hal use?

A 8

B 6

C 20

D 12

Name____________________ **Lesson 9.3**

Model with Arrays

You can use arrays to model division.

How many groups of 7 are in 28?
Use square tiles to make an array.
Use 28 tiles.

Make rows of 7

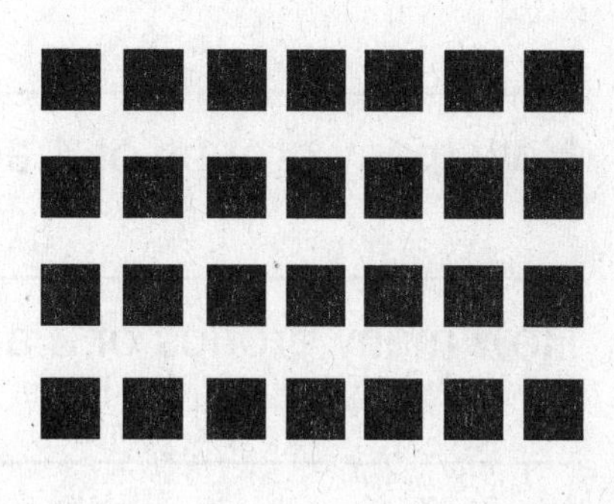

There are 28 tiles in 4 equal groups of 7.

So, there are 4 groups of 7 in 28.

Use square tiles to make an array. Solve.

1. How many groups of 2 are in 10? ____________

2. How many groups of 4 are in 12? ____________

Make an array. Write a division sentence for each one.

3. 6 groups of 36 tiles ____________

4. 8 groups of 24 tiles ____________

NS 2.0 Students calculate and solve problems involving addition, subtraction, multiplication, and division.

Name______________________________

Lesson 9.3

Model with Arrays

Use square tiles to make an array. Solve.

1. How many groups of 5 are in 25? ______
2. How many groups of 9 are in 36? ______
3. How many groups of 3 are in 12? ______
4. How many groups of 7 are in 42? ______
5. How many groups of 4 are in 16? ______
6. How many groups of 6 are in 24? ______
7. How many groups of 3 are in 18? ______
8. How many groups of 5 are in 35? ______
9. How many groups of 2 are in 14? ______
10. How many groups of 6 are in 54? ______
11. How many groups of 7 are in 21? ______
12. How many groups of 5 are in 40? ______
13. How many groups of 2 are in 18? ______
14. How many groups of 8 are in 16? ______

Make an array. Write a division sentence for each one.

15. 6 groups of 18 tiles ______
16. 7 groups of 28 tiles ______
17. 4 groups of 36 tiles ______
18. 3 groups of 30 tiles ______
19. 7 groups of 63 tiles ______
20. 4 groups of 16 tiles ______

21. George made an array with 70 tiles. He placed 7 tiles in each row. How many rows did George make in all? ______
22. Linda made an array with 48 tiles. She placed 8 tiles in each row. How many rows did Linda make in all? ______

Name ______________________ **Week 13**

Spiral Review

For 1–5, find the product.

1. $1 \times 8 =$ ______

2. $9 \times 0 =$ ______

3. $10 \times 1 =$ ______

4. $15 \times 0 =$ ______

5. $0 \times 21 =$ ______

For 6–9, find the perimeter of each figure.

6. ______ units

7. ______ units

8. ______ units

9. ______ units

For 10–12, answer the questions using the information in the line plot.

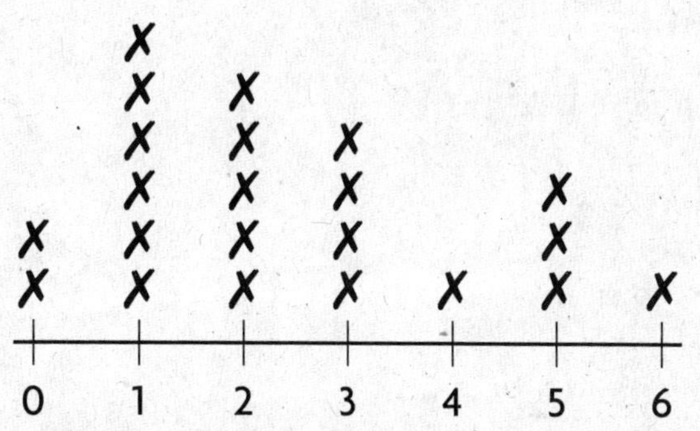

How Many Pencils Are in My Desk?

10. How many students have 2 pencils in their desk?

______

11. What is the range of the data?

______

12. What is the mode of the data?

______

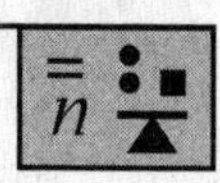

For 13–14, compare amounts to solve. Write <, >, or =.

13. Scientists found 2,508 sharks and 2,580 dolphins in the ocean. Did scientists find more sharks or more dolphins?

______

14. A male blue whale ate 6,597 krill. A female blue whale ate 6,907 krill. Which whale ate more krill?

______

Name____ **Lesson 9.4**

Algebra: Multiplication and Division

An **array** is a rectangular set of objects. An array contains *rows* and *columns*. The *rows* flow left to right and the *columns* flow up and down.

Complete the description of the array below.

There are ________ rows of counters in the array.

There are ________ columns of counters in the array.

There are ________ total counters in the array.

You can write related multiplication and division sentences to describe the array:

3 rows of 7 counters is 21 total counters ⟶ $3 \times 7 = 21$

21 counters are arranged in 3 rows of 7 counters ⟶ $21 \div 3 = 7$

So, the array is a model for $3 \times 7 = 21$ and $21 \div 3 = 7$.

For 1–5, use the array at the right.

• • • •
• • • •
• • • •
• • • •
• • • •
• • • •

1. How many rows are in the array? ____________
2. How many columns are in the array? ____________
3. How many counters are in the array? ____________
4. Complete each multiplication sentence below.

   $4 \times 6 =$ ________ $6 \times 4 =$ ________
5. Complete each division sentence below.

   $24 \div 4 =$ ________ $24 \div 6 =$ ________

Name________________________________ Lesson 9.4

Algebra: Multiplication and Division

Complete.

1.

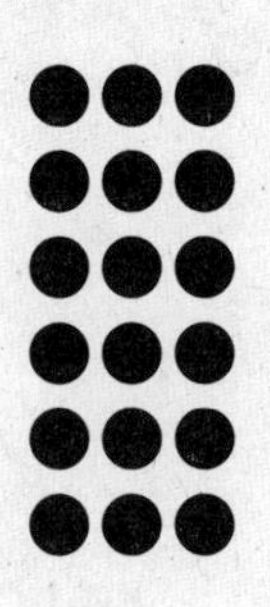

6 rows of ____ = 18

$18 \div 6 =$ ____

2.

2 rows of ____ = 12

$12 \div 2 =$ ____

3.

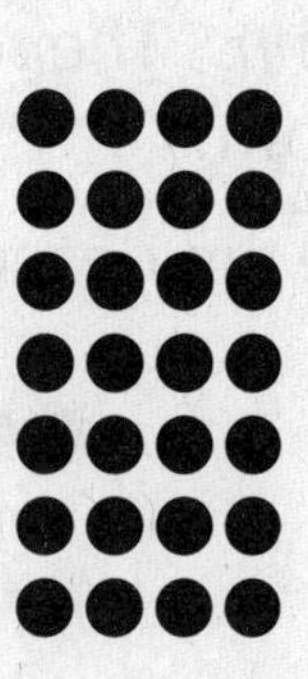

7 rows of ____ = 28

$28 \div 7 =$ ____

Complete each number sentence. Draw an array help.

4. $3 \times$ ____ $= 24$

$24 \div 3 =$ ____

5. $4 \times$ ____ $= 32$

$32 \div 4 =$ ____

6. $6 \times$ ____ $= 24$

$24 \div 6 =$ ____

7. $9 \times$ ____ $= 36$

$36 \div 9 =$ ____

Complete.

8. $3 \times 3 = 18 \div$ ____

9. $32 \div 8 =$ ____ $\times 2$

10. ____ $\times 1 = 35 \div 7$

Problem Solving and Test Prep

11. Karen has 15 tickets. A hot dog costs 5 tickets. What is the maximum number of hot dogs Karen can buy?

____

12. Molly is going to the movies with her friends. She has $40. Each ticket costs $8. What is the maximum number of tickets Molly can buy?

____

13. Tina has 30 baseball cards. She wants to divide them evenly between her 5 friends. How many cards will each friend get?

A 5 **C** 4

B 6 **D** 7

14. A big fish tank has 42 fish. The fish are divided evenly into 6 tanks. How many fish are in each tank?

A 5 **C** 7

B 6 **D** 8

Name____

Algebra: Fact Families

A **fact family** is characterized as having 4 related number sentences made from 3 numbers.

Write a fact family for the array below.

There is a total of 12 tiles.
There are 2 equal groups (rows) of tiles.
There are 6 tiles in each group (row).
The three numbers in the fact family are: 2, 6, 12.
Write 2 different multiplication and 2 different division number sentences for the set of numbers.

The largest number is 12. Use 12 as the **product** and the **dividend**.

factor × factor = **product** **dividend** ÷ divisor = quotient

☐ × ☐ = **12** **12** ÷ ☐ = ☐

☐ × ☐ = **12** **12** ÷ ☐ = ☐

So, the fact family is 2 × 6 = 12, 6 × 2 = 12, 12 ÷ 6 = 2, and 12 ÷ 2 = 6.

Complete the fact family for each set of numbers.

1. 3, 6, 18

3 × ☐ = 18 18 ÷ 3 = ______

6 × ☐ = 18 18 ÷ 6 = ______

2. 5, 10, 50

5 × ☐ = 50 50 ÷ 5 = ______

10 × ☐ = 50 50 ÷ 10 = ______

Write the fact family for each set of numbers.

3. 3, 9, 27

______ ______

______ ______

4. 4, 6, 24

______ ______

______ ______

5. 5, 8, 40

______ ______

______ ______

6. 4, 9, 36

______ ______

______ ______

Name ____________________

Lesson 9.5

Algebra: Fact Families

Write the fact family for each set of numbers.

1. 4, 6, 24 ________ ________ ________ ________

2. 2, 9, 18 ________ ________ ________ ________

3. 5, 7, 35 ________ ________ ________ ________

Write the fact family for each array.

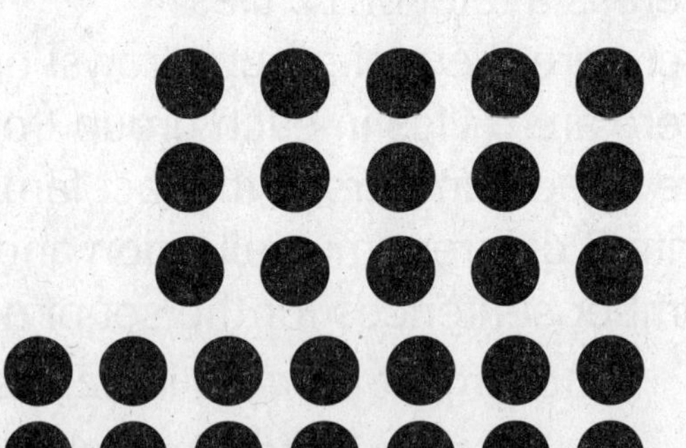

4. ________ ________

________ ________

5. ________ ________

________ ________

Complete each fact family.

6. $7 \times \_\_\_ = 42$

$6 \times 7 = \_\_\_$

$42 \div \_\_\_ = 6$

$42 \div 6 = \_\_\_$

7. $9 \times 6 = \_\_\_$

$6 \times \_\_\_ = 54$

$54 \div 9 = \_\_\_$

$54 \div \_\_\_ = 9$

8. $8 \times 3 = \_\_\_$

$\_\_\_ \times 8 = 24$

$24 \div 3 = \_\_\_$

$\_\_\_ \div 8 = 3$

9. $\_\_\_ \times 4 = 20$

$\_\_\_ \times 5 = 20$

$\_\_\_ \div 5 = 4$

$20 \div 4 = \_\_\_$

Problem Solving and Test Prep

10. Al buys a pack of watercolor paints that includes 12 colors. There are 2 colors in each of 6 rows. What is the fact family for the numbers 2, 6, and 12?

________ ________

________ ________

11. There are 18 cookies on a dish. There are 6 cookies in each of 3 rows on the dish. What is the fact family for the numbers 3, 6, and 18?

________ ________

________ ________

12. Which number sentence is not included in the same fact family as $7 \times 3 = 21$?

A $21 \div 3 = 7$

B $21 \times 3 = 7$

C $21 \div 7 = 3$

D $3 \times 7 = 21$

13. Which division sentence describes the array?

A $2 \div 3 = 6$

B $6 \div 3 = 2$

C $3 \div 2 = 6$

D $6 \div 6 = 1$

Name____ Lesson 9.6

Problem Solving Workshop Strategy: Write a Number Sentence

Jack read for 20 minutes on Monday. He read 15 minutes more on Tuesday than on Monday. For how many minutes did Jack read on Tuesday?

Read to Understand

1. What information is given?

____

____

Plan

2. Which strategy can you use to solve this problem?

____

3. Which mathematical operation can you use?

____

Solve

4. Complete the number sentence below to solve the problem

15	+	________	=	________
↑	↑	↑		↑
15 minutes	more than	Monday minutes		Tuesday minutes

5. Write your answer in a complete sentence.

____

Check

6. How can you check your answer?

____

Write a number sentence to solve. Then solve.

7. Marion played soccer 5 days this week. She played for 2 hours each day. How many hours did Marion spend playing soccer this week?

____

8. Ed spent $36 on baseball cards. Each card cost $6. How many cards did Ed buy?

____

Name____

Problem Solving Workshop Strategy: Write a Number Sentence

Problem Solving Strategy Practice

Choose the number sentence from the table at the right and solve.

$60 \div 10 = \square$	$8 + 4 = \square$
$4 \times 8 = \square$	$12 - 5 = \square$

1. Matt has 5 T-shirts. Adam has 12 T-shirts. How many more T-shirts does Adam have than Matt has?

____

2. Isabelle has 8 books in her desk. She brought 4 more books from home. How many books does Isabelle have in her desk in all?

____

3. A bag of marbles costs 60 cents. Each marble costs 10 cents. How many marbles are in the bag?

____

4. There are 4 invitations in a box. Mrs. Hannah bought 8 boxes. How many invitations did Mrs. Hannah buy?

____

Mixed Strategy Practice

USE DATA For 5, use the table.

5. Tyler spent $40 on blue tickets. Write a number sentence to show how to find the number of blue tickets Tyler bought.

____

Raffle Tickets

Color	Cost
Yellow	$2
Green	$3
Blue	$5

6. Mary spent $8 on a movie ticket, $12 on videogames, and $15 on lunch. How much money did Mary spend in all?

____

7. Randall spent $75 on tickets. Marty paid Randall $25 for tickets. Jean paid Randall $12. How much money did Randall spend of his own money?

____

Name____ Lesson 10.1

Divide by 2 and 5

You can draw a picture to show how to divide.

Divide. $14 \div 2$

Draw 14 to show the dividend.

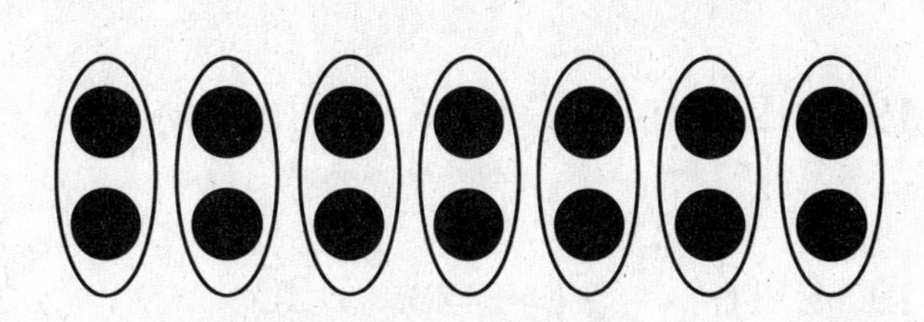

Circle groups of 2 to show the divisor.

There are 7 groups of 2. The quotient is 7.

So, $14 \div 2 = 7$

You can also use a related multiplication fact.

Divide. $20 \div 2 \longrightarrow 2 \times 10 = 20$

So, $20 \div 2 = 10$.

Find each quotient.

1. $4 \div 2 =$ ____ **2.** ____ $= 20 \div 5$ **3.** $45 \div 5 =$ ____ **4.** ____ $= 16 \div 2$

5. $8 \div 2 =$ ____ **6.** ____ $= 15 \div 5$ **7.** $12 \div 2 =$ ____ **8.** ____ $= 35 \div 5$

9. $5 \div 5 =$ ____ **10.** ____ $= 14 \div 2$ **11.** $25 \div 5 =$ ____ **12.** ____ $= 6 \div 2$

13. $10 \div 5 =$ ____ **14.** ____ $= 2 \div 2$ **15.** $40 \div 5 =$ ____ **16.** ____ $= 10 \div 2$

17. $2\overline{)18}$ **18.** $5\overline{)30}$ **19.** $2\overline{)12}$ **20.** $5\overline{)15}$

Name ____________________

Divide by 2 and 5

Find each quotient.

1. $6 \div 2 =$ ____
2. ____ $= 25 \div 5$
3. $15 \div 5 =$ ____
4. ____ $= 8 \div 2$
5. $12 \div 2 =$ ____
6. ____ $= 10 \div 5$
7. $18 \div 2 =$ ____
8. ____ $= 30 \div 5$
9. $20 \div 5 =$ ____
10. ____ $= 16 \div 2$
11. $5 \div 5 =$ ____
12. ____ $= 4 \div 2$
13. $2\overline{)14}$
14. $5\overline{)45}$
15. $2\overline{)2}$
16. $5\overline{)35}$
17. $2\overline{)10}$
18. $5\overline{)40}$
19. $2\overline{)16}$
20. $5\overline{)20}$

Problem Solving and Test Prep

21. Martin bought 40 pounds of birdseed. He bought birdseed in 5-pound packages. How many packages of birdseed did Martin buy?

____

22. **Fast Fact** A female hummingbird usually lays 2 eggs. If a researcher finds 10 eggs in one area, how many female hummingbirds are most likely in the area?

____

23. Sarah sees the same number of birds at 3 different bird feeders. She sees 12 birds in all. How many birds does Sarah see at each feeder?

A 4
B 5
C 6
D 7

24. Greg has 5 bird feeders and a 20-pound bag of bird food. He puts the same amount of bird food into each feeder. How many pounds of bird food does Greg put into each feeder?

A 3
B 4
C 5
D 6

Spiral Review

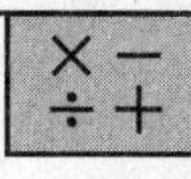

For 1–5, compare. Use <, >, or = for each ◯.

1. 1,558 ◯ 1,558
2. 7.094 ◯ 7,904
3. 969 ◯ 996
4. 6,399 ◯ 6,399
5. 3,000 ◯ 2,999

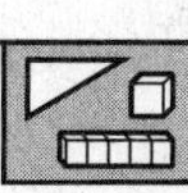

For 6–9, use the picture to answer the questions.

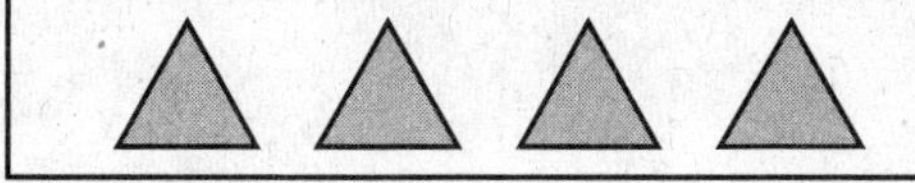

6. 1 triangle Number of sides: ____
7. 3 triangles Number of sides: ____
8. 5 triangles Number of sides: ____
9. 4 triangles Number of sides: ____

For 10–12, use the graph to answer the questions.

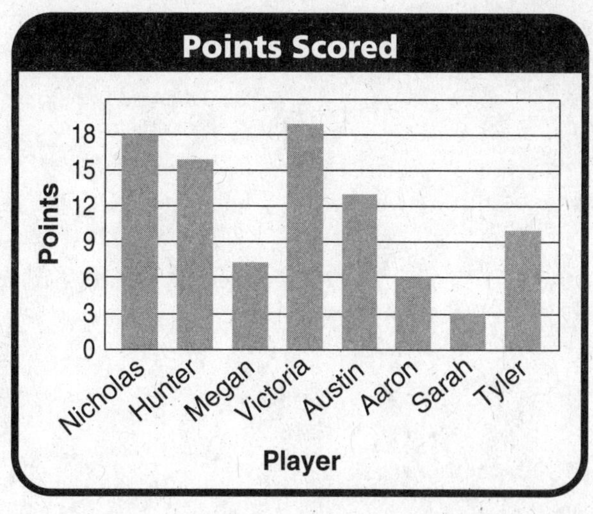

10. Which player scored the most points? ____________
11. How many points did Nicholas score? ____________
12. How many more points did Tyler score than Sarah scored? ____________

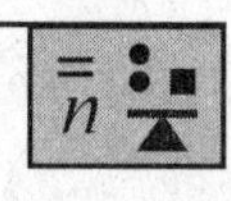

For 13–15, find the missing value.

13. ☐ $= 10 \times 4$ ______
14. $6 \times$ ☐ $= 48$ ______
15. ☐ $\times 5 = 35$ ______

Name____ Lesson 10.2

Divide by 3 and 4

There are many different ways to find the quotient.
One way is to count back on a number line.

Find: $28 \div 4$

Start at 28.

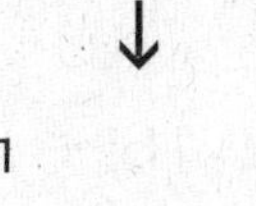

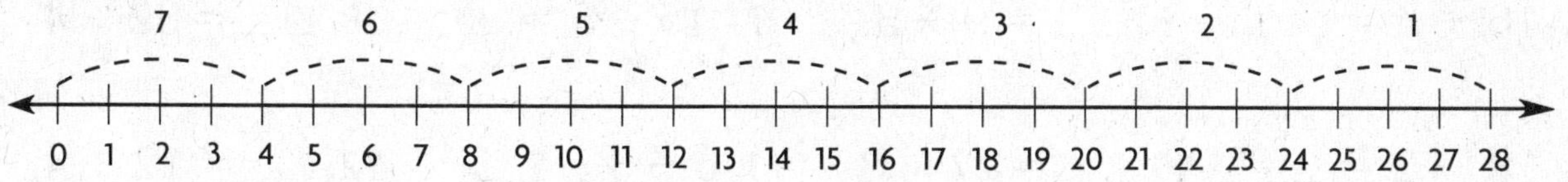

Count back by 4 on the number line until you reach 0.

Count how many times 4 was subtracted.

4 was subtracted seven times.

So, $28 \div 4 = 7$.

Find each quotient.

1. $24 \div 3 =$ ____ **2.** ____ $= 12 \div 4$ **3.** $15 \div 3 =$ ____ **4.** ____ $= 8 \div 4$

5. $36 \div 4 =$ ____ **6.** ____ $= 18 \div 3$ **7.** $20 \div 4 =$ ____ **8.** ____ $= 3 \div 3$

9. $9 \div 3 =$ ____ **10.** ____ $= 16 \div 4$ **11.** $6 \div 3 =$ ____ **12.** ____ $= 32 \div 4$

13. $4 \div 4 =$ ____ **14.** ____ $= 21 \div 3$ **15.** $12 \div 3 =$ ____ **16.** ____ $= 28 \div 4$

17. $3\overline{)27}$ **18.** $4\overline{)24}$ **19.** $3\overline{)18}$ **20.** $4\overline{)32}$

21. $3\overline{)9}$ **22.** $4\overline{)12}$ **23.** $3\overline{)21}$ **24.** $4\overline{)16}$

NS 2.3 Use the inverse relationship of multiplication and division to compute and check results.

Name________________________________ Lesson 10.2

Divide by 3 and 4

Find each quotient.

1. $12 \div 3 =$ ___ **2.** ___ $= 20 \div 4$ **3.** $21 \div 3 =$ ___ **4.** ___ $= 8 \div 4$

5. $16 \div 4 =$ ___ **6.** ___ $= 9 \div 3$ **7.** $15 \div 5 =$ ___ **8.** ___ $= 32 \div 4$

9. $18 \div 2 =$ ___ **10.** ___ $= 27 \div 3$ **11.** $6 \div 3 =$ ___ **12.** ___ $= 12 \div 4$

13. $4\overline{)28}$ **14.** $3\overline{)30}$ **15.** $3\overline{)18}$ **16.** $4\overline{)24}$

Complete.

17. $12 \div$ ___ $= 2$ **18.** $24 \div$ ___ $= 3$ **19.** $36 \div$ ___ $= 4$ **20.** $3 \div$ ___ $= 3$

21. $27 \div$ ___ $= 9$ **22.** $36 \div$ ___ $= 6$ **23.** $32 \div$ ___ $= 8$ **24.** $28 \div$ ___ $= 7$

Problem Solving and Test Prep

25. There are 24 students signed up for the relay race. Each team needs 4 students. How many teams will there be in the relay race?

____

26. Twenty-one students form a study group. If they want to sit evenly at 3 tables, then how many students will be at each table?

____

27. Jeremy has 36 crackers. He puts 4 crackers in each of his bags. How many bags does Jeremy fill?

A 6
B 7
C 8
D 9

28. Lea has 27 beads. She makes 3 bracelets, each with the same number of beads. How many beads are on each bracelet?

A 9
B 8
C 7
D 6

Name____ Lesson 10.3

Division Rules for 1 and 0

Division rules can help you understand how to divide with 1 and 0.

Rule A: When you divide any number by 1, the quotient equals that number.

$5 \div 1 = 5$ OR $1\overline{)5}$ = 5

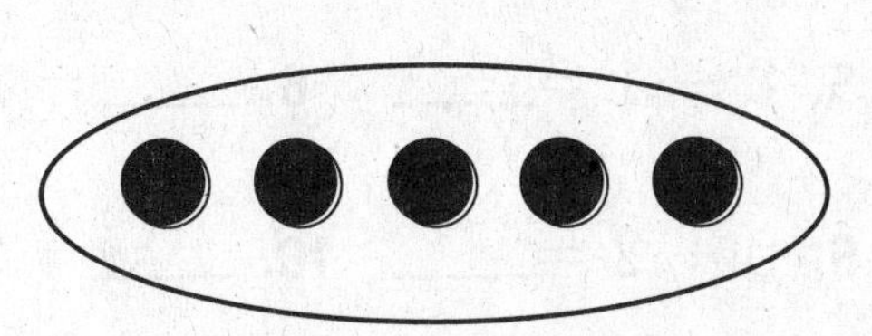

One group of 5.

Rule B: When you divide any number by itself (except 0), the quotient is 1.

$5 \div 5 = 1$ OR $5\overline{)5}$ = 1

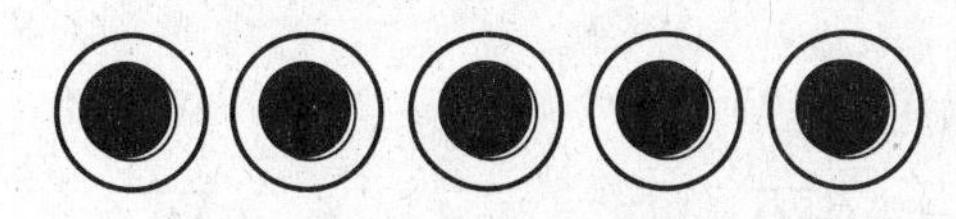

Five groups of 1.

Rule C: When you divide 0 by any number (except 0), the quotient is 0.

$0 \div 5 = 0$ OR $5\overline{)0}$ = 0

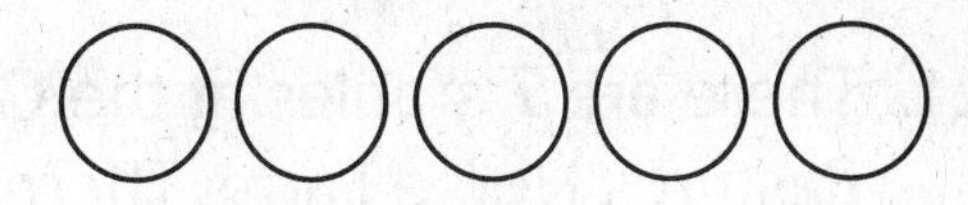

Five groups of 0.

Rule D: You cannot divide a number by 0.

$5 \div 0$ $\qquad$ $0\overline{)5}$

Find each quotient.

1. $4 \div 1 =$ ______ **2.** ______ $= 2 \div 2$ **3.** $8 \div 1 =$ ______ **4.** ______ $= 7 \div 7$

5. $3 \div 1 =$ ______ **6.** ______ $= 0 \div 9$ **7.** $10 \div 1 =$ ______ **8.** ______ $= 6 \div 1$

9. $5 \div 5 =$ ______ **10.** ______ $= 0 \div 4$ **11.** $0 \div 2 =$ ______ **12.** ______ $= 3 \div 1$

13. $7\overline{)0}$ **14.** $8\overline{)8}$ **15.** $3\overline{)3}$ **16.** $1\overline{)9}$

Name____ Lesson 10.3

Division Rules for 1 and 0

Find each quotient.

1. $5 \div 5 =$ ____
2. ____ $= 0 \div 4$
3. $3 \div 1 =$ ____
4. ____ $= 0 \div 9$
5. $8 \div 1 =$ ____
6. ____ $= 7 \div 7$
7. $10 \div 2 =$ ____
8. ____ $= 6 \div 1$
9. $0 \div 2 =$ ____
10. ____ $= 18 \div 3$
11. $4 \div 1 =$ ____
12. ____ $= 2 \div 2$
13. $1\overline{)0}$
14. $3\overline{)3}$
15. $1\overline{)9}$
16. $5\overline{)35}$
17. $8\overline{)8}$
18. $5\overline{)0}$
19. $4\overline{)36}$
20. $4\overline{)4}$

Problem Solving and Test Prep

21. There are 7 stables at the Green Pastures Horse Farm. There are 7 horses that live on the farm. How many horses are in each stable, if there are an equal number of horses per stable?

____

22. Trevor plans to give 3 grapes to each parrot in a store. There is 1 parrot in the store. How many grapes does Trevor give the parrot?

____

23. Katherine has 5 birds. She only has 1 birdcage to keep them in. How many birds are in that cage?

A 0
B 1
C 5
D 10

24. Which is the quotient?

$4\overline{)0}$

A 0
B 1
C 2
D 4

Name____________________ Lesson 10.4

Algebra: Practice the Facts

There are many different ways to find the **quotient**.
One way is to use a number line.

Write a division sentence for the number line below.

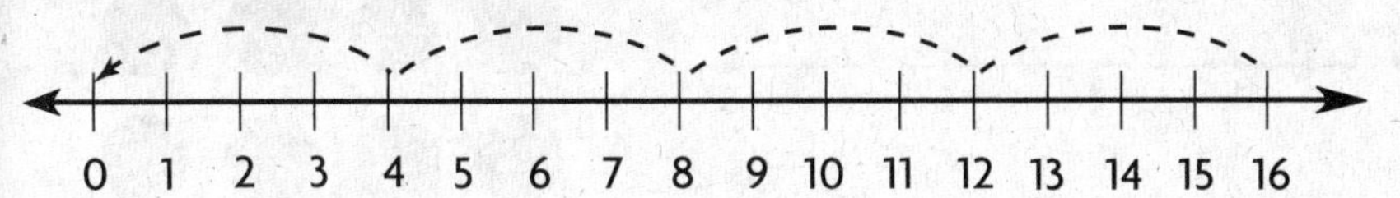

Start with the number farthest to the right on the number line.

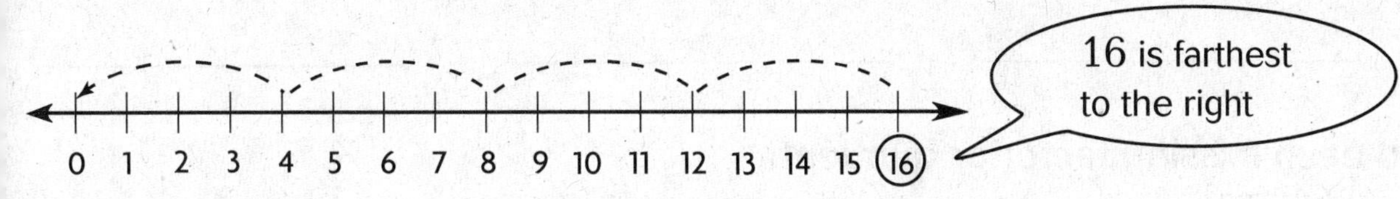

16 is the dividend.
The number line jumps numbers 4 times.
There are 4 numbers between each jump.

So, the division sentence represented by this number line is; $16 \div 4 = 4$.

Write a division sentence for each.

1. 

______

2. 

______

3. 

______

Find each missing factor or quotient.

4. $3 \times$ ___ $= 18$ $18 \div 3 =$ ___

5. $4 \times$ ___ $= 8$ $8 \div 4 =$ ___

Find each quotient.

6. $12 \div 2 =$ ___

7. $24 \div 4 =$ ___

8. $35 \div 5 =$ ___

9. ___ $= 4 \div 1$

10. $40 \div 5 =$ ___

11. $12 \div 3 =$ ___

12. $9 \div 3 =$ ___

13. ___ $= 27 \div 3$

Name ______________________

Lesson 10.4

Algebra: Practice the Facts

Write a division sentence for each.

1.

2. 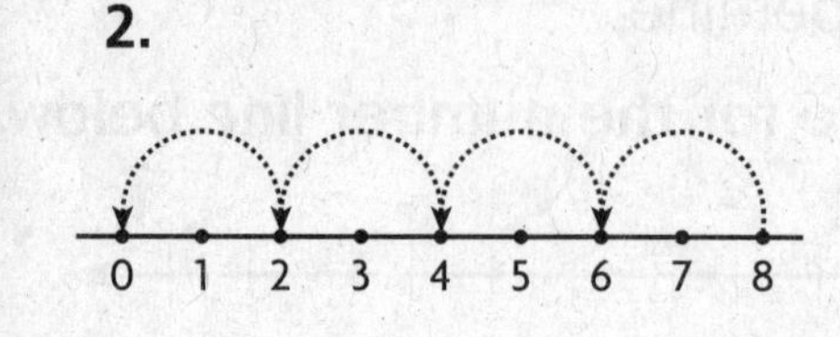

3. 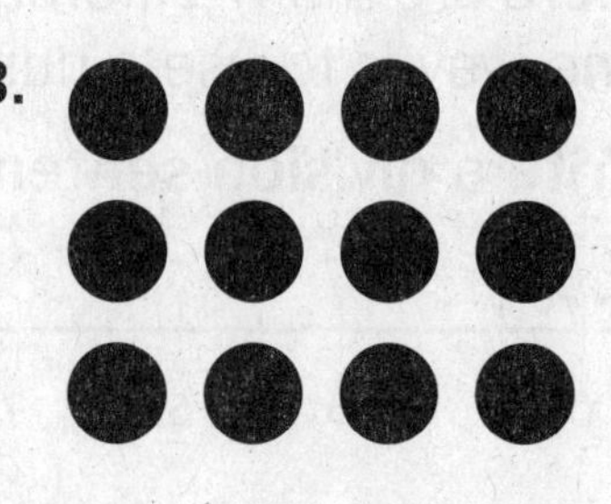

______ ______ ______

Find each missing factor or quotient.

4. $4 \times \square = 36$ $36 \div 4 =$ ____

5. $3 \times \square = 0$ $0 \div 3 =$ ____

Find each quotient.

6. $27 \div 3 =$ ____

7. $18 \div 3 =$ ____

8. $20 \div 4 =$ ____

9. ____ $= 32 \div 4$

10. $5\overline{)15}$

11. $2\overline{)2}$

12. $3\overline{)21}$

13. $2\overline{)10}$

Problem Solving and Test Prep

14. A craft store sells beads in packages of 4. Tara needs 24 beads for a project. How many packages of beads will Tara need to buy?

______

15. Two brothers sell lemonade in their neighborhood. They make $6 on Saturday. How much money should each brother receive if they split this money evenly?

______

16. Which division sentence is related to $3 \times 4 = 12$?

A $24 \div 2 = 12$

B $4 \div 2 = 2$

C $12 \div 6 = 2$

D $12 \div 3 = 4$

17. Which division sentence is related to $3 \times 8 = 24$?

A $24 \div 3 = 8$

B $24 \div 2 = 12$

C $24 \div 6 = 4$

D $24 \div 4 = 6$

Name ____________________

Problem Solving Workshop Skill: Choose the Operation

T-shirts are priced for $9 each at a gift shop. What is the cost of 3 T-shirts?

1. What are you asked to find?

____

____

2. How can you decide what operation to use to solve this problem? Name the operation you will use to solve this problem.

____

____

3. Write a number sentence for this problem. Then solve.

______ ◯ ______ ◯ ______

4. What is the cost of 3 T-shirts? ______________________________

____

5. Write a number sentence to show how you might check your answer.

____

Choose the operation. Write a number sentence. Then solve.

6. Nick buys 2 T-shirts at a gift shop. The total cost for both T-shirts is $16. He pays the same amount for each T-shirt. What is the cost of each T-shirt?

____

____

7. An amusement park has 15 rides for children under the age of 5, and 37 rides for anyone over the age of 5. How many rides are at the amusement park in all?

____

____

Name______________________

Problem Solving Workshop Skill: Choose the Operation

Problem Solving Skill Practice

Choose the operation. Write a number sentence. Then solve.

1. The Murphy family spent \$36 for 4 tickets to the nature center. how much did each ticket cost?

____

2. There were 27 children and 9 adults on the nature center tour. How many people were on the tour in all?

____

3. The nature center has a petting zoo with 5 areas. Each area has the same number of animals. There are 25 animals in the petting zoo. How many animals are in each area?

____

4. Drinks at the nature center cost \$7. Mr. Chin gave the clerk \$20 for 1 drink. how much change will Mr. Chin get back from the clerk?

____

Mixed Applications

for 5, use the table.

5. USE DATA Martha only hikes the Echo Trail. She hikes this trail 3 times each week. How many miles does Martha hike each week?

____

Nature Trails	
Trail Name	**Distance**
Echo Trail	4 miles
View Trail	12 miles
Pine Trail	47 miles
Green Trail	15 miles
Steep Trail	23 miles

6. Cora, Sal, Marty, and Jane are standing in line. Jane is first in line. Marty is behind Cora. Cora is in front of Sal. Sal is behind Marty. In what order are the four people standing in line?

____

7. Anna needs 28 balloons. They come in packages of 4, 6, or 9 balloons. How many of each package should she buy in order to have the exact number of balloons she needs?

____

Name ______________________ Week 15

Spiral Review

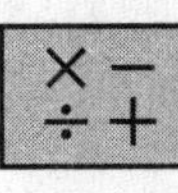

For 1–5, write the fact family for each set of numbers.

1. 7, 4, 28

______ ______

______ ______

2. 5, 7, 35

______ ______

______ ______

3. 8, 2, 16

______ ______

______ ______

4. 9, 3, 27

______ ______

______ ______

5. 6, 6, 36

______ ______

For 10–13, extend the patterns.

10. Skip count by twos: 16, 18,

______, ______, ______, ______

11. Skip count by threes: 31, 34,

______, ______, ______, ______

12. Skip count by fives: 55, 60,

______, ______, ______, ______

13. Skip count by tens: 49, 59,

______, ______, ______, ______

For 6–9, draw a line on the figure to make the given shapes.

6.  Make a triangle and a trapezoid.

7. 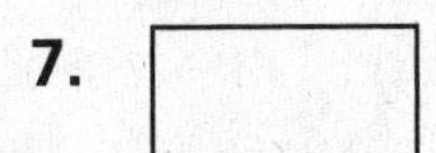 Make two triangles.

8. 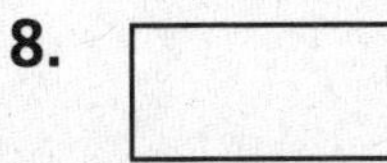 Make two rectangles.

9.  Make a triangle and a trapezoid.

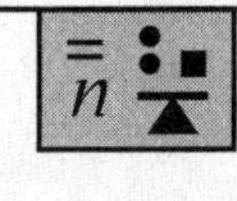

For 14–16, find the missing factor.

14. $(5 \times \square) \times 4 = 40$ ______

15. $5 \times (3 \times \square) = 45$ ______

16. $(2 \times \square) \times 7 = 56$ ______

Name ____________________

Divide by 6

You can use a multiplication chart to divide by 6.

Find the missing factor and quotient.

$6 \times \square = 42$ **$42 \div 6 = \square$**

Use a multiplication chart to find the missing factor.

Find the 6s column.

Trace down the column until you find 42.

Trace left across the row until you reach 7.

7 is the factor you multiply 6 by to get a product of 42.

So, $6 \times 7 = 42$.

Use the related multiplication sentence to find the missing quotient.

Since, $6 \times 7 = 42$, then $42 \div 6 = 7$.

×	0	1	2	3	4	5	6	7	8	9	10	11	12
0	0	0	0	0	0	0	0	0	0	0	0	0	0
1	0	1	2	3	4	5	6	7	8	9	10	11	12
2	0	2	4	6	8	10	12	14	16	18	20	22	24
3	0	3	6	9	12	15	18	21	24	27	30	33	36
4	0	4	8	12	16	20	24	28	32	36	40	44	48
5	0	5	10	15	20	25	30	35	40	45	50	55	60
6	0	6	12	18	24	30	36	42	48	54	60	66	72
7	0	7	14	21	28	35	42	49	56	63	70	77	84
8	0	8	16	24	32	40	48	56	64	72	80	88	96
9	0	9	18	27	36	45	54	63	72	81	90	99	108
10	0	10	20	30	40	50	60	70	80	90	100	110	120
11	0	11	22	33	44	55	66	77	88	99	110	121	132
12	0	12	24	36	48	60	72	84	96	108	120	132	144

Find each missing factor and quotient.

1. $6 \times \square = 30$ $30 \div 6 = \square$ **2.** $6 \times \square = 48$ $48 \div 6 = \square$

3. $6 \times \square = 18$ $18 \div 6 = \square$ **4.** $6 \times \square = 24$ $24 \div 6 = \square$

5. $6 \times \square = 42$ $42 \div 6 = \square$ **6.** $6 \times \square = 12$ $12 \div 6 = \square$

7. $6 \times \square = 6$ $6 \div 6 = \square$ **8.** $6 \times \square = 54$ $54 \div 6 = \square$

Name____________________ **Lesson 11.1**

Divide by 6

Find each missing factor and quotient.

1. $6 \times \_\_\_ = 42$ **2.** $36 \div 6 = \_\_\_$ **3.** $6 \times \_\_\_ = 24$ **4.** $\_\_\_ \times 6 = 30$

5. $42 \div 6 = \_\_\_$ **6.** $6 \times \_\_\_ = 18$ **7.** $60 \div 6 = \_\_\_$ **8.** $54 \div 6 = \_\_\_$

Find each quotient.

9. $72 \div 6 = \_\_\_$ **10.** $24 \div 3 = \_\_\_$ **11.** $\_\_\_ = 48 \div 6$ **12.** $\_\_\_ = 12 \div 6$

13. $27 \div 3 = \_\_\_$ **14.** $\_\_\_ = 35 \div 5$ **15.** $21 \div 3 = \_\_\_$ **16.** $\_\_\_ = 45 \div 9$

Problem Solving and Test Prep

17. Toni bought 24 hotdogs. They come in packages of 6. How many packages of hotdogs did Toni buy?

____

18. Kara brought 36 muffins to a picnic. Each package contains 6 muffins. How many packages of muffins did Kara bring?

____

19. There are 42 books, divided evenly among six shelves in the bookcase. How many books are on each shelf?

A 8
B 6
C 5
D 7

20. There are 30 peaches in a basket. Frank separates the peaches evenly into 6 bags. How many peaches are in each bag?

A 8
B 6
C 5
D 7

Name____ **Lesson 11.2**

Divide by 7 and 8

You can use counters to divide by 7 and 8.

Find the missing factor and quotient: $8 \times \square = 56$, $56 \div 8 = \square$

Think: what number times 8 equals 56?

Use 56 counters to make 8 groups. Put one counter into each group.

Repeat until all 56 counters are used.

There are 7 counters in each of 8 groups.

So, $8 \times 7 = 56$ or $56 \div 8 = 7$.

Find the missing factor and quotient: $7 \times \square = 63$, $63 \div 7 = \square$.

Think: what number times 7 equals 63?

Use 63 counters to make 7 groups. Put one counter into each group.

Repeat until all 63 counters are used.

There are 9 counters in each of 7 groups.

So, $9 \times 7 = 63$, and $63 \div 7 = 9$.

Find each missing factor or quotient.

1. $32 \div 8 =$ ___ **2.** $35 \div 7 =$ ___ **3.** $54 \div 6 =$ ___ **4.** $9 \times$ ___ $= 72$

5. ___ $\times 8 = 16$ **6.** ___ $\times 6 = 42$ **7.** $48 \div 6 =$ ___ **8.** $7 \times$ ___ $= 28$

9. ___ $\times 3 = 24$ **10.** ___ $\times 4 = 36$ **11.** $56 \div 7 =$ ___ **12.** $8 \times$ ___ $= 40$

Name____________________ Lesson 11.2

Divide by 7 and 8

Find each missing factor and quotient.

1. $8 \times \_\_\_ = 48$
2. $49 \div 7 = \_\_\_$
3. $7 \times \_\_\_ = 28$
4. $\_\_\_ \times 8 = 40$
5. $24 \div 8 = \_\_\_$
6. $7 \times \_\_\_ = 56$
7. $63 \div 7 = \_\_\_$
8. $32 \div 8 = \_\_\_$

Find each quotient.

9. $56 \div 8 = \_\_\_$
10. $21 \div 7 = \_\_\_$
11. $\_\_\_ = 35 \div 7$
12. $\_\_\_ = 16 \div 2$
13. $7 \div 1 = \_\_\_$
14. $\_\_\_ = 70 \div 7$
15. $32 \div 8 = \_\_\_$
16. $\_\_\_ = 45 \div 9$

Problem Solving and Test Prep

17. The Williams family went camping at a lake. There are 56 members in the Williams family. Each cabin holds 8 people. How many cabins did the Williams family rent?

____

18. Juana bought juice boxes for a camping trip. She needed 40 juice boxes. They come in packages of 8. How many packages of juice boxes did Juana buy?

____

19. There were 56 apples in a cart. Don emptied the cart and put 7 apples into each of his bags. How many bags did Don fill?

A 12
B 7
C 8
D 6

20. Eva has 24 flowers. She arranges them into bunches of 8. How many bunches does Eva arrange?

A 6
B 24
C 8
D 3

ame______________________________ Lesson 11.3

ʼroblem Solving Workshop Strategy: Work Backward

yan collects model cars. He gave half his model car collection to a ʼiend. Then he bought 6 more model cars. Now Ryan has 14 model cars. ɪow many cars did Ryan have to start?

ʀead to Understand

. What are you asked to find?

____

____

ʼlan

. Which strategy can you use to solve this problem?

____

____

·olve

First, subtract the number of cars Ryan bought from the total number of cars Ryan ended up with.

$14 - 6 = \_\_\_$

Then multiply your answer by 2.

How many model cars did Ryan have to start with? ______________

ʼheck

How can you check your answer?

____

____

ʼork backward to solve.

Sarah was making a fruit salad. She divided the fruit evenly into 2 bowls. Sarah ate 6 pieces of fruit from one of the bowls, which then had 18 pieces of fruit left in it. How many pieces of fruit did Sarah have to start with?

____

7. Patricia spent half of her allowance on ice cream. Her dad gave her $4.00. Now Patricia has $9.00. How much is Patricia's allowance?

____

MR 1.1 Analyze problems by identifying relationships, distinguishing relevant from irrelevant information, sequencing and prioritizing informaiton, and observing patterns.

Name________________________________ **Lesson 11.3**

Problem Solving Workshop Strategy: Work Backward

Problem Solving Strategy Practice

Work backward to solve.

1. Rachel cut a ribbon in half. She then cut 7 inches off of one of these halves, leaving 5 inches of ribbon on this half. What was the length of the original ribbon?

____

2. Abby cut a piece of construction paper into 2 equally long pieces. She then cut off 5 inches in length from one piece. This piece is now 4 inches long. What was the length of the original piece of construction paper?

____

Mixed Strategy Practice

USE DATA For 3–4 use the table.

3. Trent wants to buy a total of 10 shirts of either blue or green color. Are there enough shirts in inventory for Trent to buy the shirts he wants? Show your work.

____

4. Frank bought 4 red T-shirts and 2 yellow T-shirts on Tuesday. If Frank wants to buy 16 more T-shirts on Wednesday, are there enough left in inventory for him to do so?

____

T-Shirt Inventory

Color	Number of Shirts
Blue	5
Green	4
Yellow	7
Red	6

5. Greg collected $81 selling 9 boxes of candy bars. How much did Greg charge for each box of candy bars?

____

6. Anna has 2 tambourines and 4 guitars. How many musical instruments does Anna have in all? Show your work.

____

Name____ Lesson 11.4

Divide by 9 and 10

You can use counters to divide by 9 and 10.

Find the quotient for $63 \div 9 = \square$.

Place 63 counters into 9 equal groups.

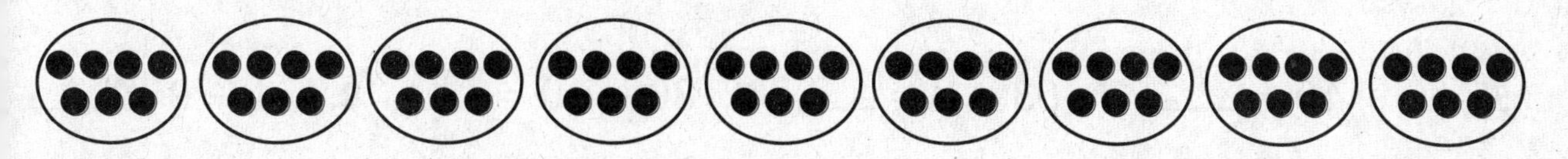

There are 7 counters in each group.

So, $63 \div 9 = 7$.

Find the quotient for $30 \div 10 = \square$.

Place 30 counters into 10 equal groups.

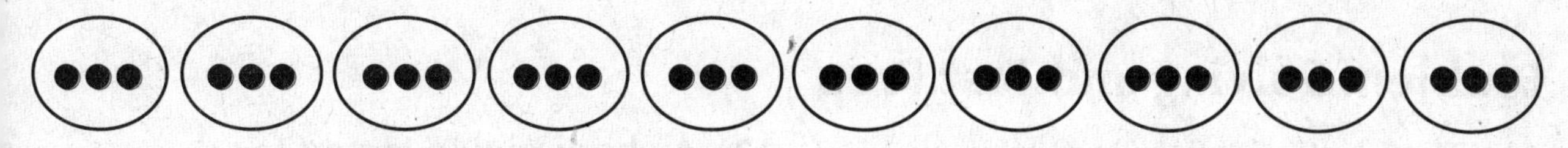

There are 3 counters in each group.

So, $30 \div 10 = 3$.

Find each quotient.

1. $18 \div 9 =$ ____
2. $20 \div 10 =$ ____
3. $60 \div 10 =$ ____
4. $72 \div 9 =$ ____
5. $90 \div 10 =$ ____
6. $54 \div 9 =$ ____
7. $45 \div 9 =$ ____
8. $100 \div 10 =$ ____
9. $36 \div 6 =$ ____
10. $27 \div 9 =$ ____
11. $90 \div 9 =$ ____
12. $20 \div 10 =$ ____

Name ______________________________

Divide by 9 and 10

Find each quotient.

1. $70 \div 10 =$ ____ 2. $36 \div 9 =$ ____ 3. $40 \div 10 =$ ____ 4. $27 \div 9 =$ ____

5. $50 \div 10 =$ ____ 6. $63 \div 9 =$ ____ 7. $30 \div 10 =$ ____ 8. $54 \div 9 =$ ____

Complete each table.

9.

÷	40	60	80	100
10				

10.

÷	27	45	72	81
9				

Problem Solving and Test Prep

11. There are 54 fish, in 9 tanks, at an aquarium. Each tank contains an equal number of fish. How many fish are in each tank?

____

12. A shark movie lasted for 50 minutes. The movie spent 10 minutes featuring each shark. How many sharks were featured in the movie?

____

13. There are 40 people waiting in lines at an aquarium. There are 10 people in each line. How many lines are there?

A 1

B 4

C 40

D 400

14. Nine fish in a tank display a total of 36 stripes. If they each display an equal number of stripes, how many stripes does each fish display?

A 9

B 5

C 6

D 4

Spiral Review

For 1–5, find the product.

1. $5 \times 4 =$ ______
2. $8 \times 6 =$ ______
3. $9 \times 4 =$ ______
4. $10 \times 10 =$ ______
5. $2 \times 7 =$ ______

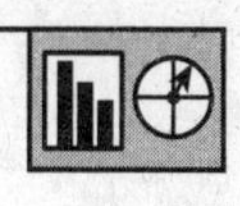

For 10–12, use the pattern in the table to answer the questions.

number of cars	1	2	3	4		6
number of wheels	4	8	12		20	

10. How many wheels are on 4 cars?

______

11. How many cars have 20 wheels in all?

______

12. How many wheels are on 6 cars?

______

For 6–9, draw a line to match the shape with the number of faces it has.

6. 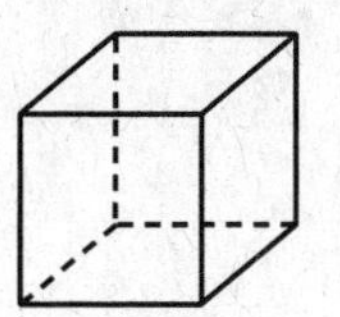 5

7. 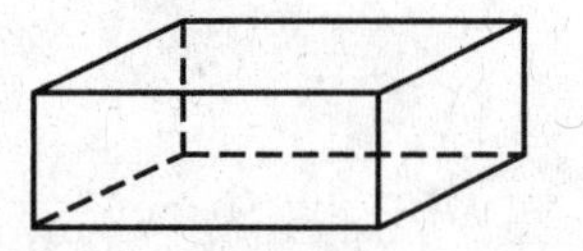 6

8. 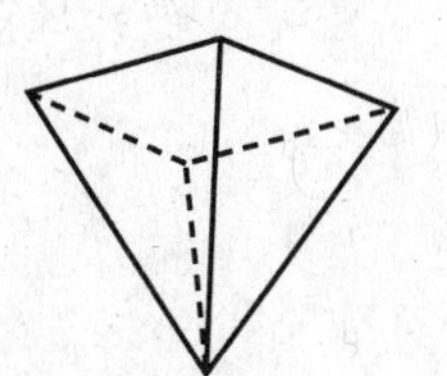 6

9. 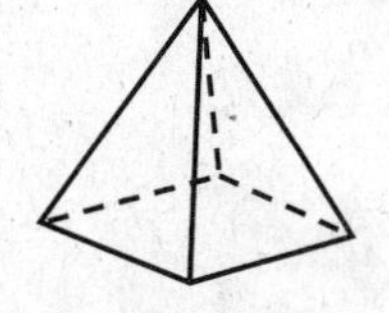 5

For 13–15, predict the next number in each pattern. Explain.

13. 1, 2, 4, 8, ☐

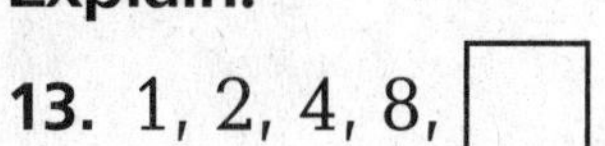

____

14. 19, 23, 27, 31, ☐

____

15. 900, 800, 700, 600, ☐

____

Name________________________________ **Lesson 11.5**

Algebra: Practice the Facts

You can use repeated subtraction to divide.

Write the division sentence that has been represented by the repeated subtraction model below.

$$\begin{array}{r}40\\-10\\\hline 30\end{array}\nearrow\begin{array}{r}30\\-10\\\hline 20\end{array}\nearrow\begin{array}{r}20\\-10\\\hline 10\end{array}\nearrow\begin{array}{r}10\\-10\\\hline 0\end{array}$$

Start with the number 40 and subtract 10. Continue to subtract 10 from the difference until you get to 0. Then count the number of times you subtracted 10.

Since you subtracted 10 4 times, 4 is the quotient.

4 is the quotient.

So, the division sentence for this problem is: **$40 \div 10 = 4$.**

$4 \times \square = 40$

Since $40 \div 10 = 4$, that means $4 \times 10 = 40$.

Find each missing factor and quotient.

1. $3 \times$ _____ $= 27$ $27 \div 3 =$ _____ **2.** $4 \times$ _____ $= 20$ $20 \div 4 =$ _____

3. $6 \times$ _____ $= 42$ $42 \div 6 =$ _____ **4.** $3 \times$ _____ $= 9$ $9 \div 3 =$ _____

5. $5 \times$ _____ $= 45$ $45 \div 5 =$ _____ **6.** $7 \times$ _____ $= 56$ $56 \div 7 =$ _____

Name ____________________

Algebra: Practice the Facts

Write a division sentence for each.

1. 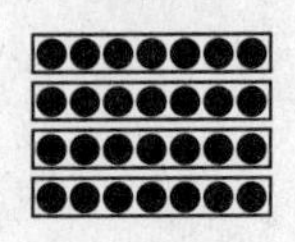

2. 

3. $\begin{array}{r} 18 \\ -6 \\ \hline 12 \end{array}$ ↙ $\begin{array}{r} 12 \\ -6 \\ \hline 6 \end{array}$ ↙ $\begin{array}{r} 6 \\ -6 \\ \hline 0 \end{array}$

________ ________ ________

Find each missing factor and quotient.

4. $8 \times$ ____ $= 40$ $\quad 40 \div 8 =$ ____

5. $4 \times$ ____ $= 36$ $\quad 36 \div 4 =$ ____

6. $9 \times$ ____ $= 63$ $\quad 63 \div 9 =$ ____

7. $3 \times$ ____ $= 24$ $\quad 24 \div 3 =$ ____

Compare. Write <, >, or = for each ◯.

8. 3×4 ◯ $24 \div 2$

9. $40 \div 5$ ◯ $15 - 5$

10. 7×4 ◯ 9×3

Problem Solving and Test Prep

11. Thomas goes hiking in a Colorado state park. The entire trail takes 54 minutes to complete. Each section of the trail takes 9 minutes to complete. How many sections does the trail have?

____

12. Carrie took 40 pictures on her nature walk. She took 4 pictures of each flower that she saw. How many flowers did Carrie see?

____

13. Hal walked 27 miles, evenly over 9 days. How many miles did Hal walk each day?

A 7 **C** 3

B 4 **D** 5

14. Nancy bought 4 new flashlights. Each flashlight cost $6.00. How much money did Nancy spend?

A $18 **C** $30

B $24 **D** $10

Name______________________________

Algebra: Find the Cost

You can use multiplication and division to find the cost of items.

Write a number sentence for the following problem. Then solve.

Josie buys 3 markers for $2 each. How much does Josie spend?

Each marker costs $2.

To find how much 3 markers will cost you can multiply.

3	×	$2	= $6
Markers	×	Cost per Marker	

So, 3 × $2 = $6; Josie spends $6.

Write a number sentence. Then solve.

1. Karen spent $18 on 3 stamps. How much did each stamp cost?

____

2. Charlotte needs 4 pages to fill her photo album. Each page costs $4. How much will it cost Charlotte to buy the pages she needs?

____

3. Bob bought 5 new pens. Each pen cost $3. How much did Bob spend?

____

4. Christine bought 6 paint sets. She spent $42. How much did each paint set cost?

____

5. The notebooks cost $3 each. How much do 7 notebooks cost?

____

6. Michael paid $27 for 3 sets of watercolors. How much did each set of watercolors cost?

____

NS 2.7 Determine the unit cost when given the total cost and number of units.

Name____

Lesson 11.6

Algebra: Find the Cost

Write a number sentence. Then solve.

1. Alex buys 2 sets of paint. Each set costs $7. How much does Alex spend on paint sets?

____

2. Mr. Walker buys 5 boxes of erasers. Each box costs $4. How much does Mr. Walker spend on erasers?

____

Find the cost of the total number of items.

3. Packages of pens cost $3 each.

5 packages ________

7 packages ________

4. Notebooks cost $7 each.

4 notebooks ________

6 notebooks ________

5. Paint brushes cost $4 each

3 brushes ________

9 brushes ________

Find the cost of one of each item.

6. 8 pencils cost $24

____

7. 5 folders cost $10

____

8. 6 glue sticks cost $18

____

Problem Solving and Test prep

9. Riley buys 9 sheets of paper for her scrapbook. She spends a total of $18. How much money does each sheet of paper cost?

____

10. Each package of colored pencils costs $7. Kylie has $20. What is the maximum number of packages of pencils Kylie can buy?

____

11. If 8 yards of ribbon costs $40, what is the cost of 1 yard of ribbon?

A $4

B $5

C $6

D $7

12. Ellen buys 5 packages of stickers. Each package costs $3. How much does Ellen spend?

A $2

B $8

C $10

D $15

Name____ **Lesson 11.7**

Algebra: Expressions and Equations

You can write expressions and equations to help you solve problems.

There were 7 children, who each paid $4 to bowl. How much did the children pay in all?

Decide which operation to use.

Since each of 7 children paid $4, and you need to find a total amount paid, use multiplication.

Write an expression to show how much the children paid to bowl.

7	×	$4
Number of Children	×	*Amount Paid Per Child*

Use the expression to write an **equation**. An **equation** is a number sentence that uses an equal sign to show that two amounts are equal.

7	×	$4	=	$28
7 children	×	$4 to bowl	=	total amount the children paid

So, the children paid $28 in all.

Write an expression. Then write an equation and solve.

1. Michelle has 4 scarves. She received 5 more from her mother. How many scarves does Michelle have now?

   ____

2. Each backpack comes with 6 zippers. How many zippers are on 6 backpacks?

   ____

3. Billy walked 4 miles a day. How many days did it take Billy to walk 28 miles?

   ____

4. Hector had 32 newspapers. He delivered 23. How many more newspapers does Hector have to deliver?

   ____

Name ______________________________

Algebra: Expressions and Equations

Write an expression. Then write an equation to solve.

1. There are 4 tables in the cafeteria. Eight children sit at each table. How many children are in the cafeteria?

____

____

2. Sam brought 28 cookies. There are 22 students in his class. If each student had one cookie, how many cookies were left?

____

____

3. Francesca collected 72 shells from the beach. She collected 9 shells each day. How many days did Francesca collect shells at the beach?

____

____

4. There were 27 adults and 15 children at a party. How many people were at the party in all?

____

____

Problem Solving and Test Prep

5. Freddy organizes a dodge ball tournament with 6 equal teams. A total of 42 students sign up to play in the tournament. Write an expression to help Freddy find how many students should play on each team. Then write an equation to solve.

____

____

6. Kathie played 6 games each week for 4 weeks. Write the expression that shows how many games Kathie played in 4 weeks.

____

____

7. Rob needs 3 sheets of paper to make a paper airplane. What is the maximum number of paper airplanes Rob can make with 30 sheets of paper?

A 10 **C** 3
B 27 **D** 15

8. Alex spent \$6 on a game of bowling and \$3 to rent shoes. Which equation shows how much money Alex spent in all?

A $\$6 \times \$3 = \$18$ **C** $\$6 + \$3 = \$9$
B $\$6 - \$3 = \$4$ **D** $\$6 \div \$3 = \$2$

Name____

Lesson 12.1

Line Segments and Angles

The terms below can help you describe figures.

A **point** is an exact position or location.	A **line** is straight and does not end. It continues forever in both directions.	A **line segment** is straight and is part of a line. It has an endpoint at each end.	A **ray** is straight and is part of a line. It has one endpoint and continues in one direction.

Name this figure. 

It shows an arrow at each end.

So, the figure is a line.

Angles are formed by two rays or two line segments that share an endpoint.

A **right angle** is a special angle that forms a square corner.

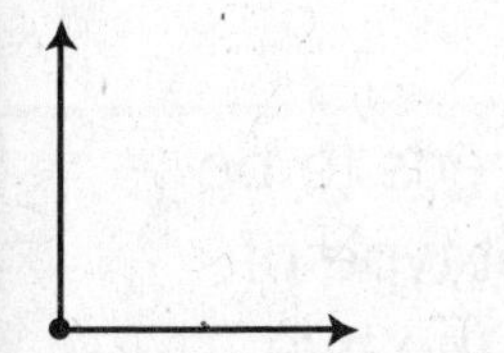

Some angles are less than a right angle.

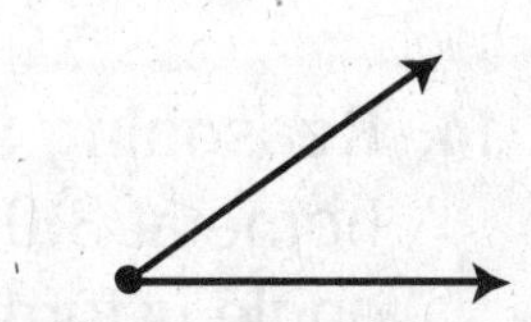

Some angles are greater than a right angle.

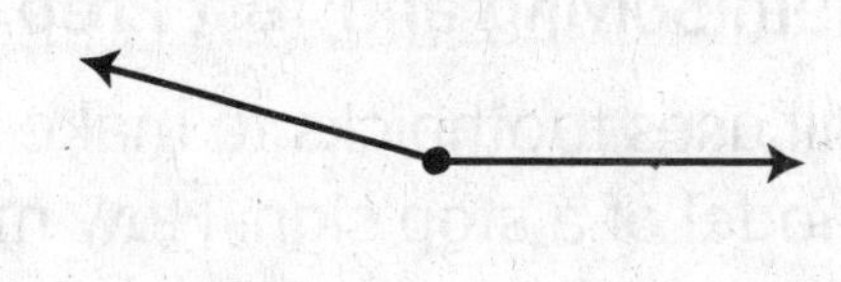

For 1-4 name each figure. Write *point, line, ray,* or *line segment.*

1.

2. 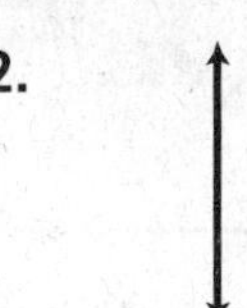

3.

4.

__________ __________ __________ __________

For 5-7 use the corner of a sheet of paper to tell if each angle appears to be a right angle, greater than a right angle, or less than a right angle.

5. 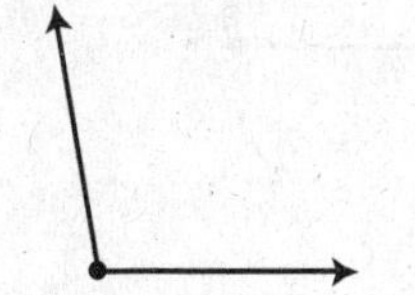

6. 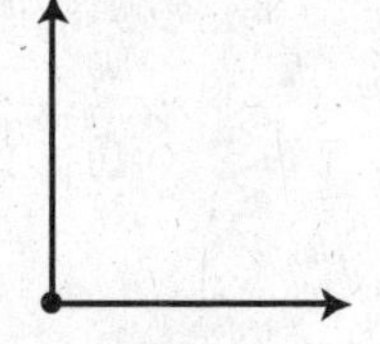

7. 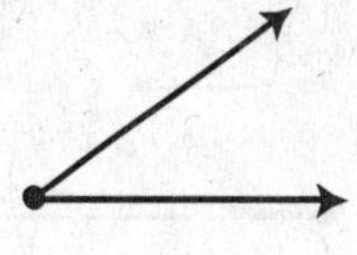

__________ __________ __________

Name______________________________________ **Lesson 12.1**

Line Segments and Angles

Tell whether each is a *line*, a *line segment*, or a *ray*.

1. 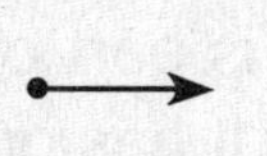 ____________

2. 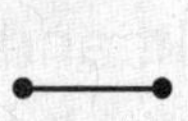 ____________

3.  ____________

4. ____________

5. ____________

6.  ____________

7. 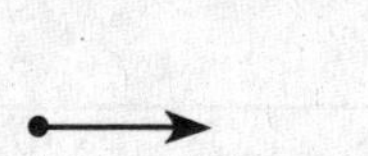 ____________

8.  ____________

Use the corner of a sheet of paper to tell whether each angle is a *right angle, greater than* a right angle, or *less than* a right angle.

9. 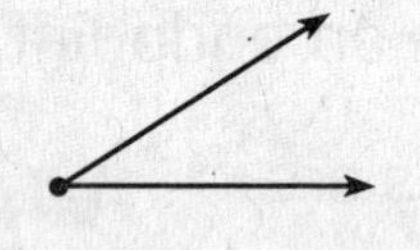 ____________

10. 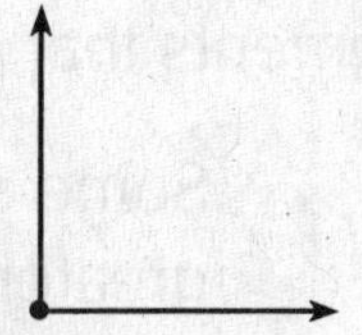 ____________

11. 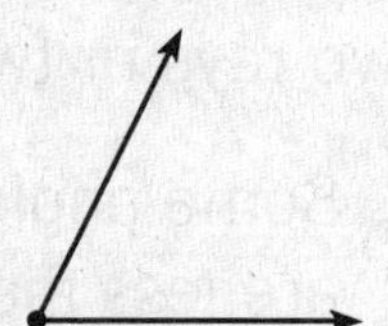 ____________

12. 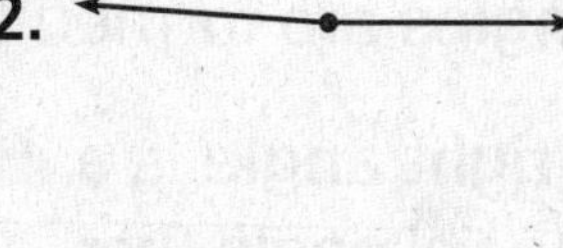 ____________

Problem Solving and Test Prep

13. Bill uses toothpicks to make a model of a stop sign. How many line segments did Bill use for the sides of the stop sign?

____

14. **Reasoning** Sally needs to be home at 3:00. What type of angle is formed by the two hands on a clock at 3:00?

____

15. Which shows an angle greater than a right angle?

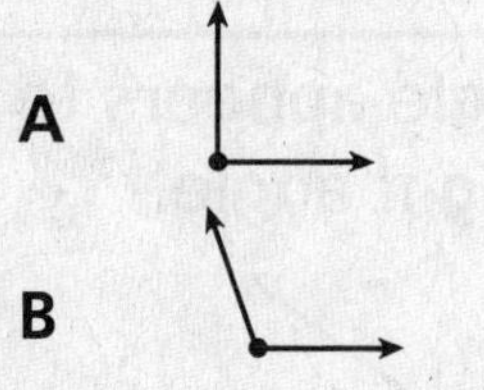

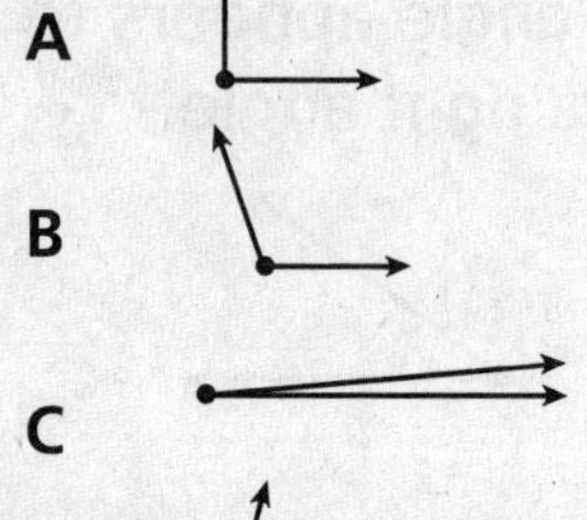

A

B

C

D

16. Which shows a line segment?

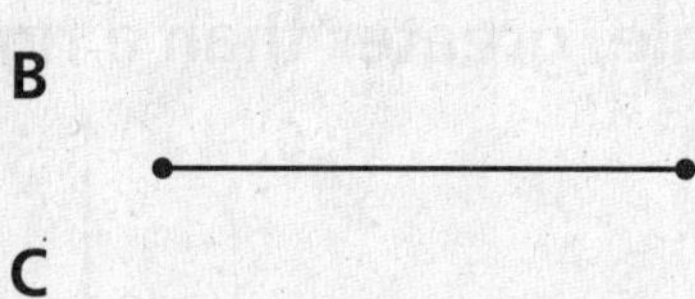

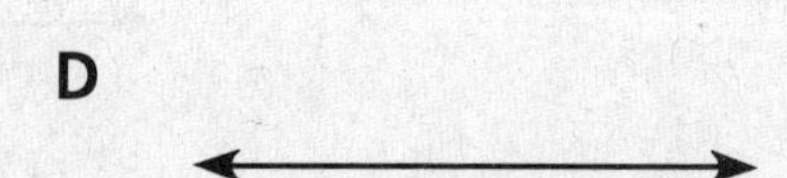

A

B

C

D

Name ______________________ **Week 17**

Spiral Review

For 1–5, find each quotient.

1. $50 \div 5 =$ ____

2. $36 \div 6 =$ ____

3. $54 \div 6 =$ ____

4. $42 \div 7 =$ ____

5. $27 \div 9 =$ ____

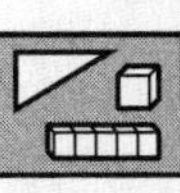

For 6–9, use the picture to answer the questions.

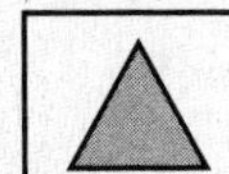    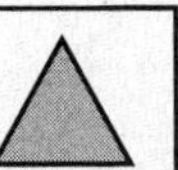

6. 1 triangle   Number of angles: ____

7. 3 triangles   Number of angles: ____

8. 5 triangles   Number of angles: ____

9. 4 triangles   Number of angles: ____

For 10–12, a class takes a survey about favorite colors. Write the results as tally marks.

10. 13 students chose red.

____

11. 9 students chose blue.

____

12. Look at the table at the right. How many more students wore white shirts than wore green shirts?

Shirts Worn	
Color	**Students**
White	卌 |
Orange	||
Green	||||

____

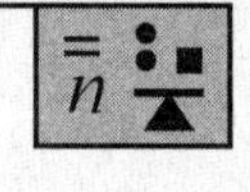

For 13–15, find the missing factor.

13. $(2 \times \square) \times 9 = 90$ ____

14. $8 \times (4 \times \square) = 64$ ____

15. $(2 \times \square) \times 6 = 36$ ____

Name ____

Types of Lines

Pairs of lines can be described according to how the lines cross, or don't cross.

Intersecting lines cross to form angles.

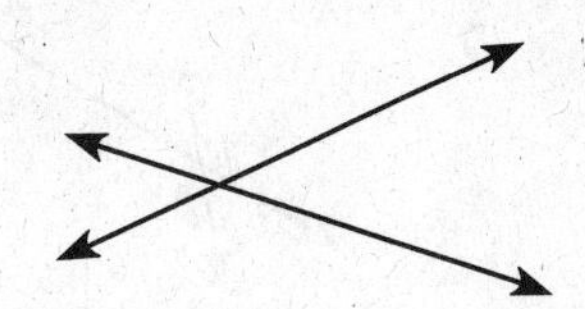

Some intersecting lines cross to form right angles.

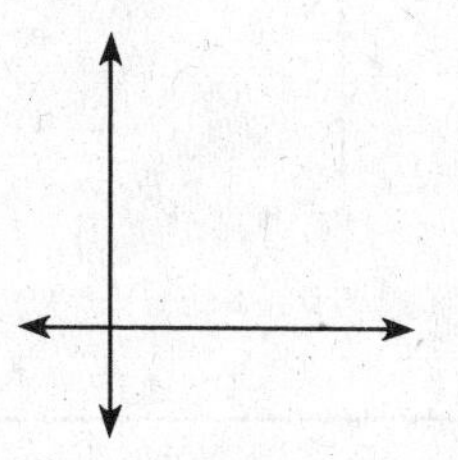

Parallel lines never cross. They are always the same distance apart. They do not form any angles.

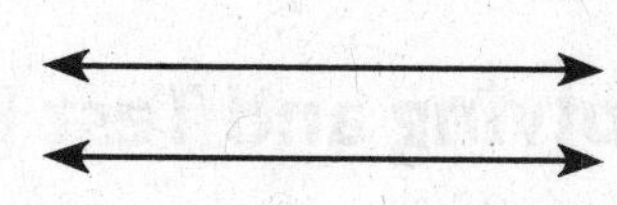

Describe the lines. Write intersecting or parallel.

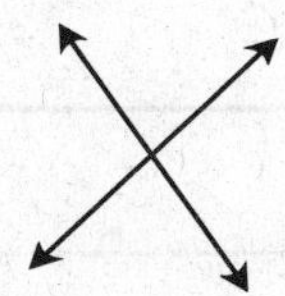

The two lines cross so they cannot be parallel lines. So, the lines are intersecting.

Describe the lines. Write *intersecting* or *parallel*.

1.

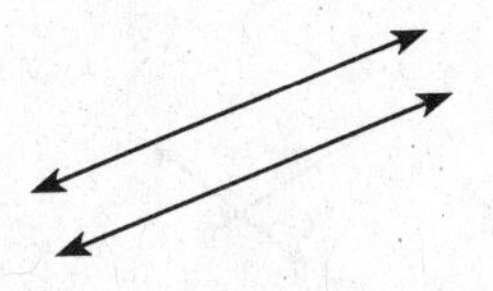

____

2.

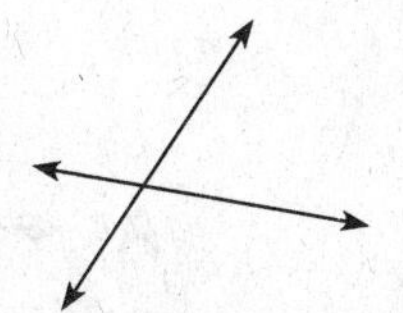

____

3.

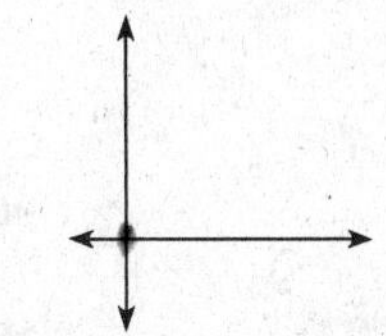

____

4.

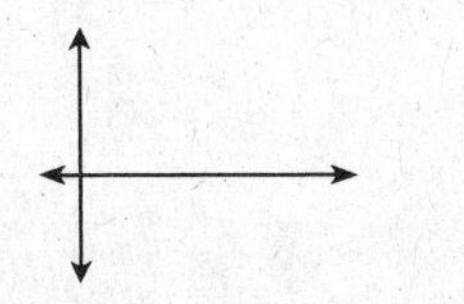

____

5.

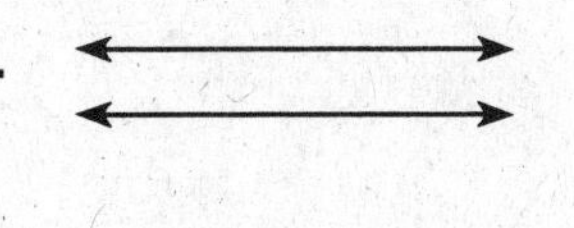

____

6.

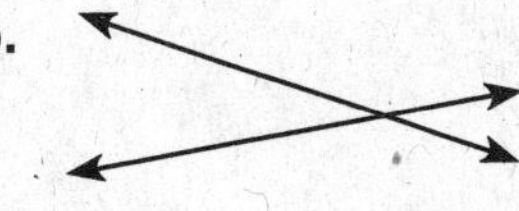

____

Name____

Lesson 12.2

Types of Lines

Describe the lines. Write *intersecting*, or *parallel*.

1. 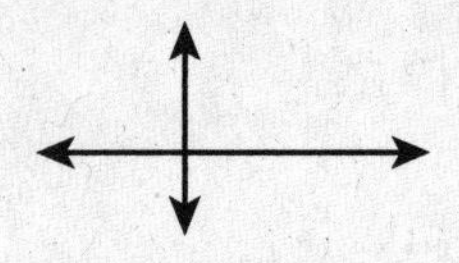

____

2. 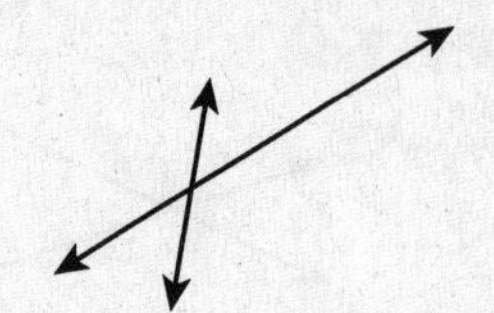

____

3. 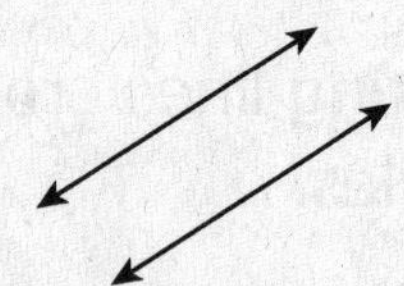

____

4. 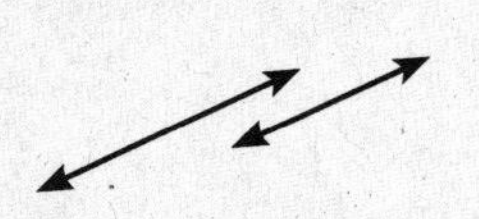

____

5. 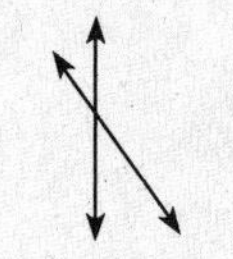

____

6. 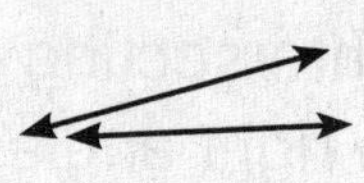

____

Problem Solving and Test Prep

7. Marc wonders if every intersecting pair of lines forms right angles. What would you tell him?

____

____

____

8. Can parallel lines be intersecting lines as well? Why or why not?

____

____

____

9. Which of these pairs of lines appear to be parallel?

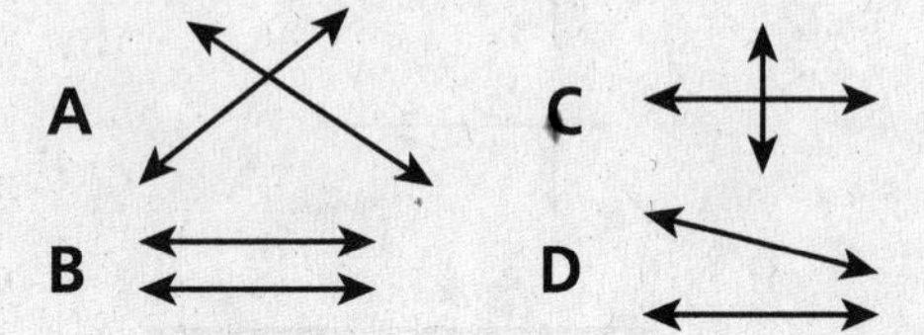

10. Which of these pairs of lines intersect at right angles?

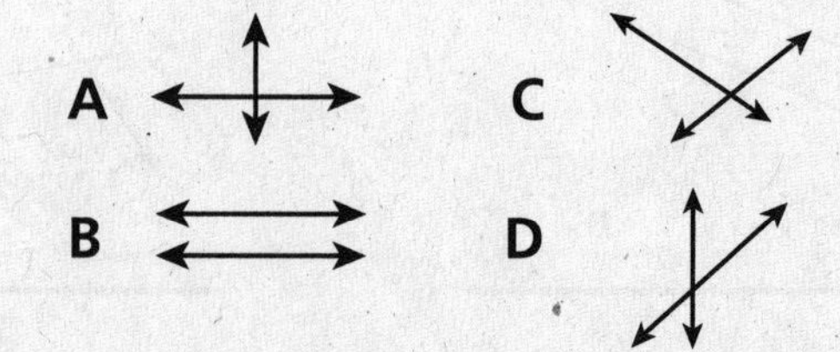

Name______________________________________ **Lesson 12.3**

Identify Plane Figures

Plane figures are figures on a flat surface that can be formed by curved lines, straight lines, or both. A special kind of plane figure is a **polygon.** A polygon is a closed plane figure that has line segments as sides. Polygons are named for the number of sides they have.

Type of Polygon	Triangle	Quadrilateral	Pentagon	Hexagon	Octagon
number of sides	3	4	5	6	8
number of angles	3	4	5	6	8
example					

Tell if the figure below is a polygon.

- The figure is closed.
- A polygon is a closed plane figure.
- The figure does not have sides.
- A polygon must have straight sides.

The figure is not a polygon.

Tell if each figure is a polygon. Write *yes* or *no*.

1. 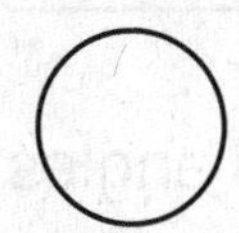
____

2. 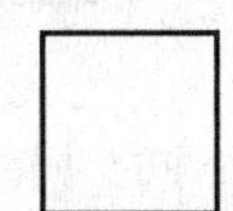
____

3. 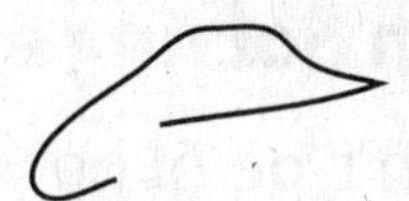
____

4. 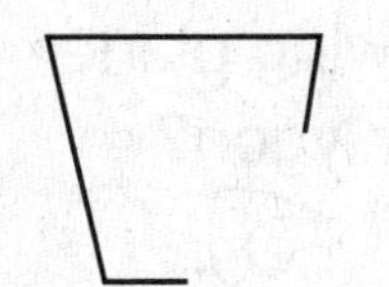
____

5. 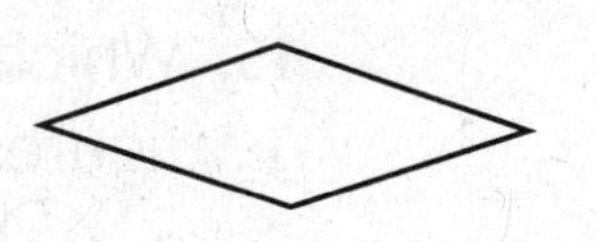
____

6. 
____

Name____________________

Identify Plane Figures

Tell whether each figure is a polygon. Write *yes* or *no*.

1. 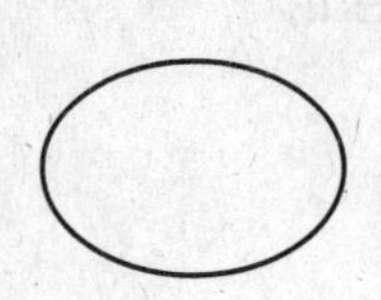
______

2. 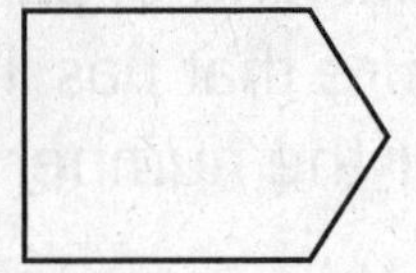
______

3. 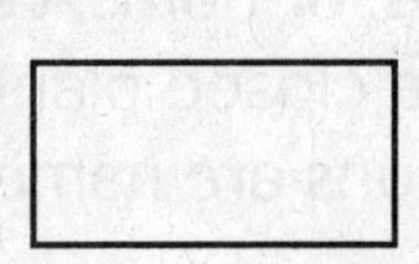
______

4. 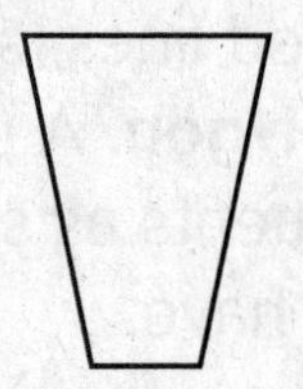
______

Name each polygon. Tell how many sides.

5. 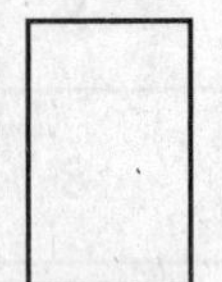
______

6. 

______

7. 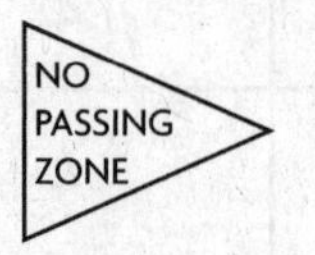

______

8. 

______

For 9–11, use figures A–E.

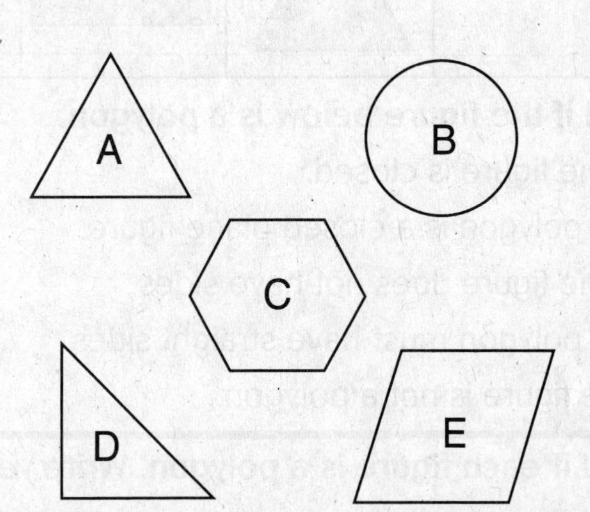

9. Which of the figures are closed?

______

10. Which figures are polygons?

______

11. Which figure is a plane figure but not a polygon?

______

Problem Solving and Test Prep

12. Which type of polygon has 6 sides and 6 angles?

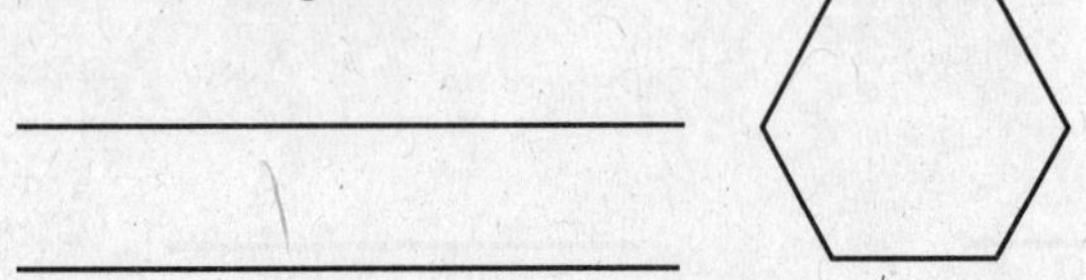

______

______

13. How many sides and angles does the polygon below have?

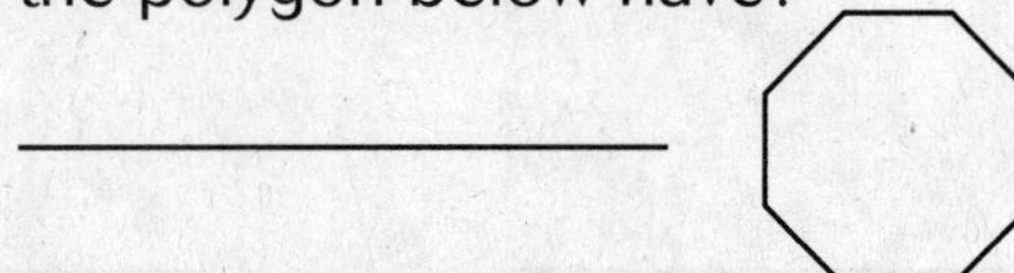

______

______

14. How many sides does a quadrilateral have?

A 4 **C** 6

B 5 **D** 8

15. Which of the following plane figures is also a polygon?

A ○ **C** 

B ⬭ **D** □

Name ____________________

Triangles

Triangles can be named by their equal sides.

Triangles Named by Their Sides	
An **equilateral triangle** has 3 sides that are equal in length.	
An **isosceles triangle** has two sides that are equal in length.	
A **scalene triangle** has no sides that are equal in length.	

Name the triangle below. Write equaliteral, isosceles, or scalene.

How long are the sides?

- Measure each side.
- The sides of this triangle are all equal, each measuring 4 centimeters in length.

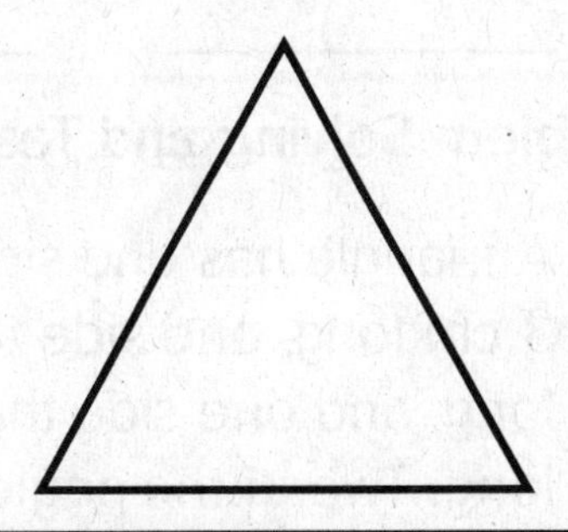

So, this is an equilateral triangle.

Name each triangle. Write equilateral, isosceles, or scalene.

1.
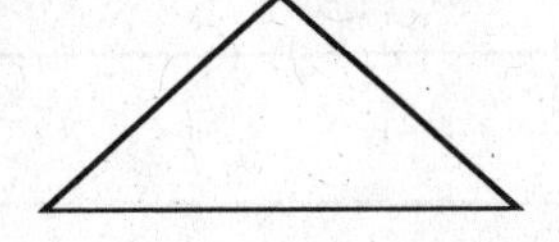
____

2.
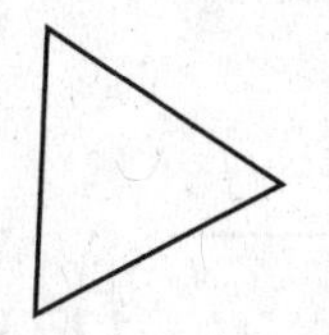
____

3.
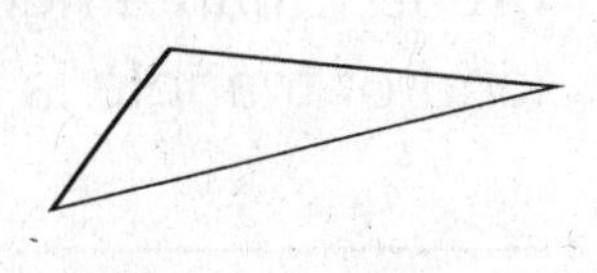
____

4.
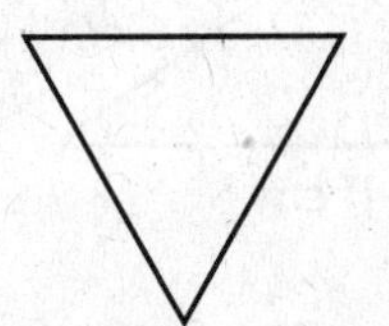
____

5.
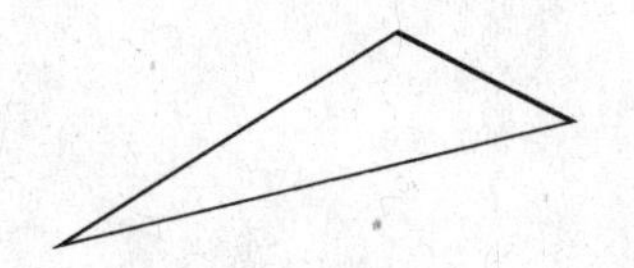
____

6.
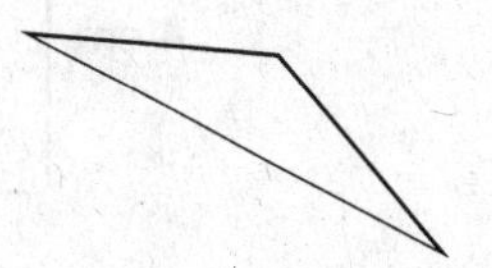
____

MG 2.2 Identify attributes of triangles (e.g., two equal sides for the isosceles triangle, three equal sides for the equilateral triangle, right angle for the right triangle).

Name ____________________

Triangles

Name each triangle. Write *equilateral*, *isosceles*, or *scalene*.

1. 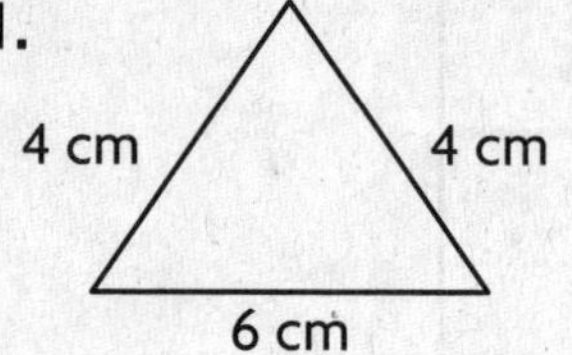

____

2. 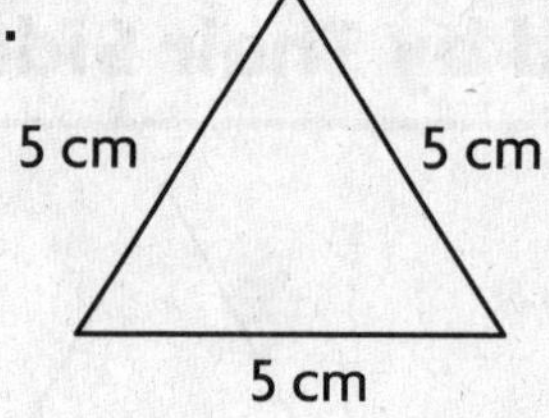

____

3. 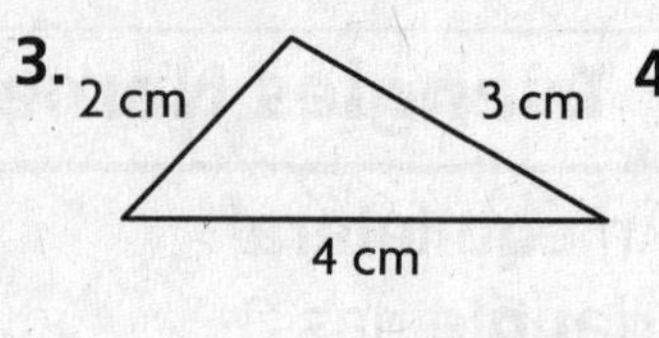

____

4. 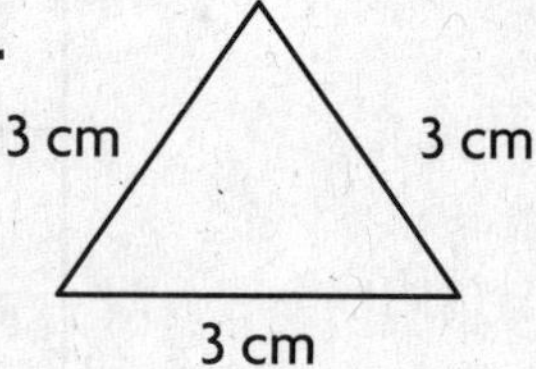

____

For 5–8, write *a*, *b*, or *c*.

a. It has 1 right angle

b. It has 1 angle greater than a right angle.

c. It has 3 angles less than a right angle.

5. 6 cm, 8 cm, 10 cm
____

6. 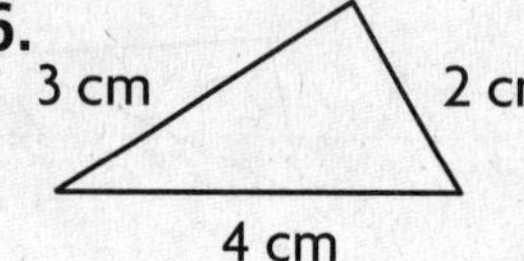

____

7. 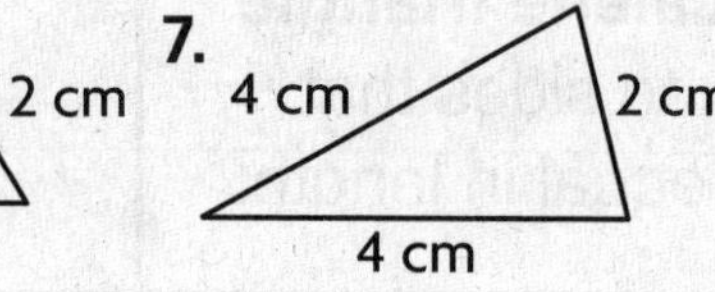

____

8. 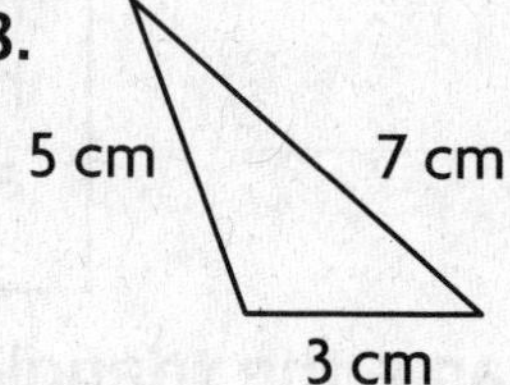

____

Problem Solving and Test Prep

9. A triangle has one side that is 3 cm long, one side that is 2 cm long, and one side that is 4 cm long. Two of the angles are less than a right angle and one angle is greater than a right angle. What kind of triangle is it?

____

10. Can a triangle with a right angle also be an isosceles triangle? Explain.

____

____

11. Which correctly names this triangle?

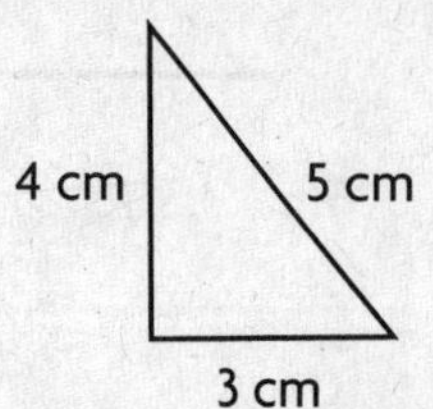

A scalene **C** isosceles

B equal **D** equilateral

12. Which correctly names this triangle?

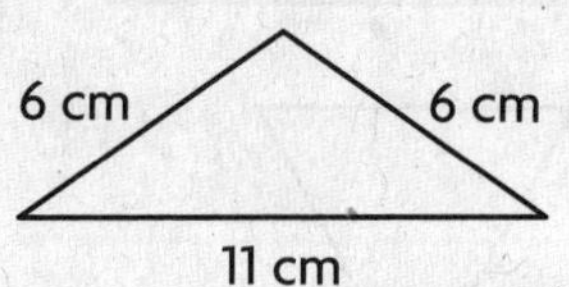

A equilateral **C** right

B scalene **D** isosceles

Name____________________

Quadrilaterals

Quadrilaterals are four-sided polygons.
Some quadrilaterals are named for their sides and their angles.

Trapezoids have one pair of parallel sides. The side lengths and the sizes of the angles may not be the same.

Parallelograms have 2 pairs of parallel sides and 2 pairs of equal sides.

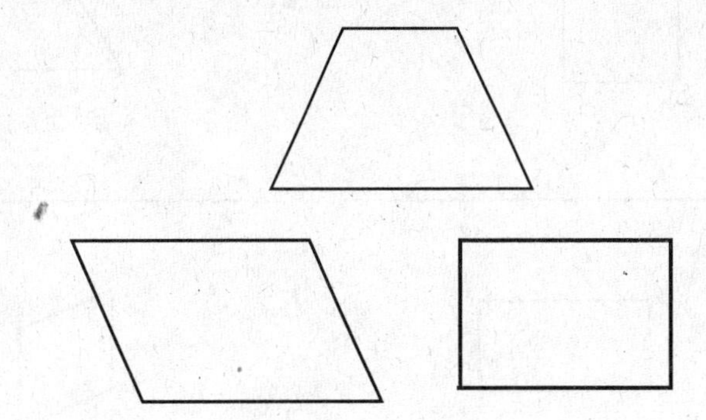

Some parallelograms have special names.

A **rhombus** has 2 pairs of parallel sides and 4 equal sides.

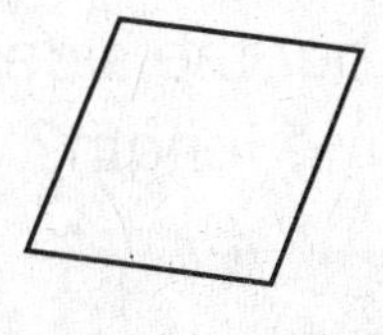

A **rectangle** has 4 right angles, 2 pairs of parallel sides, and 2 pairs of equal sides.

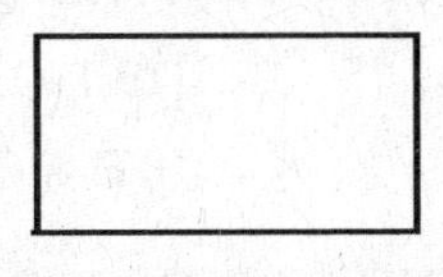

A **square** has 4 right angles, 4 equal sides, and 2 pairs of parallel sides.

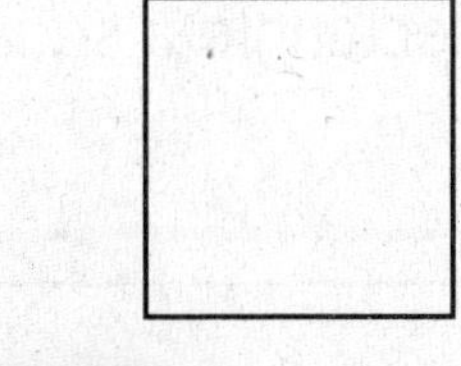

Write as many names for the quadrilateral below as you can.

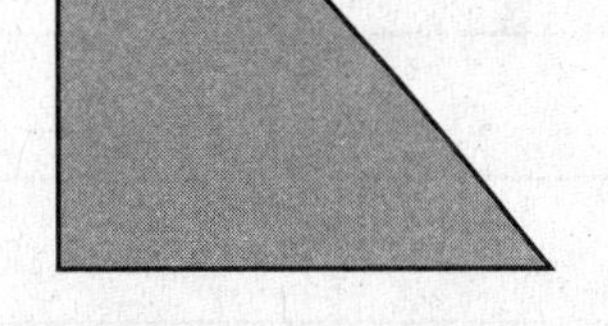

The quadrilateral has only 1 pair of parallel sides.

So, the quadrilateral is a trapezoid.

Write as many names for each quadrilateral as you can.

1. 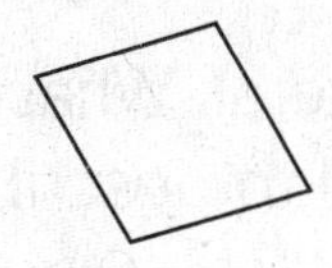

____

2. 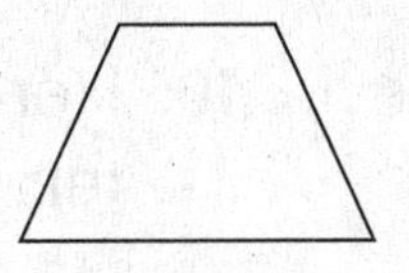

____

3. 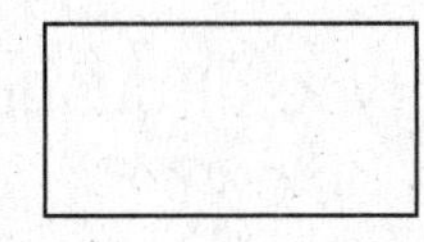

____

4. 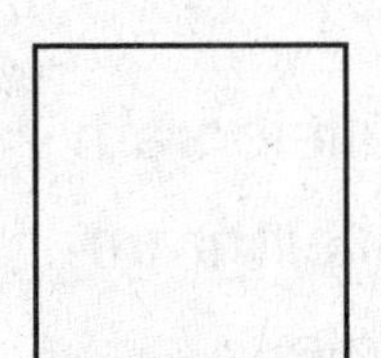

____

5. 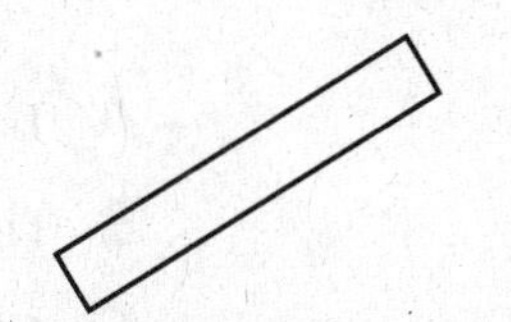

____

6. 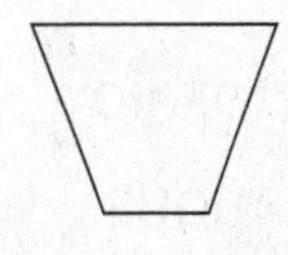

____

Quadrilaterals

Write as many names for each quadrilateral as you can.

1. 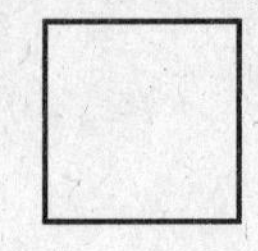

____

2. 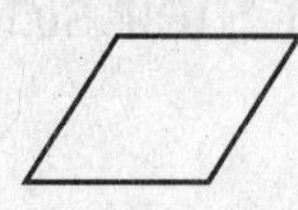

____

3. 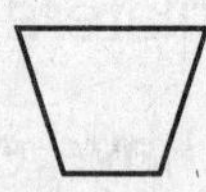

____

4. 

____

5. 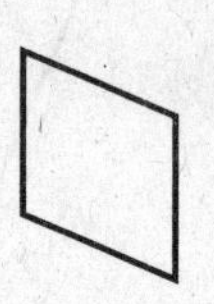

____

6. 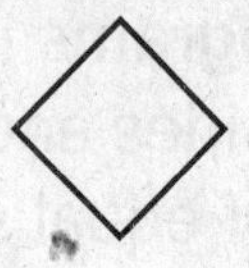

____

Problem Solving and Test Prep

7. Reasoning A square is a rectangle. Is a rectangle a square? Explain.

____

____

____

____

8. What type of quadrilateral has one pair of parallel sides, but the sides are not always the same length?

____

9. What type of quadrilateral is the figure below?

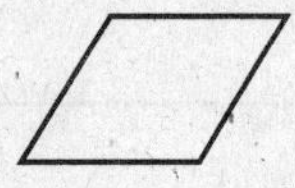

A trapezoid
B rhombus
C square
D rectangle

10. Here is a quadrilateral. What two terms can be used to describe it?

A rectangle, parallelogram
B rhombus, parallelogram
C square, rectangle
D rhombus, square

Spiral Review

For 1–3, use the Lunch Menu to solve the problems.

Lunch Menu	
Chili $8 Soup $4 Fruit $3	Sandwich $6 Salad $2 Drink $2

1. Tom ordered chili, a salad, and a drink. He paid with two $10 bills. How much change did Tom receive? ______
2. There are 9 players on the team. They each ordered a sandwich and fruit. What was the total cost of the team's order? ______
3. Kyla and Blake share a sandwich, a soup, and two drinks. If they each pay an equal amount, how much do they each pay? ______

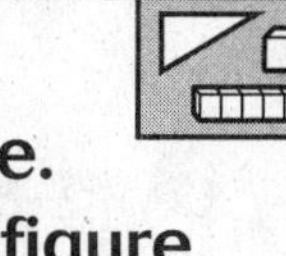
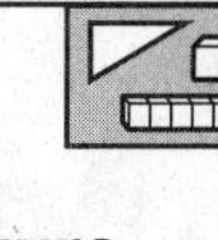

For 4–5, combine the given figures to make a new figure. Draw an outline of the new figure.

4.  

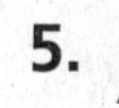
5.    

For 6–8, a class takes a survey about their favorite subject. Let stand for 1 person. Use to replace the tally marks for each subject in the tally table below.

Favorite Subject	
Subject	**Students**
Math	𝍸 |
Science	𝍸 ||
Reading	𝍸

6. Math ______________
7. Science ______________
8. Reading ______________

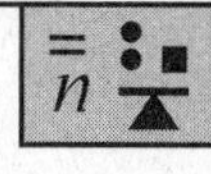
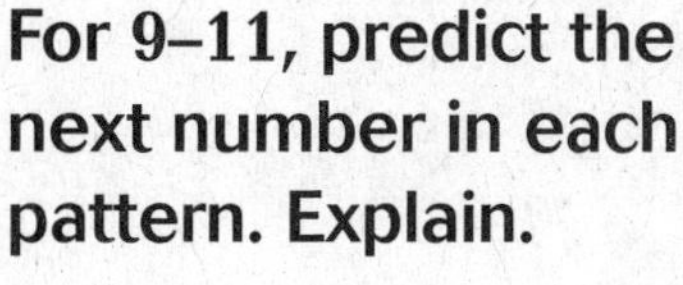

For 9–11, predict the next number in each pattern. Explain.

9. 22, 26, 30, 34, 38, ☐

____

10. 99, 88, 77, 66, ☐

____

11. 300, 350, 400, 450, ☐

____

Name____ Lesson 12.6

Compare Plane Figures

You can compare plane figures by looking at the number of sides, types of angles, and pairs of parallel sides for each figure.

Look at the table below.

Figure	A	B	C	D
Number of Sides	4	4	4	6
Types of Angles	2 right angles, 1 angle greater than a right angle, 1 angle less than a right angle.	4 right angles	4 right angles	6 greater than right angles
Parallel Sides	1 pair	2 pairs	2 pairs	3 pairs

Which of the figures below have 4 sides?

Count the sides of each figure.

Each figure has 4 sides except for figure D.

So, figures A, B, and C each have 4 sides.

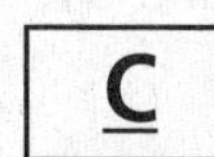

For 1-4, use the figures below.

1. Which figures have three sides? ________

2. Which figures have four sides? ________

3. Which figures have eight angles? ________

4. Which figures have parallel sides? ________

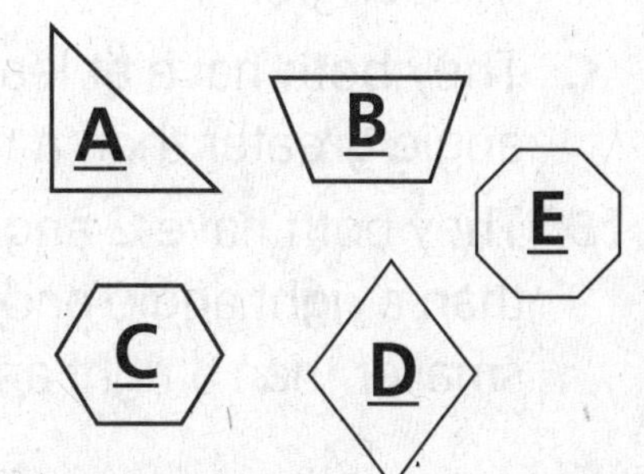

Name________________________________

Lesson 12.6

Compare Plane Figures

For 1–3, use the plane figures at the right.

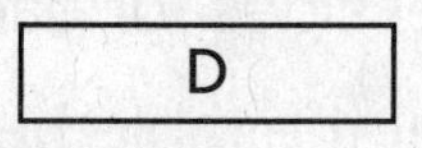

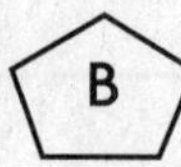

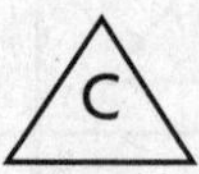

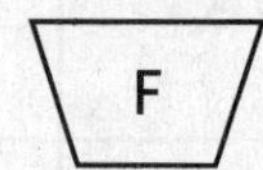

1. Which figures have only 3 sides?

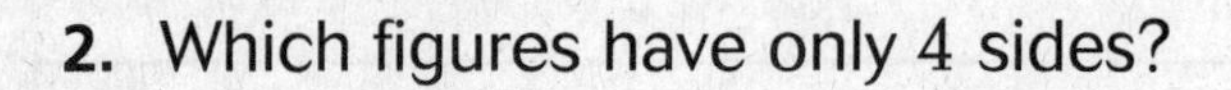

2. Which figures have only 4 sides?

____

3. Which figures have parallel sides?

____

Problem Solving and Test Prep

4. How are a Stop sign and a No Passing Zone sign alike?

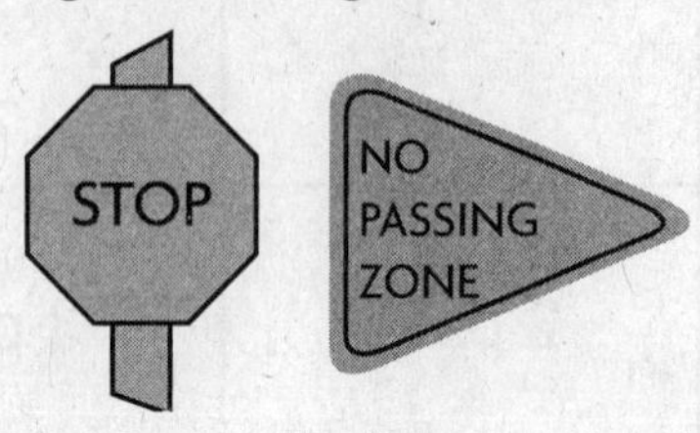

____

5. How might a square and a rectangle be different?

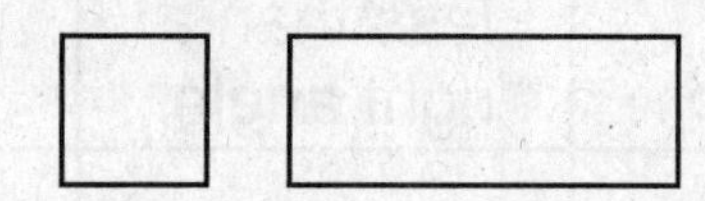

____

6. How are these figures alike?

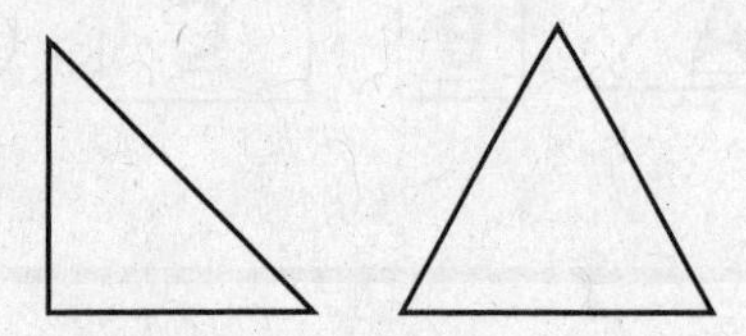

A They both have 3 sides.

B They both have at least one right angle.

C They both have at least one angle greater than a right angle.

D They both have 2 angles greater than a right angle and one angle smaller than a right angle.

7. How are these figures different?

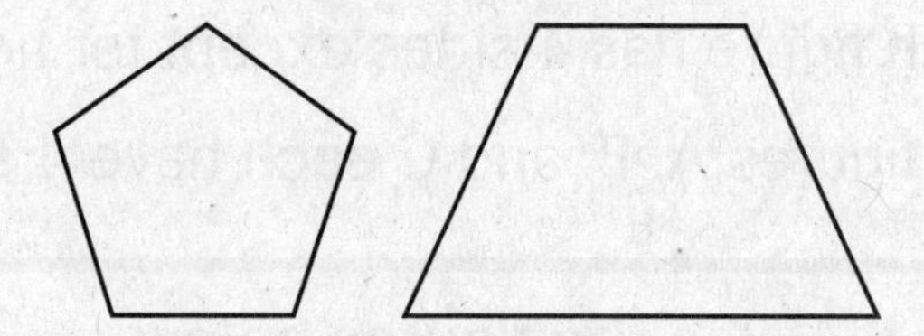

A They have a different number of sides.

B One is a closed plane figure, the other is not.

C They both have at least one right angle.

D They both have at least one angle greater than a right angle.

Name____ **Lesson 12.7**

Problem Solving Workshop Strategy: Find a Pattern

Steven painted a border around a picture frame. His pattern unit was 3 circles, and 1 triangle. He painted a total of 24 figures. What shape was the twelfth figure?

Read to Understand

1. What are you asked to find?

____

____

Plan

2. What strategy can you use to solve the problem?

____

____

Solve

3. Solve the problem. Describe the strategy you used.

____

____

____

____

Check

4. Is there another strategy you could use to solve the problem? Explain.

____

____

Find a pattern to solve.

5. Shelby drew the figures below on her notebook:

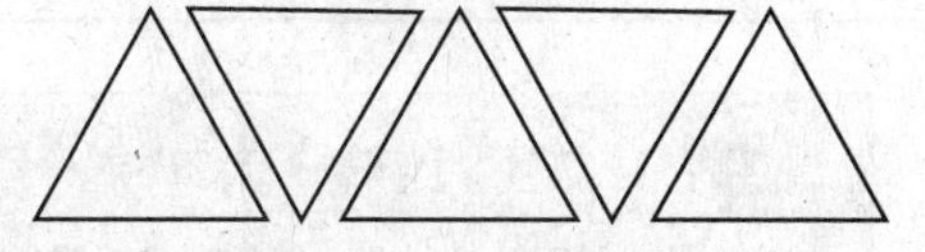

If the pattern continues, what will the twelfth figure be?

____

6. Mr. Carlson wrote the numbers below on the chalkboard:

17 23 29 35 41

If the pattern continues, what will the next 2 numbers be?

____

Name______________________________ Lesson 12.7

Problem Solving Workshop Strategy: Find a Pattern

Problem Solving Strategy Practice

Find a pattern to solve.

1. What are the next three numbers in this pattern?

5, 10, 15, 20, ______, ______, ______.

2. What are the next three numbers in this pattern?

100, 92, 84, ______, ______, ______.

3. What is the next shape in the pattern?

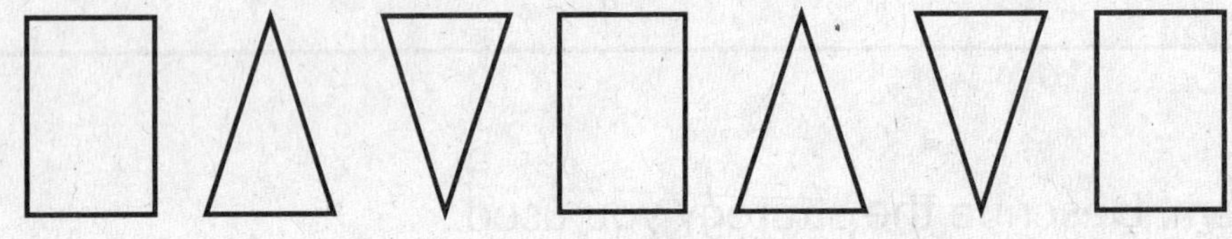

____

Mixed Strategy Practice

4. There were 4 different colored cars sitting at a stop light. Lar's car is not blue. Teresa's car is red. Vincent's car is not blue. Uri's car is blue. Lar's car is not green. There is a yellow car sitting at the stop light. What color is Lar's car?

____

5. Florence saw a total of 19 books on Tuesday. In the morning she saw one book on the coffee table. In the afternoon she saw books at her office. In the evening she saw 5 books in her friend's car. How many books did Florence see at her office?

____

6. **Use Data** How many students voted in all for their favorite type of movie? Show your work.

____

Favorite Type of Movie

Type	Number of Votes
Western	5
Comedy	10
Cartoon	7

Name________________________________ **Lesson 13.1**

Identify Solid Figures

A solid figure is called a **three-dimensional** figure because it has three dimensions: *length, width,* and *height.*

Name the solid figure that the baseball is shaped like.

- The baseball has no line segments, so it is not a polygon.
- A sphere has no line segments, so it is not a polygon.
- The baseball is a round object with no flat surfaces.
- A sphere is also round with a curved surface.

Solid Figures

cone

cube

cylinder

rectangular prism

sphere

square pyramid

So, the baseball is shaped like a **sphere.**

Name the solid figure that each object is shaped like.

1. 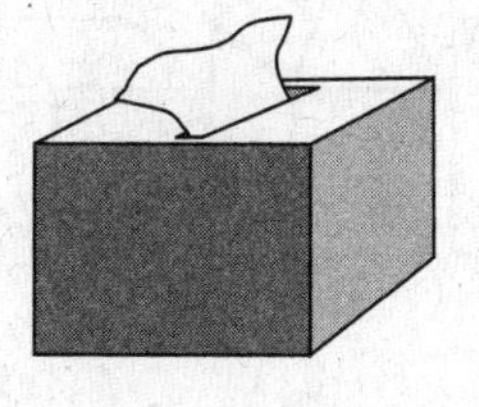

____

2. 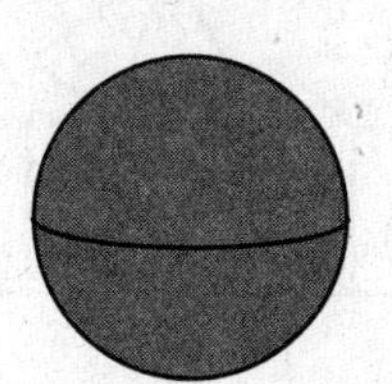

____

3. 

____

4. 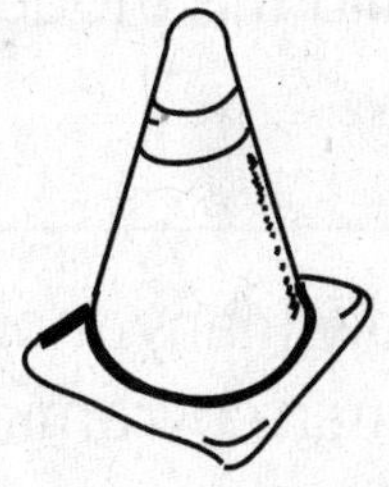

____

5. 

____

6. 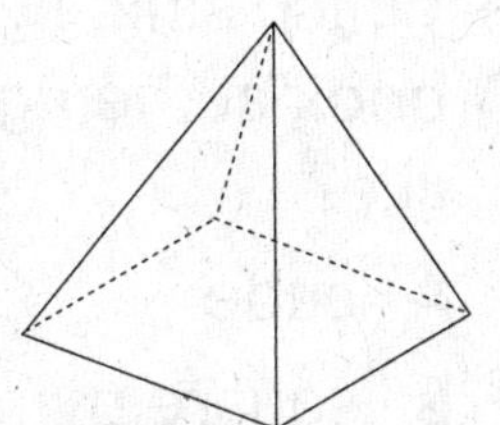

____

7. 

____

8. 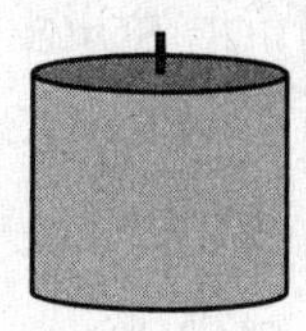

____

Name______________________________

Identify Solid Figures

Name the solid figure that each object is shaped like.

1. 

______

2. 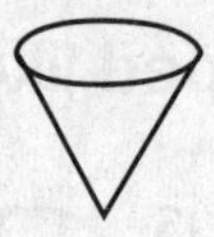

______

3. 

______

4. 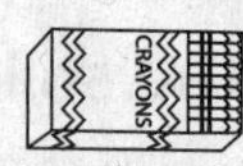

______

5. 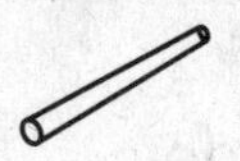

______

6. 

______

7. 

______

8. 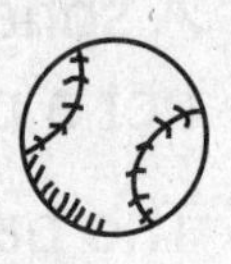

______

Problem Solving and Test Prep

USE DATA For 9–10, use the model below.

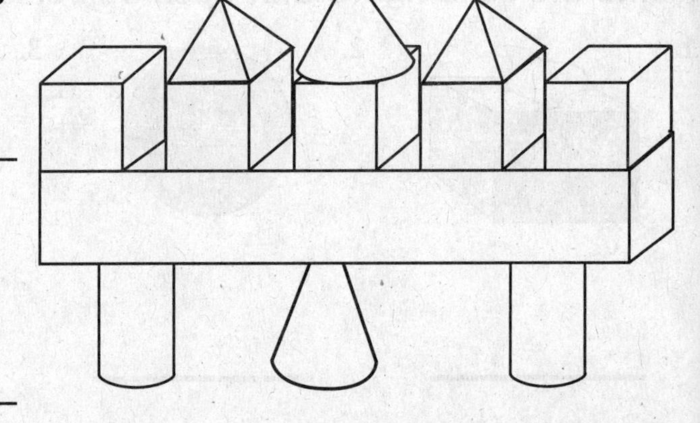

9. How many cylinders were used to build the model?

______

10. How many more cubes than square pyramids were used to build the model?

______

11. Which solid figure is the tent below shaped like?
    - **A** cone
    - **B** cube
    - **C** square pyramid
    - **D** rectangular prism

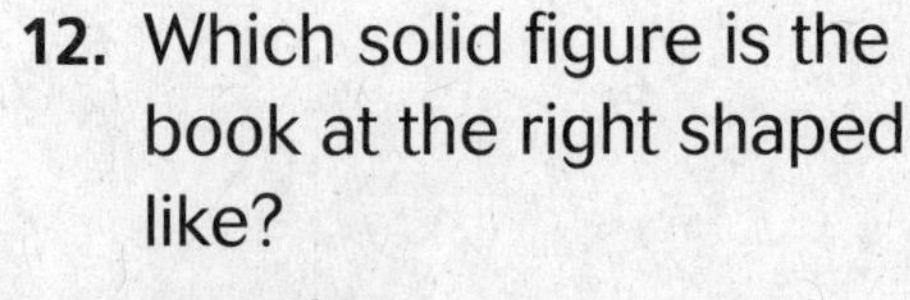

12. Which solid figure is the book at the right shaped like?
    - **A** cube
    - **B** square pyramid
    - **C** rectangular prism
    - **D** triangular prism

Name____

Faces, Edges, and Vertices

Solid figures have faces, edges, and vertices.

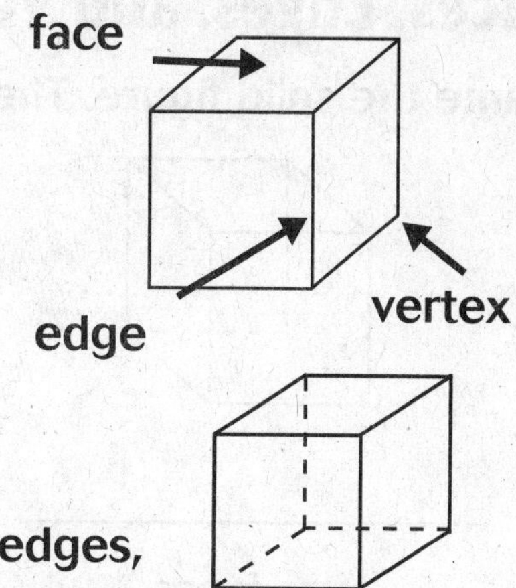

A **face** is a flat surface of a solid figure.

An **edge** is the line segment formed where two faces meet.

A **vertex** is formed when 3 or more edges meet at a point. The plural for vertex is vertices.

Name the solid figure. Then tell how many faces, edges, and vertices the solid figure has.

To count the *faces* of the cube, first draw the faces.

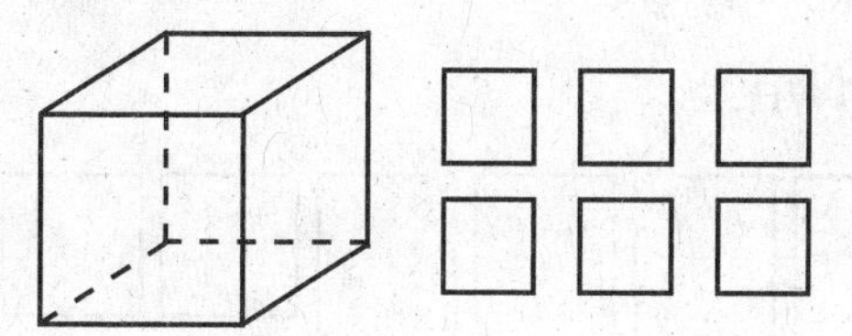

The cube has 6 faces.

Find the edges of the cube.

The top face has 4 edges.

The bottom face has 4 edges.

The top and bottom faces are connected by 4 edges.

So, $4 + 4 + 4 = 12$ edges

To count the *vertices* of the cube, count the corners.
There are 8 corners on the cube. So there are 8 vertices.

So, a cube has 6 faces, 12 edges, and 8 vertices.

Name the solid figure. Then tell how many faces, edges, and vertices.

1.

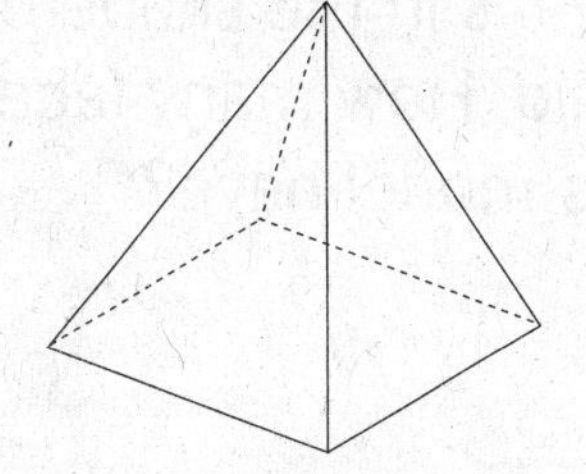

____

________ faces

________ edges

________ vertices

2.

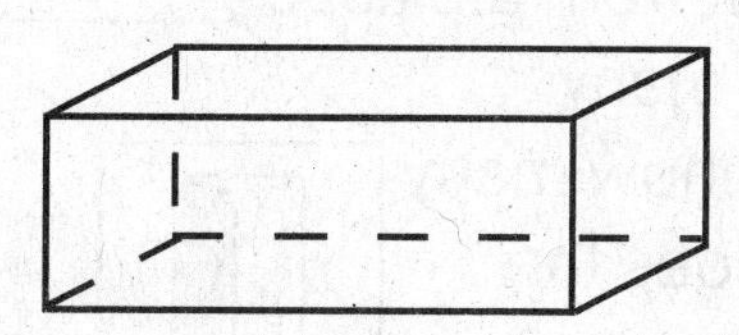

____

________ faces

________ edges

________ vertices

Name______________________________ **Lesson 13.2**

Faces, Edges, and Vertices

Name the solid figure. Then tell how many faces, edges, and vertices.

1.

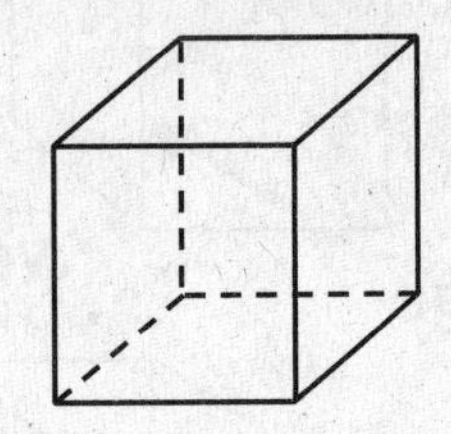

____

_______ faces

_______ edges

_______ vertices

2.

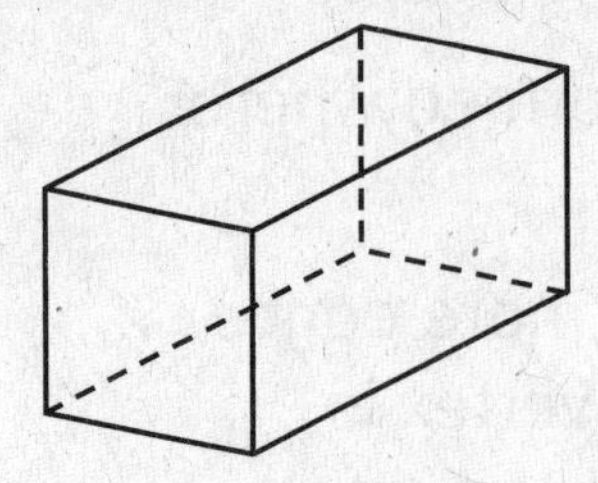

____

_______ faces

_______ edges

_______ vertices

3.

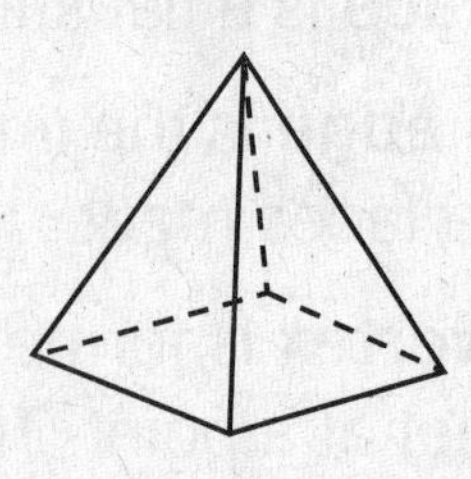

____

_______ faces

_______ edges

_______ vertices

Name the solid figure that has the faces shown.

4.

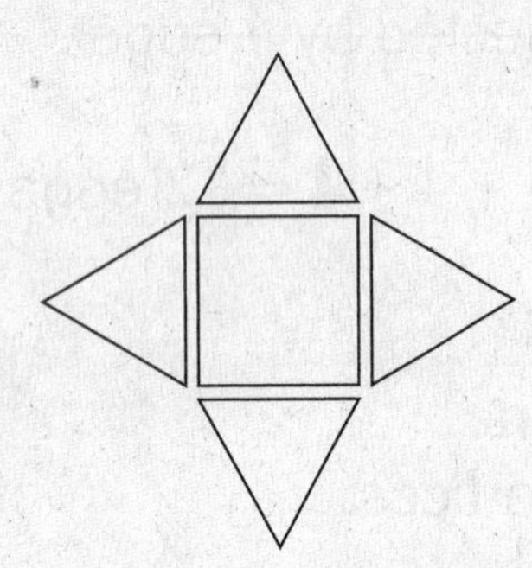

____

5.

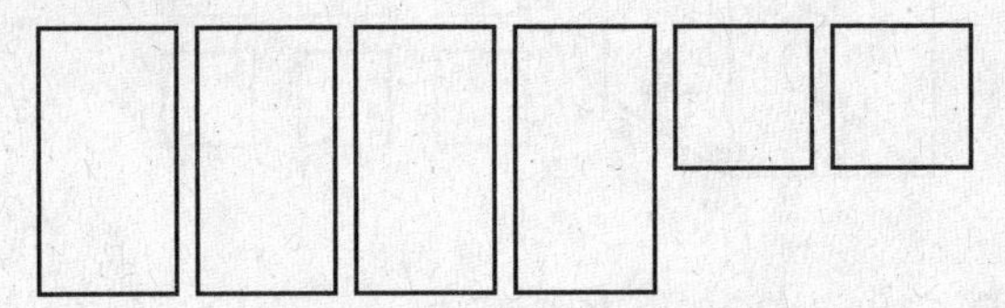

____

Problem Solving and Test Prep

6. Rene made the birdfeeder at the right from a plastic box. How many faces and how many vertices does the birdfeeder have?

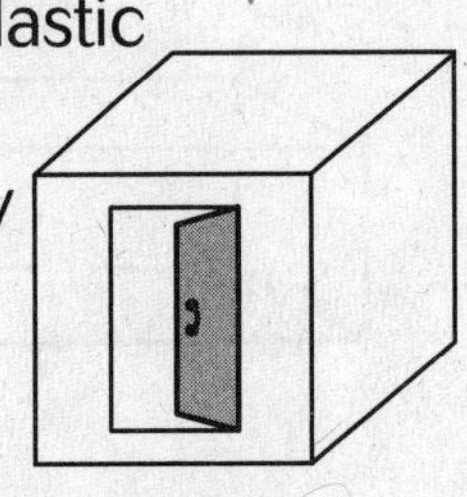

____

7. Gwynn makes a wooden model of a tent. The tent is in the shape of a square pyramid. How many faces does Gwynn's model have?

____

8. Which solid figure is shaped like a drinking straw?

A cone
B cube
C cylinder
D sphere

9. Which represents the number of edges that a small cube has?

A 8
B 6
C 4
D 12

Name ______________________________ **Week 19**

Spiral Review

For 1–3, solve the problems.

1. Morrie bought 8 goldfish. All the fish were the same price. The total cost was $56. How much money did each fish cost?

   ______

2. Paco bought 6 bullfrog tadpoles. All the tadpoles were the same price. The total cost was $36. How much money did each tadpole cost?

   ______

3. Renée bought 4 hermit crabs. All the crabs were the same price. The total cost was $20. How much money did each crab cost?

   ______

For 8–10, answer the questions using the information in the line plot.

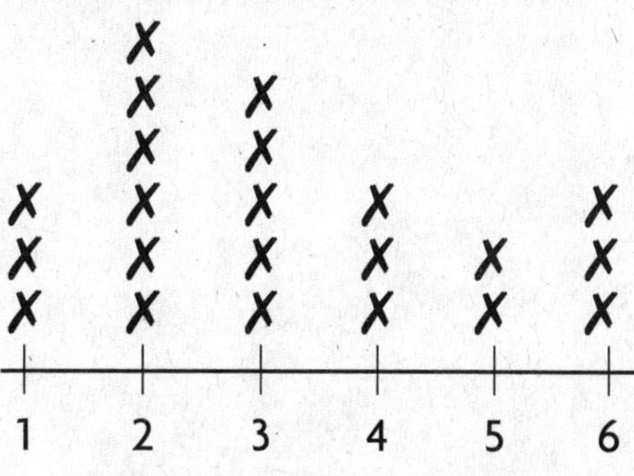

Number Cube Experiment

8. How many times was 5 rolled?

   ______

9. What is the range of the data?

   ______

10. What is the mode of the data?

   ______

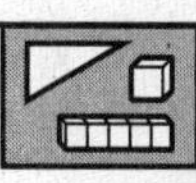

For 4–7, find the perimeter of each figure.

4. 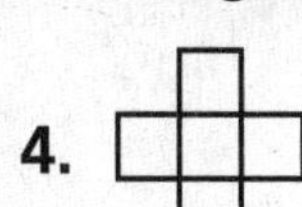 ______ units

5. ______ units

6. ______ units

7. ______ units

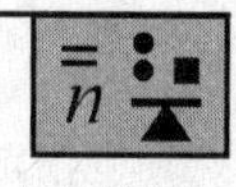

For 11–15, write the symbol that makes the statement true.

11. $100 \bigcirc 10 = 10$

12. $10 + 11 \bigcirc 21$

13. $8 \bigcirc 9 = 72$

14. $16 \bigcirc 9 = 7$

15. $0 \bigcirc 1 = 1$

Name____

Lesson 13.3

Model Solid Figures

A **net** is a flat two-dimensional pattern of a three-dimensional figure that folds to make a model of a figure.

Identify the solid figure that can be made from the net at the right.

net

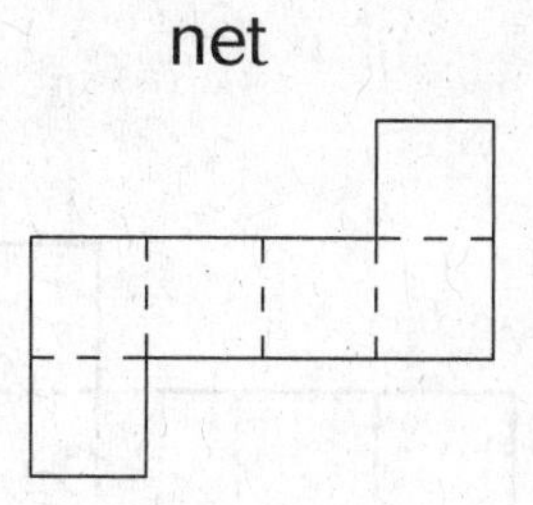

- The net has 6 square-shaped faces of equal size.

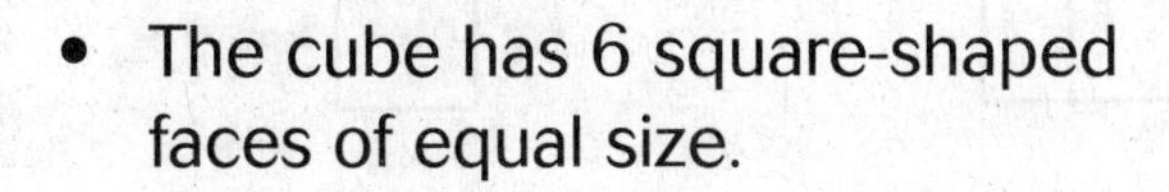

- The cube has 6 square-shaped faces of equal size.

cube

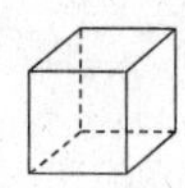

So, the net can be made into a cube.

Identify the solid figure that can be made from each net.

1.

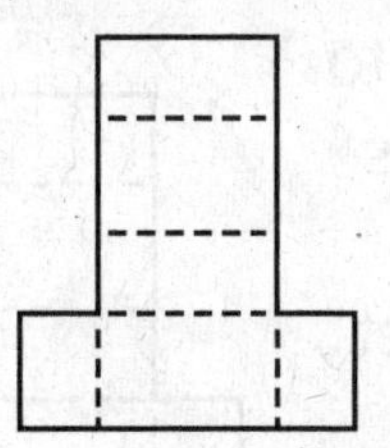

____

2.

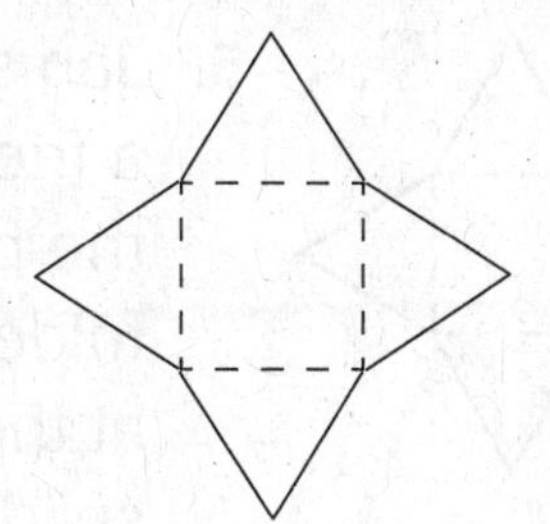

____

3.

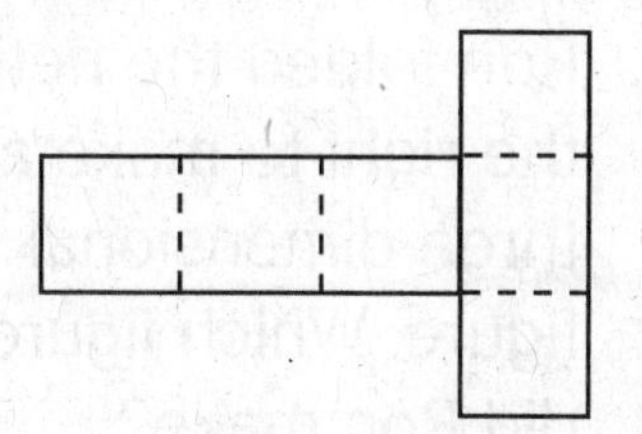

____

4.

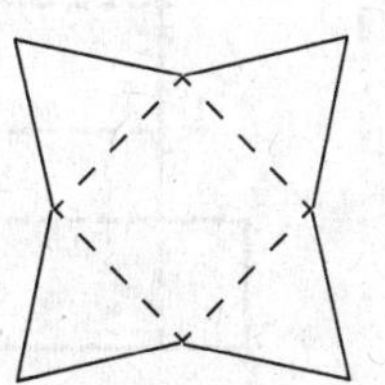

____

5.

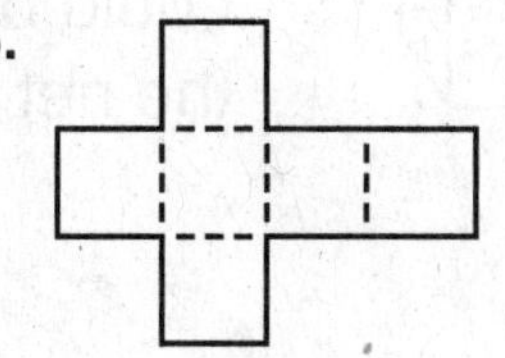

____

6.

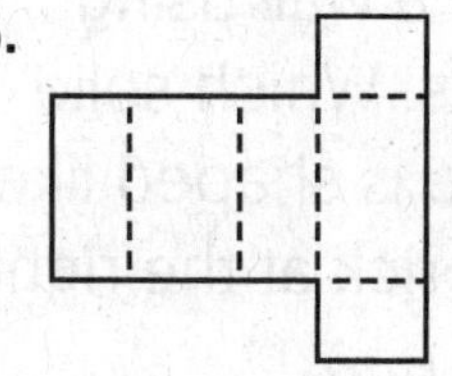

____

Name ______________________________

Model Solid Figures

Identify the solid figure that can be made from each net.

1.

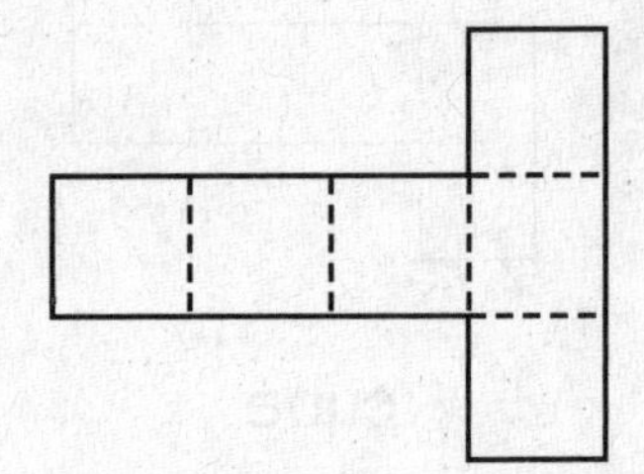

____

2.

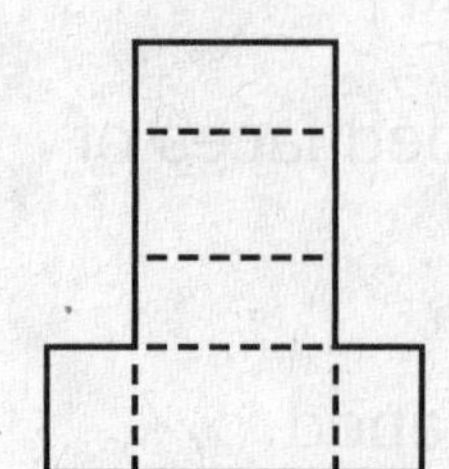

____

3.

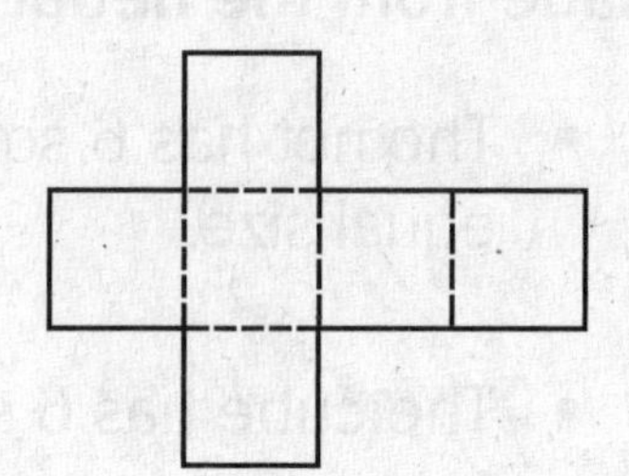

____

Problem Solving and Test Prep

4. Ron folded the net at the right to make a three-dimensional figure. Which figure did Ron make?

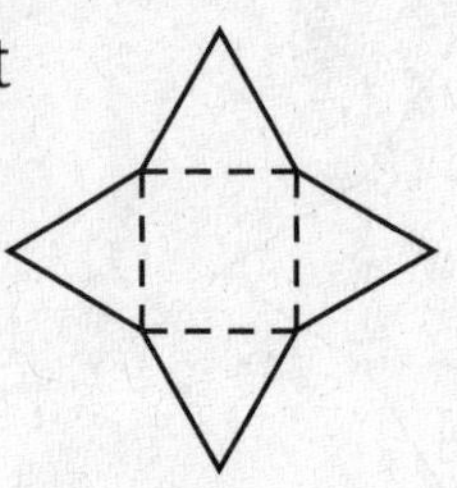

____

5. Jan sends a book to a friend. She packs the book in a box folded from the net at the right. Which solid figure does the net become?

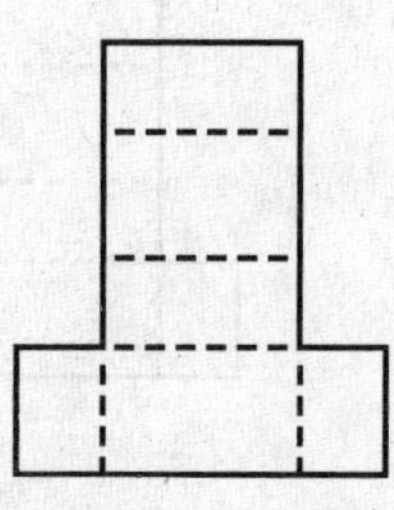

____

6. Jimmy saw workers build a wall using bricks. Which solid figure is shaped like the brick at the right?

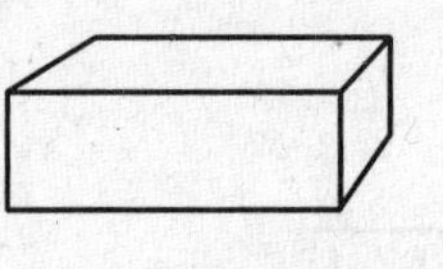

A sphere

B cylinder

C rectangular prism

D square pyramid

7. Which solid figure could be made with the net at the right?

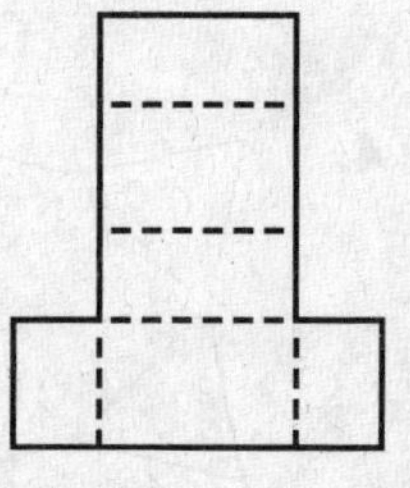

A cone

B cylinder

C sphere

D rectangular prism

Name____

Combine Solid Figures

You can combine solid figures to make new objects.

Name the solid figures used to make this object.

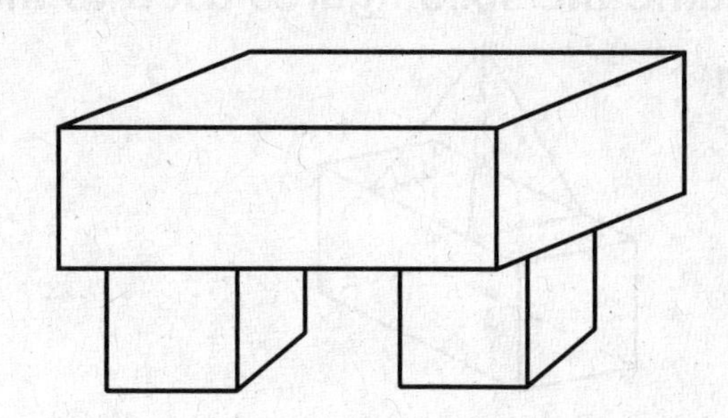

- Look at each part of the object separately.

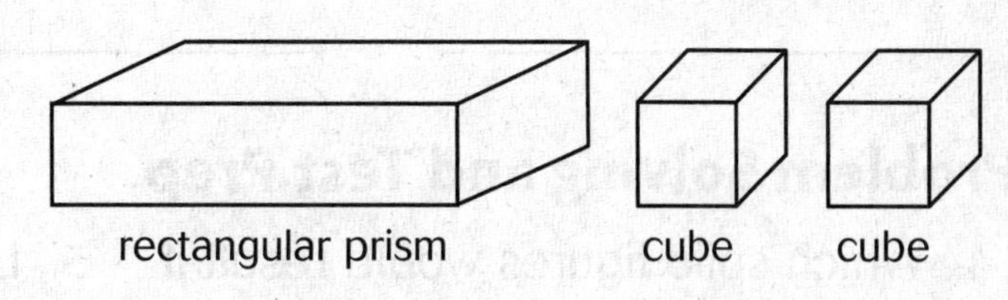

So, one rectangluar prism and two cubes are used to make the object.

Name the solid figures used to make each object.

1. 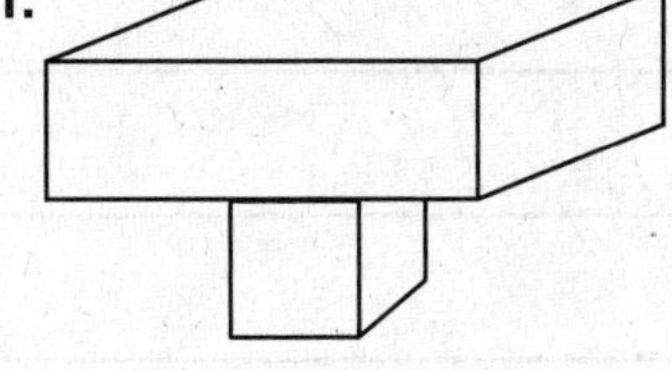

____

2. 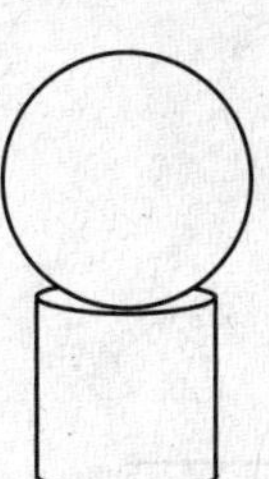

____

3. 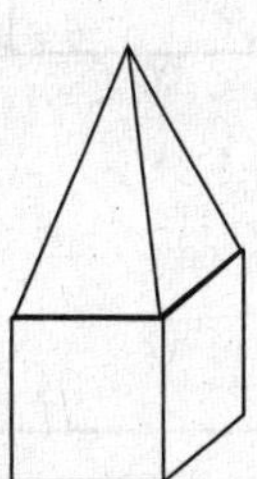

____

4. 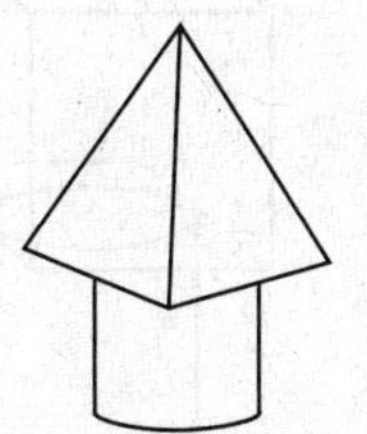

____

5. 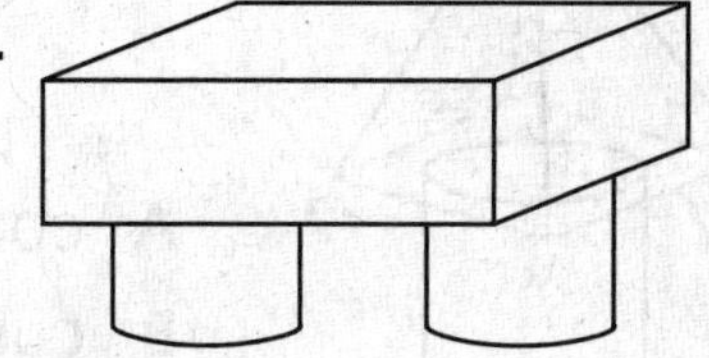

____

6. 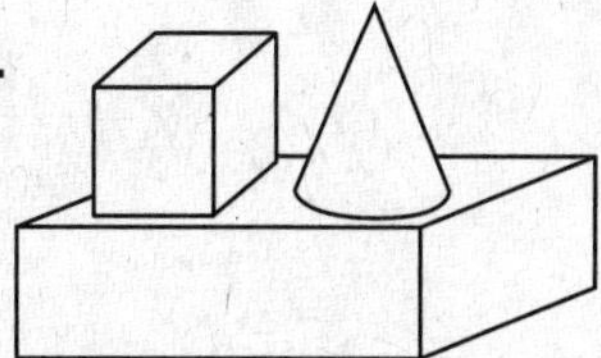

____

MG 2.6 Identify common solid objects that are the components needed to make a more complex solid object (e.g., cube, rectangular solid, sphere, prism, pyramid, cone, cylinder).

Name______________________________

Combine Solid Figures

Name the solid figures used to make each object.

1. 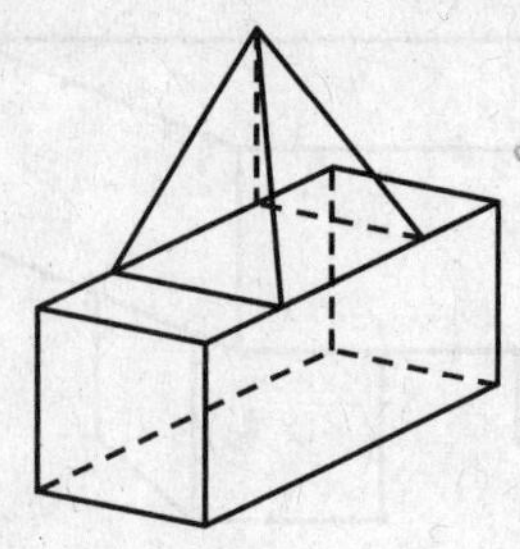

____

2. 

____

____

3. 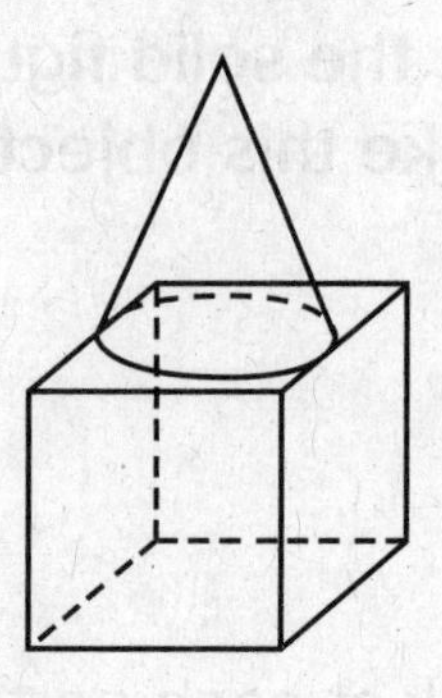

____

Problem Solving and Test Prep

4. Which solid figures would result if a rectangular prism was cut in half as shown below?

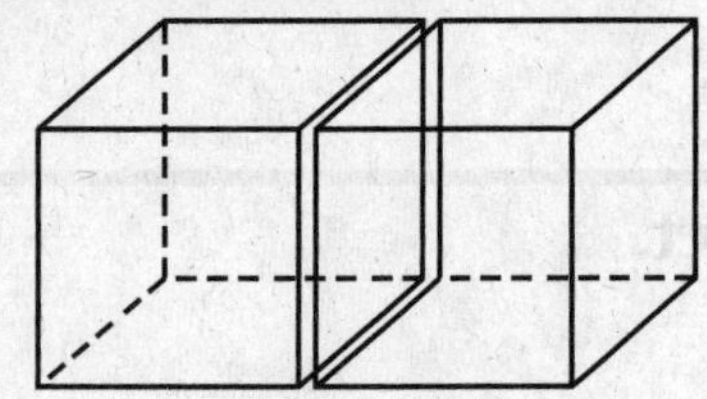

____

5. Lisa is building a birdhouse. She wants to have a point at the top of her birdhouse. Which solid figures can Lisa use?

____

____

____

6. Which solid figures are shown?

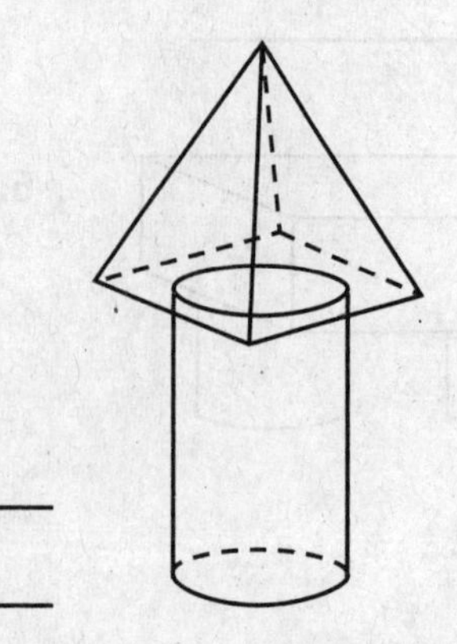

____

____

7. Which solid figures are combined to make the object below?

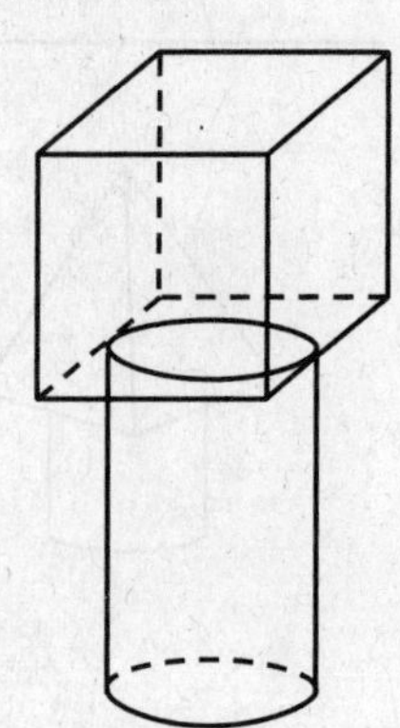

A cone, cube

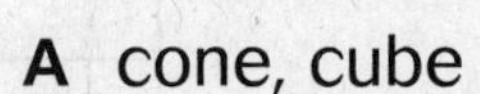

B cube, cylinder

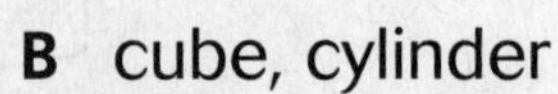

C cylinder, cone

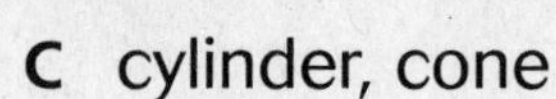

D triangle, cube

Name ______________________________ **Lesson 13.5**

Problem Solving Workshop Skill: Identify Relationships

Casey made a sponge paint border of triangles around a poster. Was his sponge in the shape of a cylinder, a sphere, or a square pyramid?

1. What are you asked to find? ______________________________

____

2. Does a cylinder, a sphere, or a square pyramid have triangular shaped faces or surfaces?

____

3. What is the answer to the question?

____

4. How can you check your answer?

____

Solve.

5. Ronaldo made a sponge paint border of circles around a poster. Was Ronaldo's sponge in the shape of a rectangular prism, a cone, or a square pyramid?

____

6. Suzie made a sponge paint border of squares around a poster. Was Suzie's sponge in the shape of a cylinder, a cube, or a cone?

____

Name____________________ Lesson 13.5

Problem Solving Workshop Skill: Identify Relationships

Problem Solving Skill Practice

Solve.

1. Skip used a sponge to make a border around his paper. He had access to 3 different sponges in the shape of a cube, a square pyramid, and a cylinder. Which sponge did Skip use to make the border at the right?

____

2. Julie used a sponge to make a border around her paper. She had access to 3 different sponges in the shape of a cube, a square pyramid, and a cylinder. Which sponge did Julie use to make the border at the right?

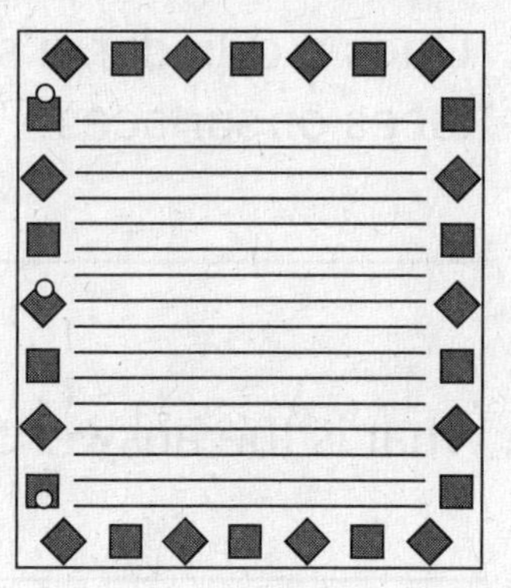

____

Mixed Applications

USE DATA For 3–4, use the Store Price List below.

3. Alice is told to spend exactly $13 at the store on two items. What will the shape of the two items be?

____

Store Price List

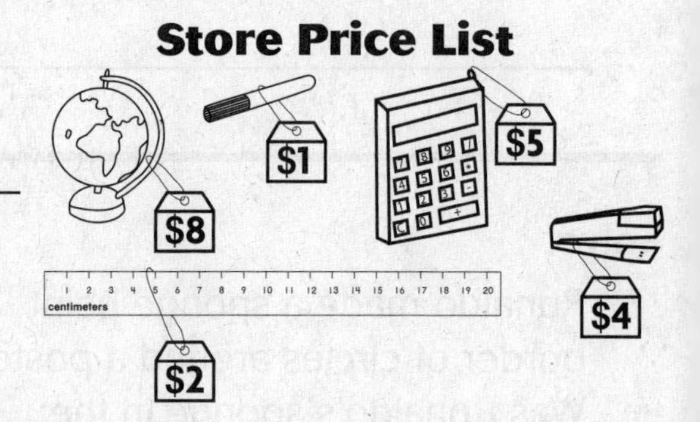

4. Cindy went to the movies on Saturday and spent $8. She went to the store on Monday and spent $8 on one item. What did Cindy buy at the store?

____

5. Bobby collected 8 baseball cards and 2 basketball cards. He put the cards evenly in each of 5 cylinder shaped canisters. How many cards were in each canister? Show your work.

____

____

Name____

Lesson 14.1

Perimeter

Perimeter is the distance around a figure.

Find the perimeter of the figure below.

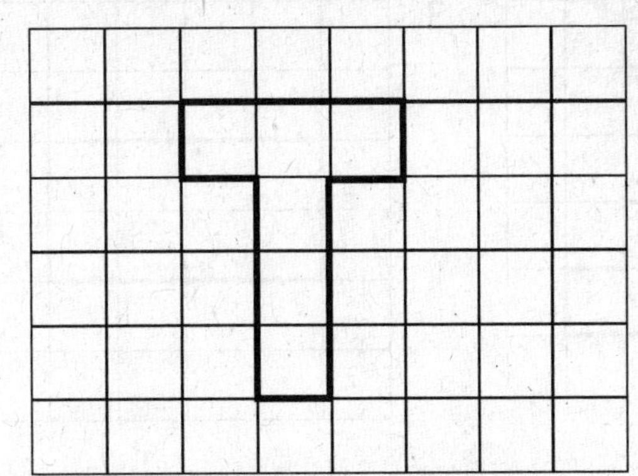

- You can count units on a grid to find perimeter.
- Each square represents one unit.

So, the figure has a perimeter of 14 units.

Find the perimeter of each figure.

1.

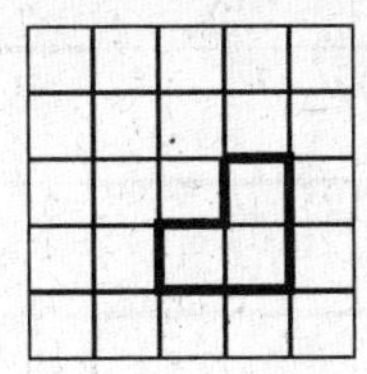

____

2.

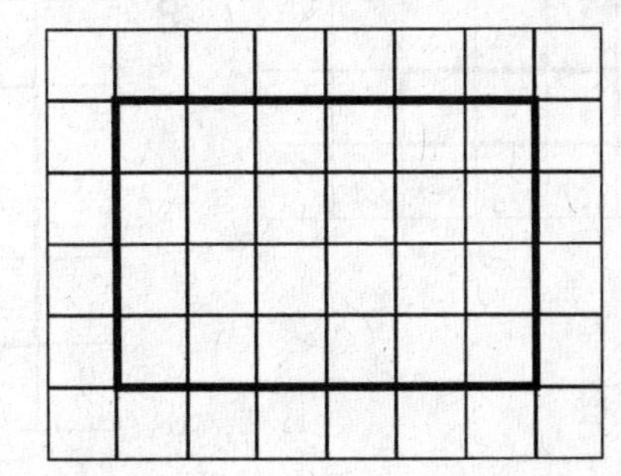

____

3.

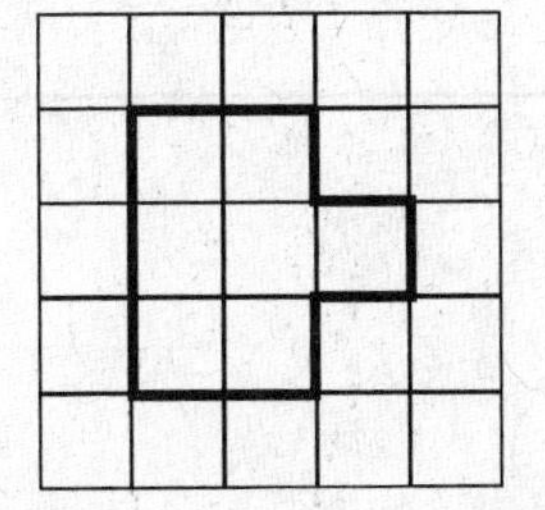

____

4.

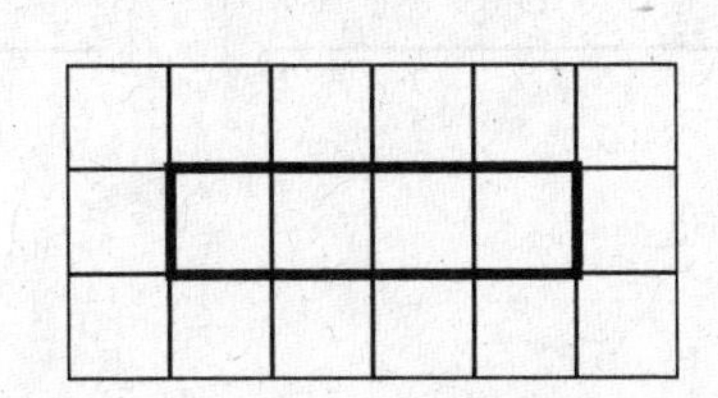

____

Name______________________________

Perimeter

Find the perimeter of each figure.

1.

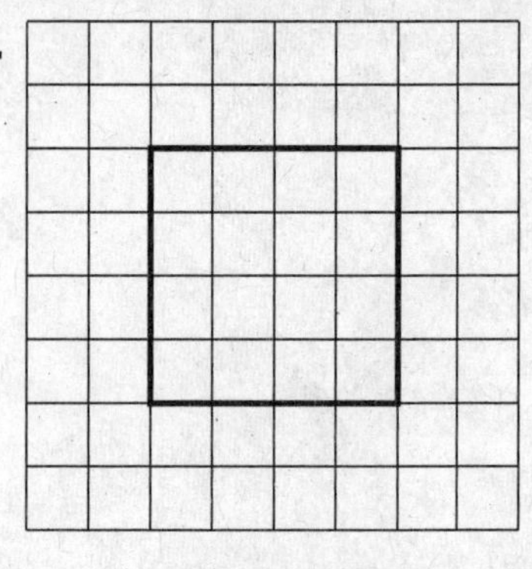

____

2.

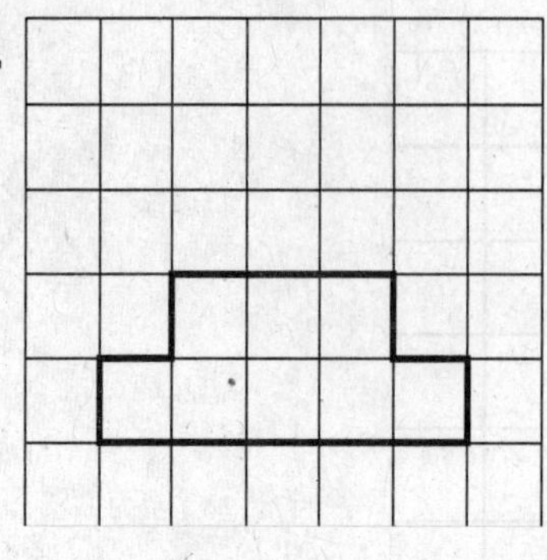

____

3.

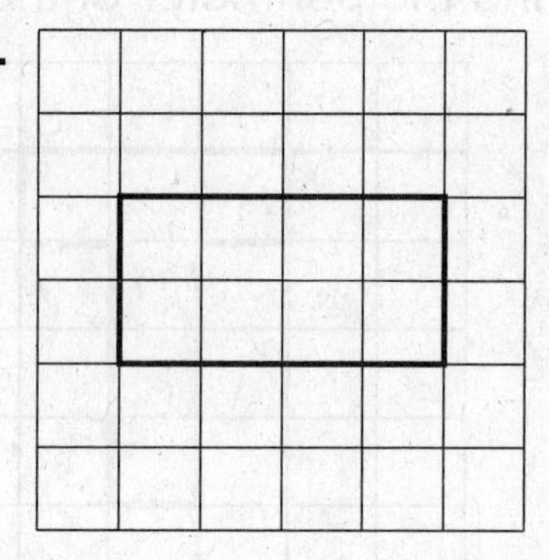

____

4.

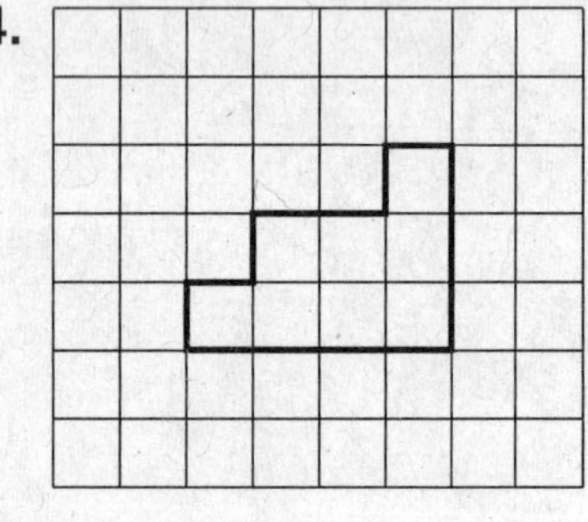

____

5.

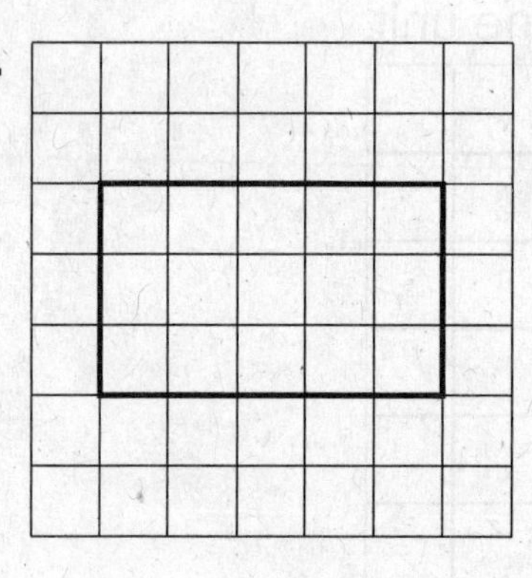

____

6.

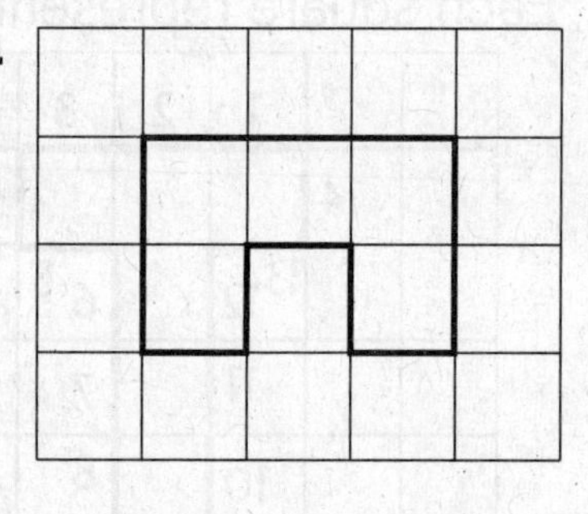

____

7.

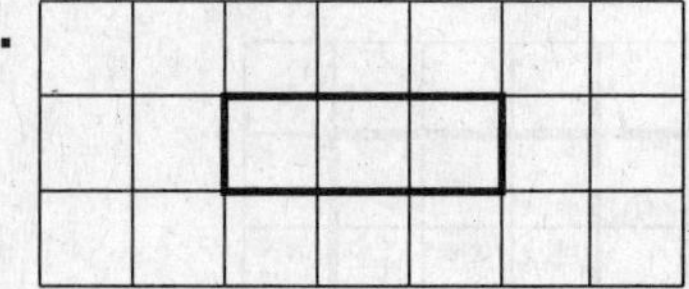

____

8.

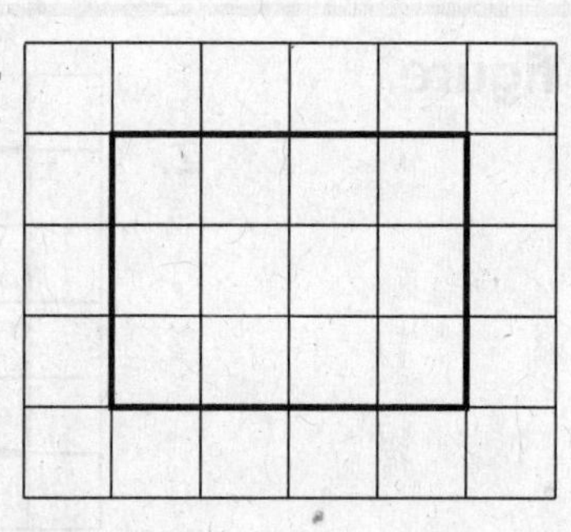

____

9.

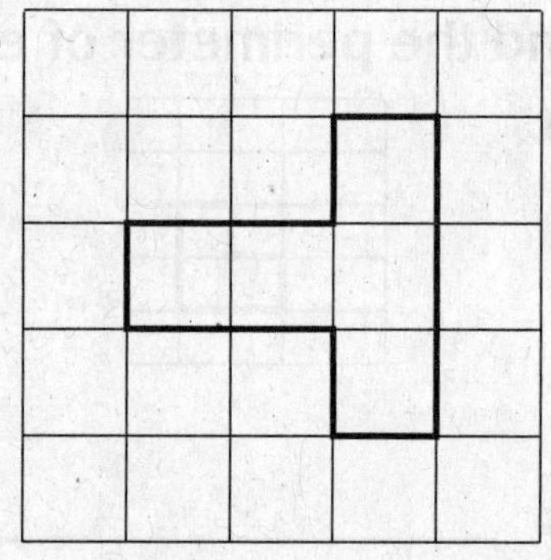

____

Spiral Review

For 1–5, find each product.

1. $3 \times 2 =$ ______
2. $7 \times 2 =$ ______
3. $9 \times 6 =$ ______
4. $8 \times 4 =$ ______
5. $5 \times 4 =$ ______

For 9–12, extend the patterns.

9. Skip count by twos: 16, 18,
______, ______, ______, ______

10. Skip count by threes: 33, 36,
______, ______, ______, ______

11. Skip count by fives: 100, 105,
______, ______, ______, ______

12. Skip count by tens: 200, 210,
______, ______, ______, ______

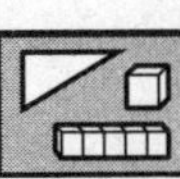

For 6–8, use the table to answer the questions.

Library Schedule	
Duration	Event
9:00 – 11:00	Story Time
11:00 – 1:00	Learn at Lunch
1:00 – 2:00	Play for Preschoolers
2:00 – 5:00	Homework Help

6. Which event lasts for 3 hours?

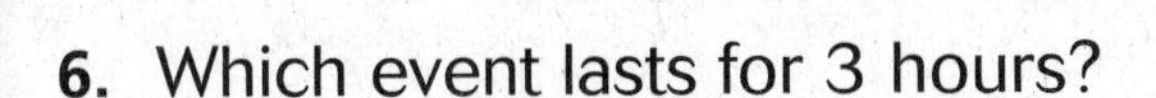

____

7. How long does Story Time last?

____

8. Which is longer, Play for Preschoolers or Learn at Lunch?

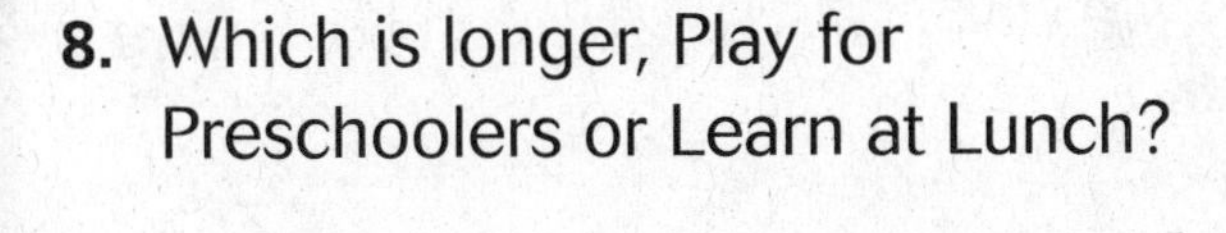

____

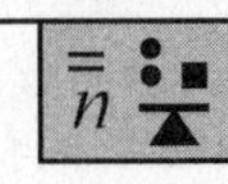

For 13–17, compare using >, <, or =.

13. $9 \div 9 \bigcirc 9 \div 1$
14. $16 \div 2 \bigcirc 24 \div 3$
15. $90 \div 10 \bigcirc 72 \div 9$
16. $56 \div 8 \bigcirc 63 \div 9$
17. $64 \div 8 \bigcirc 27 \div 3$

Name____

Estimate and Measure Perimeter

Perimeter is the distance around a figure.

When you estimate perimeter, it helps to visualize the length of the unit you are measuring in your head. Then, guess how many of these units form each side of the figure.

Estimate. Then use a centimeter ruler to find the perimeter.

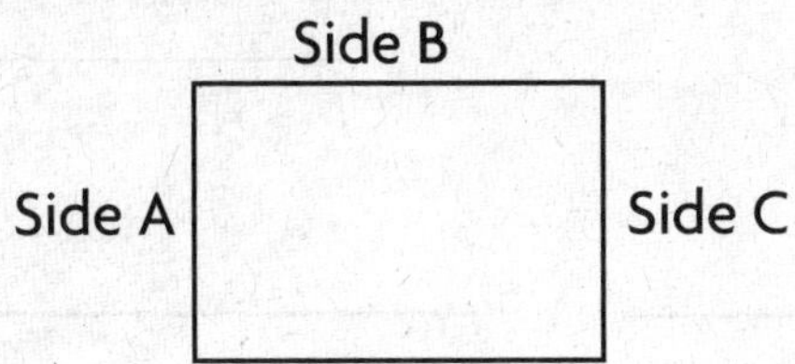

First, estimate the perimeter based on your experience with centimeter length.

Estimate: 8 cm

Find the length of Side A. Side A is 2 cm.

Find the length of Side B. Side B is 3 cm.

Find the length of Side C. Side C is 2 cm.

Find the length of Side D. Side D is 3 cm.

Write the length of each side. Then add the lengths.

3 cm + 2 cm + 3 cm + 2 cm = 10 cm

So, the perimeter of the figure is 10 cm.

Estimate. Then use a centimeter ruler to find the perimeter.

1.

Estimate Perimeter ____________

Perimeter ____________

2.

Estimate Perimeter ____________

Perimeter ____________

Name ____________________

Estimate and Measure Perimeter

Estimate. Then use a centimeter ruler to find the perimeter.

1.

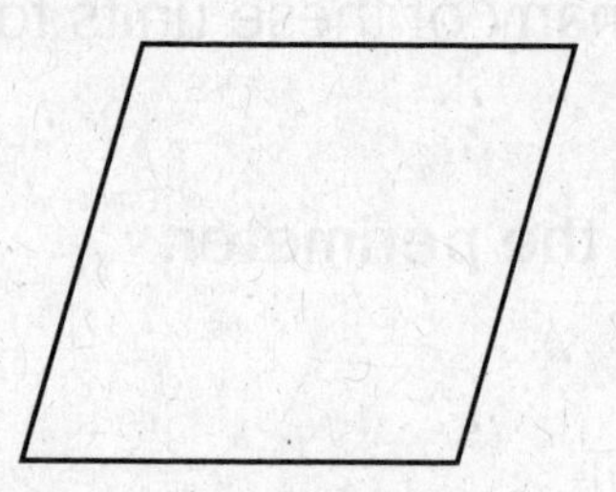

____

2.

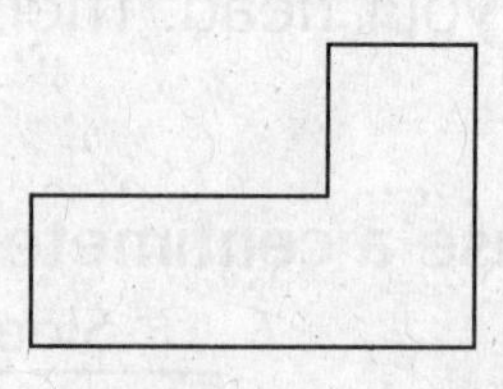

____

Estimate. Then use an inch ruler to find the perimeter.

3.

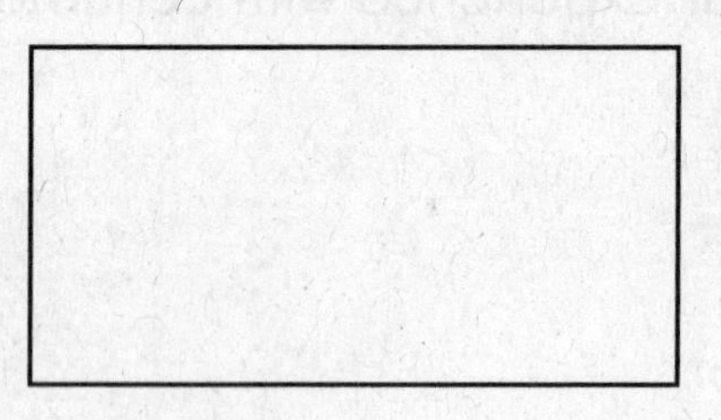

____

4.

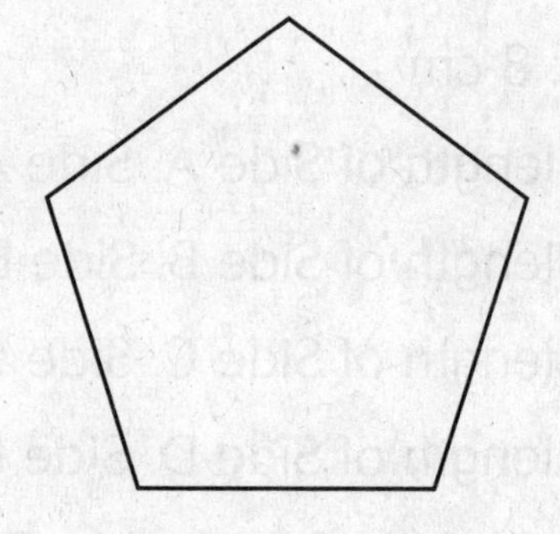

____

Problem Solving and Test Prep

5. John has a 5-inch by 7-inch picture frame and a 4-inch by 6-inch picture frame. Which picture frame has a greater perimeter?

____

6. Brian has an 8-inch by 10-inch picture frame. He wants to add 1 inch to both the width and length. Find the perimeter of the new picture frame.

____

7. What is the perimeter of this triangle?

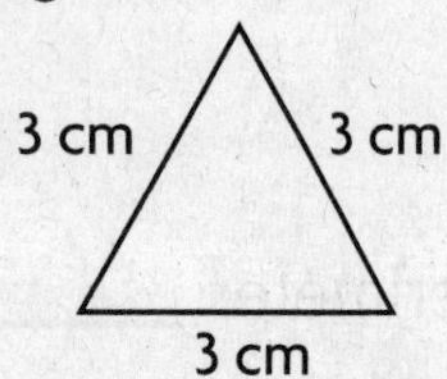

A 6 cm **C** 9 cm

B 8 cm **D** 12 cm

8. This figure has a perimeter of 20 cm. How long is the fourth side?

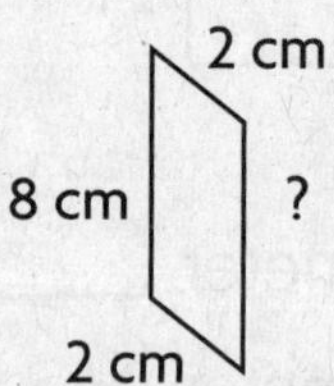

A 2 cm **C** 10 cm

B 8 cm **D** 12 cm

Area of Plane Figures

Area is the number of square units needed to cover a flat surface.
To find the area of a figure, find the number of square units that cover its surface without overlapping.

**Count or multiply to find the area of the figure below.
Write the answer in square units.**

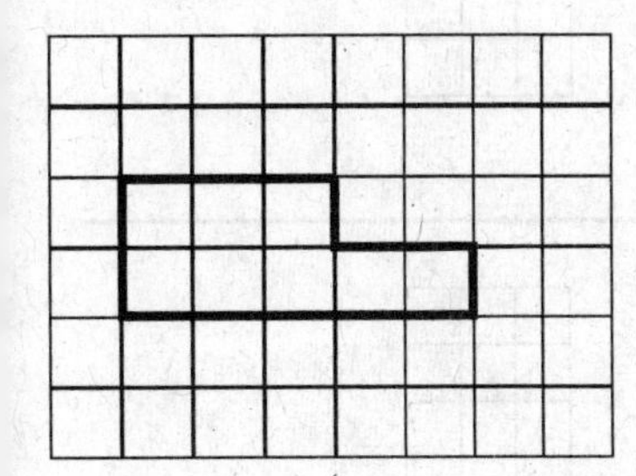

- A **square unit** is a square with a side lenth of 1 unit.
- To find the area, count how many square units the figure covers.

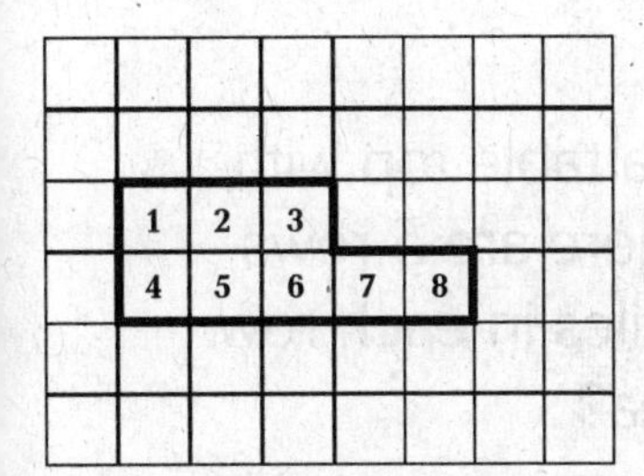

So, the area of the figure is 8 square units.

**Count or multiply to find the area of the figures below.
Write the answer in square units.**

1. 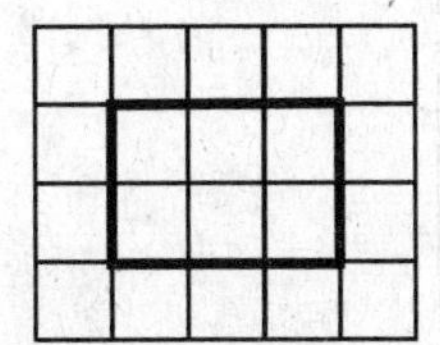

____

2. 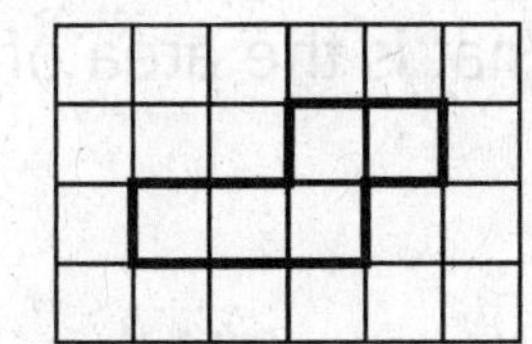

____

3. 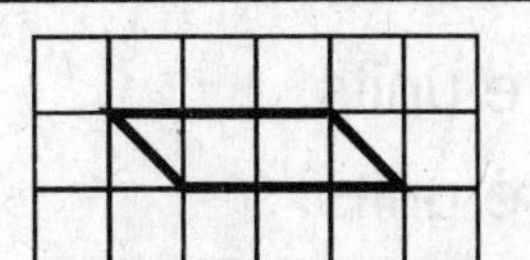

____

4.

____

Name________________________________

Area of Plane Figures

Count or multiply to find the area of each figure. Write the answer in square units.

1. 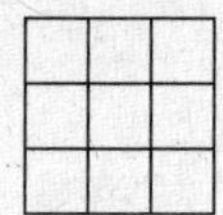

2. 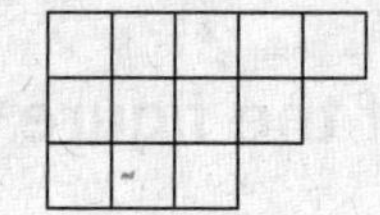

3. 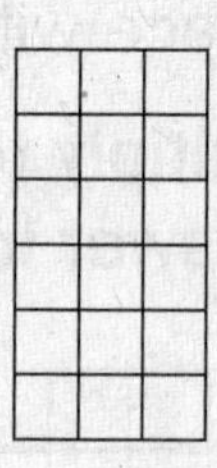

4. 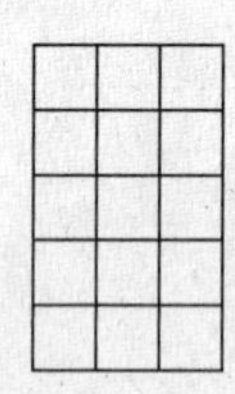

5. 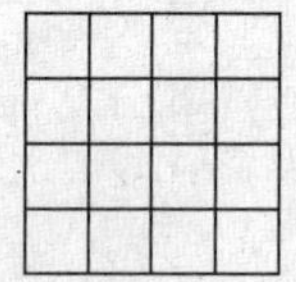

6. 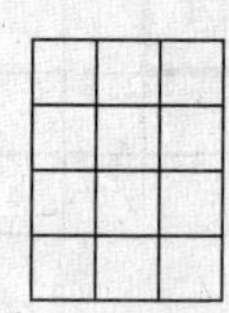

Problem Solving and Test Prep

7. Look at the figures below. Which figure has the greater area?

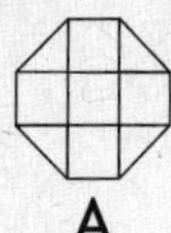

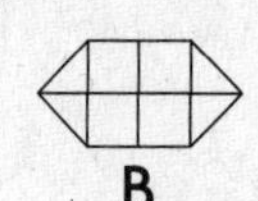

A B

8. Brian covered a table top with square tiles. There are 5 rows with 5 square tiles in each row. What is the area?

9. Julie is making a hot plate of 5 rows of square tiles with 6 tiles in each row. What is the area of the hot plate?

A 11 square units
B 12 square units
C 30 square units
D 36 square units

10. What is the area of this rectangle?

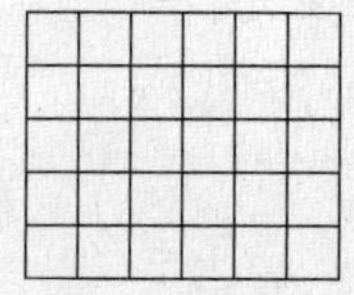

A 8 square units
B 17 square units
C 30 square units
D 72 square units

Area of Solid Figures

Use a net to find the area of solid figures.

Find the area of this solid figure.

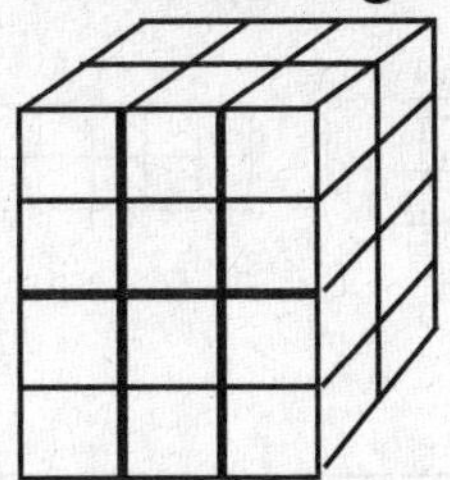

		E	E	E		
		E	E	E		
C	C	A	A	A	D	D
C	C	A	A	A	D	D
C	C	A	A	A	D	D
C	C	A	A	A	D	D
		F	F	F		
		F	F	F		
		B	B	B		
		B	B	B		
		B	B	B		
		B	B	B		

To find the area of the solid figure, follow these steps.

Find the area of face A. Count each unit.

A	A	A
A	A	A
A	A	A
A	A	A

There are 12 units. The area of face A is 12 square units.

Now, find the area of faces B, C, D, E, and F by counting units.

Face B – 12 square units.

Face C – 8 square units.

Face D – 8 square units.

Face E – 6 square units.

Face F – 6 square units.

Did you notice that opposite sides of a box have the same area?

Add the areas of each face to find the total area of the box.

12	+	12	+	8	+	8	+	6	+	6	=	52 square units
Face A		Face B		Face C		Face D		Face E		Face F		total area

So, the area of the figure is 52 square units.

Find the total area that covers each solid.

1.

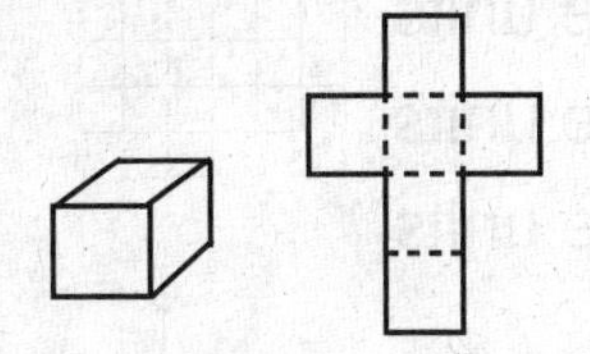

____

2.

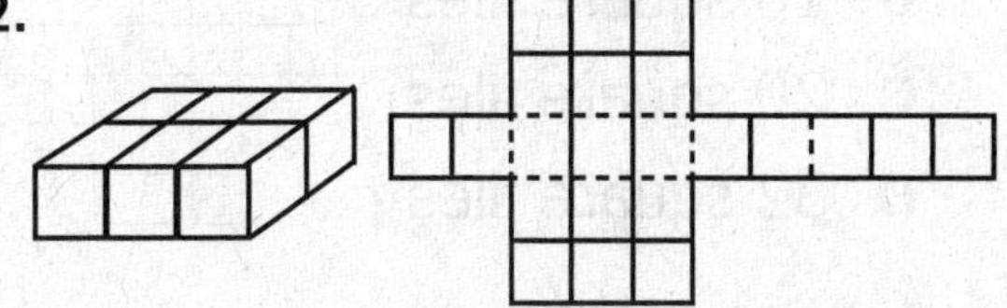

____

Name ____________________ Lesson 14.4

Area of Solid Figures

Find the total area that covers each solid figure.

1. 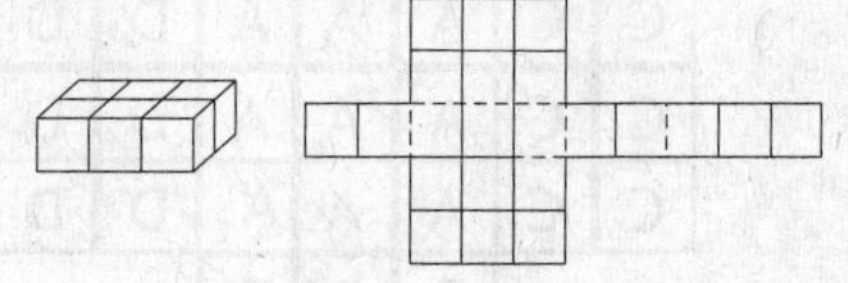

____

2. 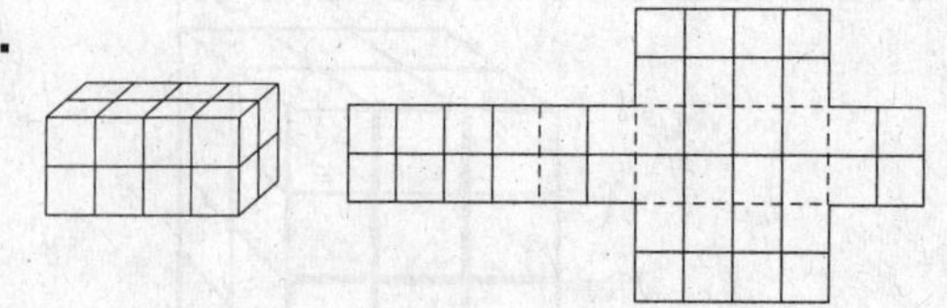

____

3. 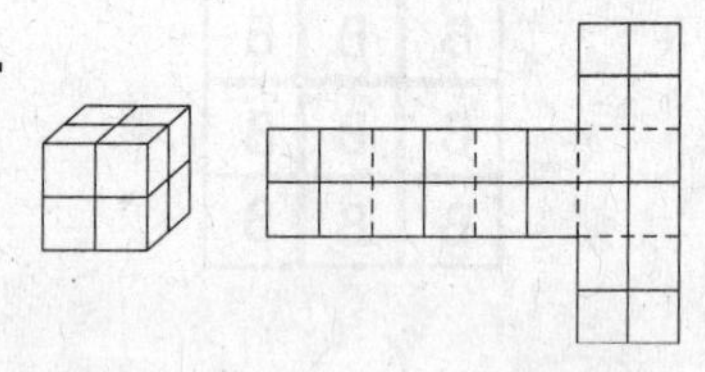

____

4. 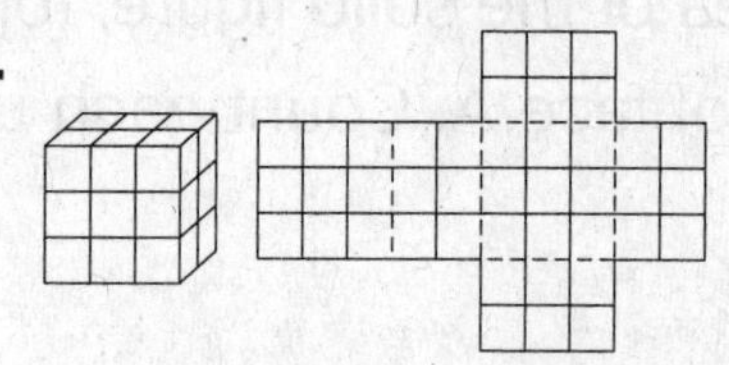

____

Problem Solving and Test Prep

5. Hon is decorating a box with square ceramic tiles. How many square units of tile will he need to cover the box?

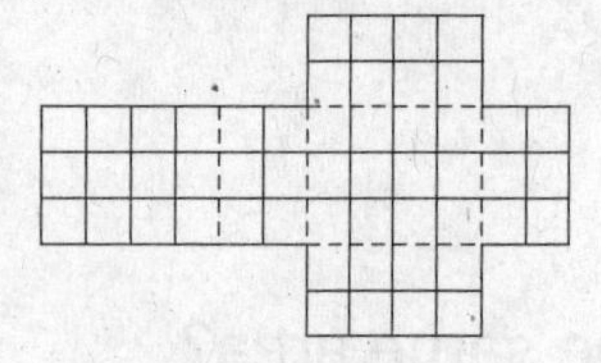

____

6. Lucy is making a mosaic box. She is covering all but the bottom with square tiles. How many tiles will she use to cover the top and sides of the box?

____

7. How many square tiles are needed to cover this box?

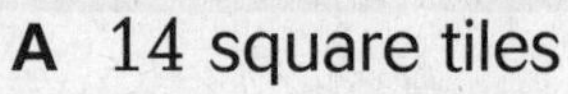
A 14 square tiles
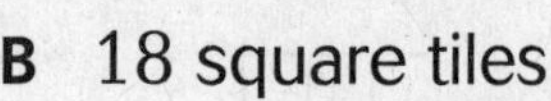
B 18 square tiles
C 20 square tiles
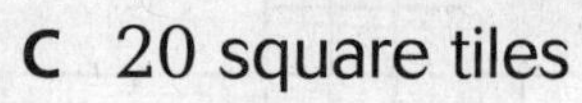
D 32 square tiles

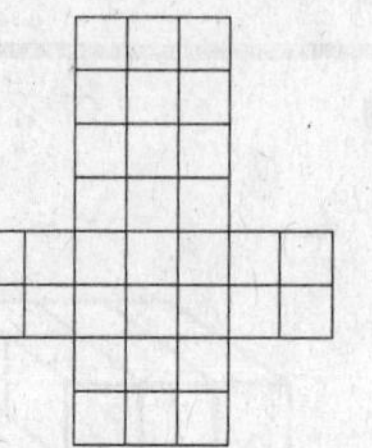

8. How many units of square tile are needed to cover this box?

A 22 square units
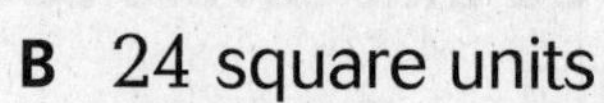
B 24 square units
C 32 square units
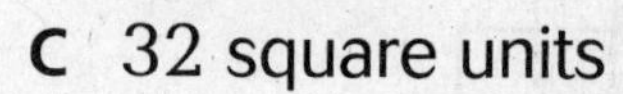
D 34 square units

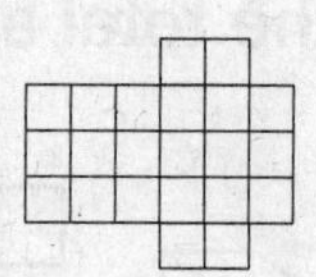

Name____

Estimate and Find Volume

Volume is the amount of space a solid figure takes up. 1 cubic unit

A **cubic unit** is used to measure volume. A cubic unit is a cube with a side length of 1 unit.

Write the volume of the solid figure in cubic units.

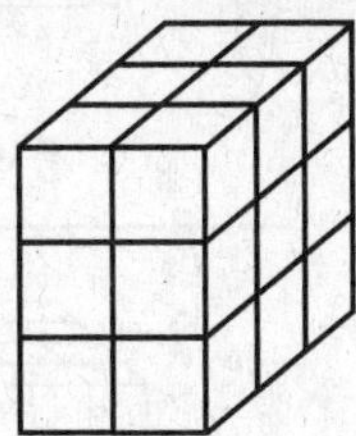

Count the number of cubes in the top layer.

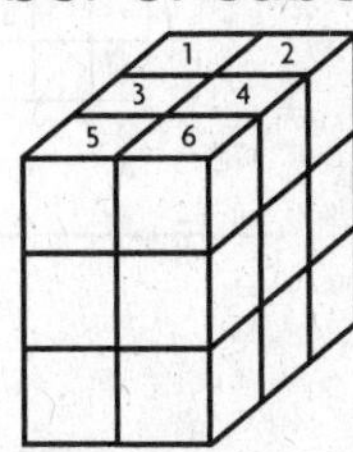

There are 6 cubes in the top layer.

Count the number of layers in the solid figure.

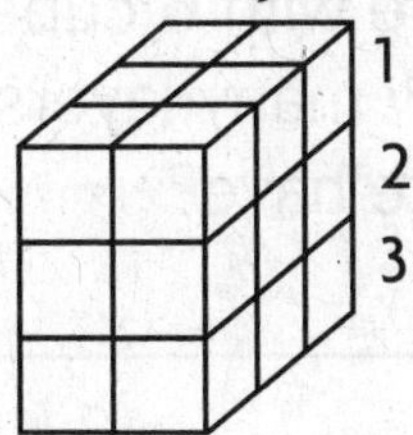

There are 3 layers.

Multiply the number of layers by the number of cubes per layer.

$6 \times 3 = 18$

So, the volume of this solid figure is 18 cubic units.

Write the volume of the solid figure in cubic units.

1.

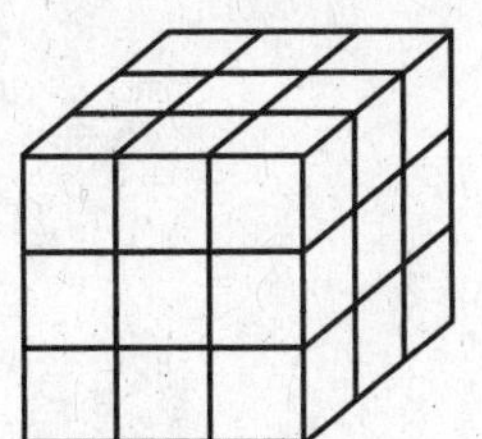

____

2.

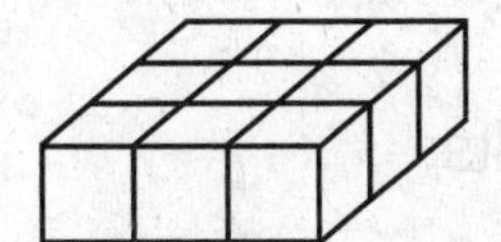

____

MG 1.2 Estimate or determine the area and volume of solid figures by covering them with squares or by counting the number of cubes that would fill them.

Name____________________

Estimate and Find Volume

Use cubes to make each solid. Then write the volume in cubic units.

1.

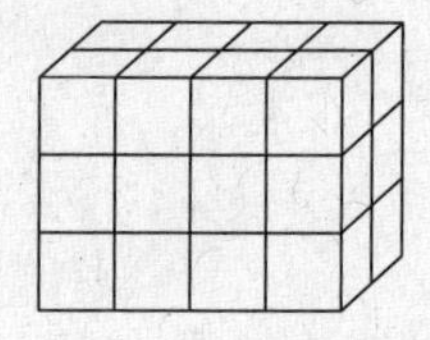

______

2.

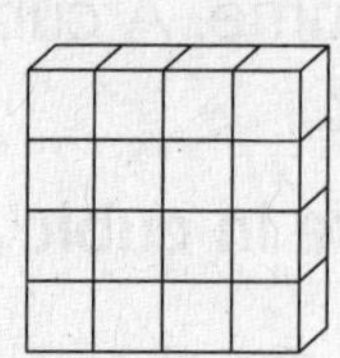

______

3.

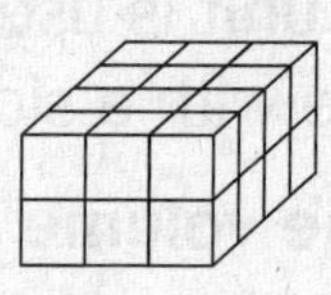

______

4.

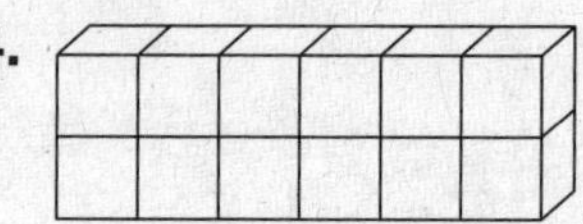

______

5.

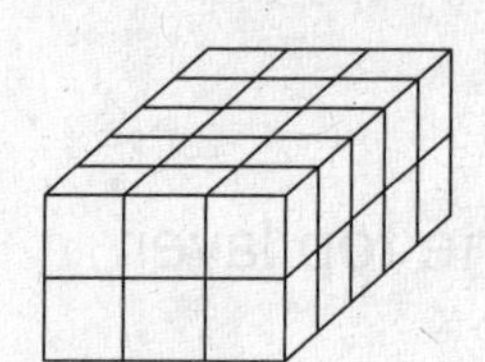

______

6.

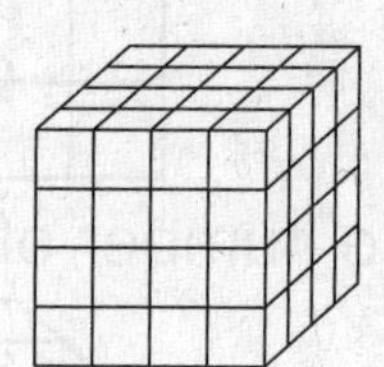

______

Problem Solving and Test Prep

7. Each layer of a rectangular prism is 4 cubic units. The volume is 8 cubic units. How many layers are in the prism?

______

8. Teresa has 18 cubes to make a solid figure with 6 cubes in each layer. How many layers will the solid figure have?

______

9. What is the volume of this solid figure?

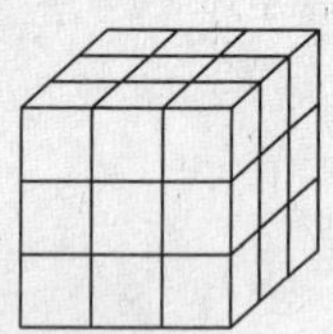

A 2 cubic units

B 8 cubic units

C 27 cubic units

D 30 cubic units

10. What is the volume of this solid figure?

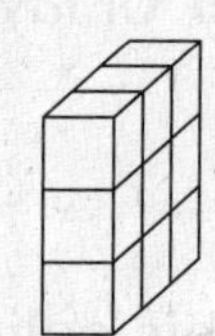

A 3 cubic units

B 6 cubic units

C 9 cubic units

D 12 cubic units

Spiral Review

For 1–5, solve the problems.

1. $3.00
   −$1.75

2. $12.98
   +$9.57

3. $21.78
   $15.65
   +$54.84

4. $ 8.45
   ×7

5. $15.57
   ×3

For 9–10, use the graph to answer the questions.

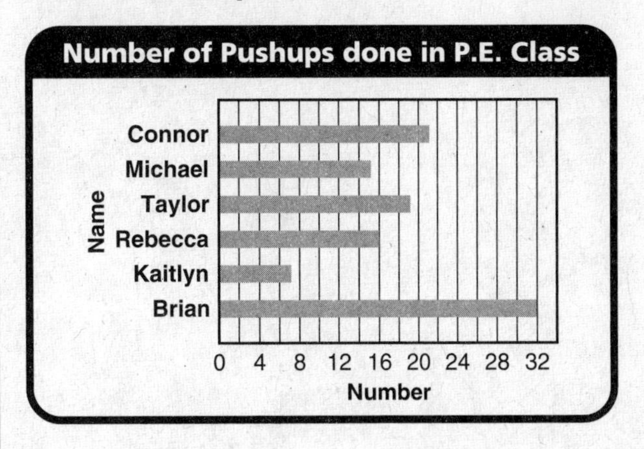

9. Which student did exactly 16 pushups? ____________

10. How many students did more than 20 pushups? ____________

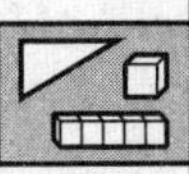

For 6–8, use the calendar to answer the following questions.

NOVEMBER

S	M	T	W	T	F	S
			1	2	3	4
5	6	7	8	9	10	11
12	13	14	15	16	17	18
19	20	21	22	23	24	25
26	27	28	29	30		

6. How many Mondays are in November? ____________

7. Rehearsals start on November 7 and last for 15 days. On what day do rehearsals end?

____

8. How many days are in November? ____________

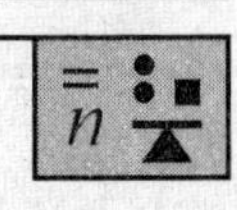

For 11–13, find the missing factor.

11. $(5 \times 2) \times 3 = 5 \times (\square \times 3)$

____

12. $4 \times (3 \times \square) = (3 \times 7) \times 4$

____

13. $(2 \times \square) \times 7 = (8 \times 2) \times 7$

Name______________________________

Problem Solving Workshop Skill: Use a Model

Ms. Liverte bought 16 mugs. The mugs come in cube-shaped boxes. She wants to put all the mugs into one large box. Which box can Ms. Liverte use to hold the 16 mugs?

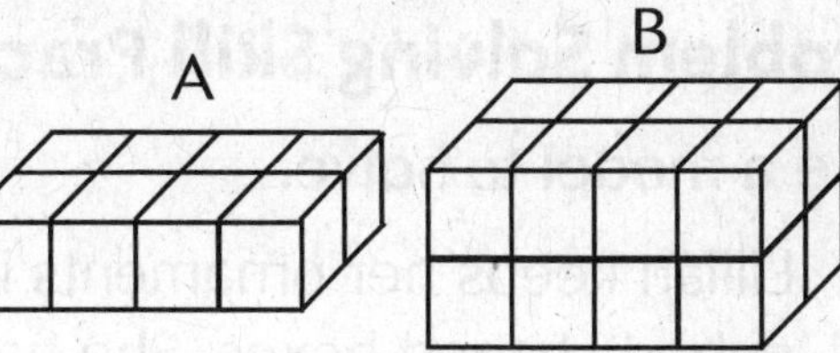

Read to Understand

1. What are you asked to find?

____

Plan

2. What can you use to help solve? ______________________________

Solve

3. How many cubes are in each layer of box A and B? ______________
4. Which box can Ms. Livetter use to hold the 16 mugs? ______________

Check

5. How many layers do the 2 boxes have each? ______________

Make a model to solve.

6. A carton has 8 picture cubes in each layer. If there are 2 layers, how many picture cubes are there in all?

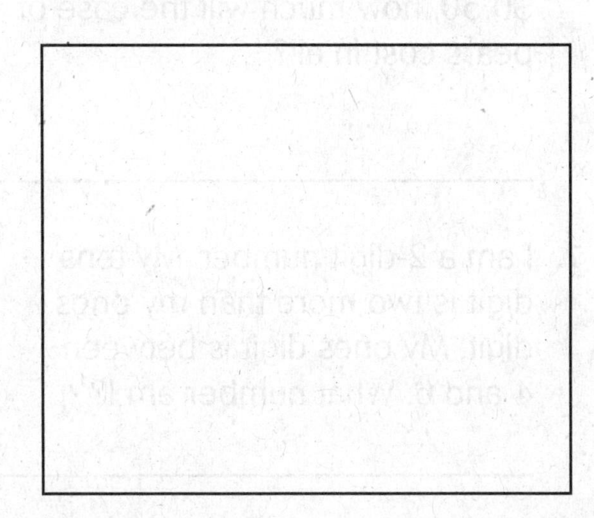

____

7. A carton has 20 gift box cubes in it. There are 4 layers with 5 boxes in each layer. If you add 1 layer, how many gift box cubes will there be in all?

____

MG 1.2 Estimate or determine the area and volume of solid figures by covering them with squares or by counting the number of cubes that would fill them.

Name____ **Lesson 14.6**

Problem Solving Workshop Skill: Use a model

Problem Solving Skill Practice

Use a model to solve.

1. Lillian keeps her ornaments in cubed-shaped boxes. She has 2 larger boxes of ornaments. She is looking for a special ornament that is in a box that holds 40 ornaments. In which box should she look?

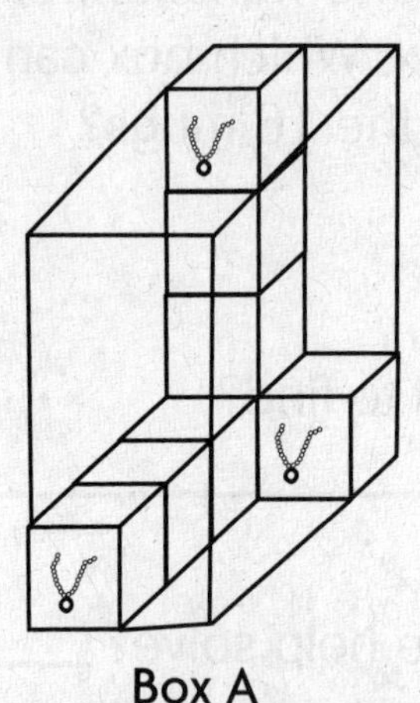

Box A

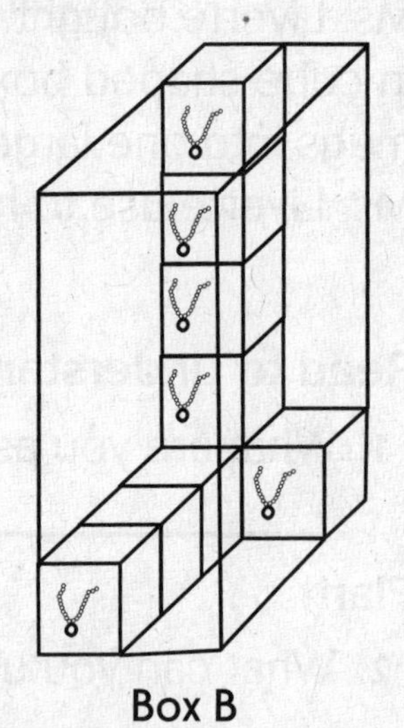

Box B

____

2. What if Box B could hold only 1 layer of cube-shaped ornament boxes? What would be the volume of Box B in cubic units?

____

3. What if Box A could hold 3 layers of cube-shaped ornament boxes? What would be the volume of Box A in cubic units?

____

Mixed Applications

4. Tom has two cartons of golf balls. Carton A has 3 layers with 15 golf balls in each layer. Carton B has 4 layers with 12 golf balls in each layer. Which carton holds the greater amount of golf balls?

____

5. Ella is buying a case of pears. Each row has 10 pears and there are 3 rows. If the cost of a pear is $0.50, how much will the case of pears cost in all?

____

6. Wesley has 4 more hockey cards than baseball cards. If he has 28 cards in all, how many hockey cards does he have?

____

7. I am a 2-digit number. My tens digit is two more than my ones digit. My ones digit is between 4 and 6. What number am I?

____

Name______________________________ Lesson 15.1

Algebra: Multiples of 10 and 100

You can use basic facts and patterns to multiply multiples of 10 and 100.

Use a basic fact and pattern to find each product.

$8 \times 5 =$ ______

$80 \times 5 =$ ______

$800 \times 5 =$ ______

Find the basic fact 8×5: $\begin{array}{r} 8 \\ \times\ 5 \\ \hline 40 \end{array}$

Adding a zero to a factor will add a zero to the product: $\begin{array}{r} 80 \\ \times\ 5 \\ \hline 400 \end{array}$

Adding two zeros to a factor will add two zeros to the product: $\begin{array}{r} 800 \\ \times\ 5 \\ \hline 4{,}000 \end{array}$

So , $8 \times 5 = 40$; $80 \times 5 = 400$; and $800 \times 5 = 4{,}000$

Use a basic fact and patterns to find each product.

1. $6 \times 3 =$ ______
$6 \times 30 =$ ______
$6 \times 300 =$ ______

2. $9 \times 3 =$ ______
$9 \times 30 =$ ______
$9 \times 300 =$ ______

3. $4 \times 8 =$ ______
$4 \times 80 =$ ______
$4 \times 800 =$ ______

4. $2 \times 3 =$ ______
$2 \times 30 =$ ______
$2 \times 300 =$ ______

5. $6 \times 5 =$ ______
$6 \times 50 =$ ______
$6 \times 500 =$ ______

6. $8 \times 7 =$ ______
$8 \times 70 =$ ______
$8 \times 700 =$ ______

Name______________________________

Algebra: Multiples of 10 and 100

Use a basic fact and patterns to find each product.

1. $5 \times 3 =$ ________ **2.** $2 \times 7 =$ ________ **3.** $9 \times 8 =$ ________

$50 \times 3 =$ ________ $2 \times 70 =$ ________ $9 \times 80 =$ ________

$500 \times 3 =$ ________ $2 \times 700 =$ ________ $9 \times 800 =$ ________

Find the product.

4. $50 \times 7 =$ ________ **5.** $4 \times 500 =$ ________ **6.** ________ $= 3 \times 200$

7. $40 \times 8 =$ ________ **8.** ________ $= 6 \times 90$ **9.** $4 \times 600 =$ ________

10. ________ $= 1 \times 60$ **11.** ________ $= 700 \times 3$ **12.** $200 \times 5 =$ ________

Problem Solving and Test Prep

13. A case contains 5 boxes of toy cars. Each box holds 100 cars. How many toy cars are in 4 cases?

____

14. Mandy has 2 boxes of earrings with 20 earrings in each box. Carla has 3 boxes of earrings with 10 earrings in each box. Who has the most earrings?

____

15. Which answer shows the product of 4×60?

A 24
B 240
C 2,400
D 24,000

16. The art club is inviting people to view their work. There are 3 exhibits with 300 paintings displayed for each exhibit. How many paintings are on display at the art club?

A 9
B 90
C 900
D 9,000

Name ______________________________

Arrays with Tens and Ones

An array shows objects in rows and columns. It can be used to show multiplication.

Find the product of 3 × 17. Show your mulitiplication and addition.

Multiply the tens place first: $3 \times 10 = 30$

Then multiply the ones: $3 \times 7 = 21$

Add the products: $30 + 21 = 51$

3 rows × 17 columns = 51 counters

So, $3 \times 17 = 51$.

Find the product. Show your multiplication and addition.

1.

$4 \times 12 =$

____

____

2.

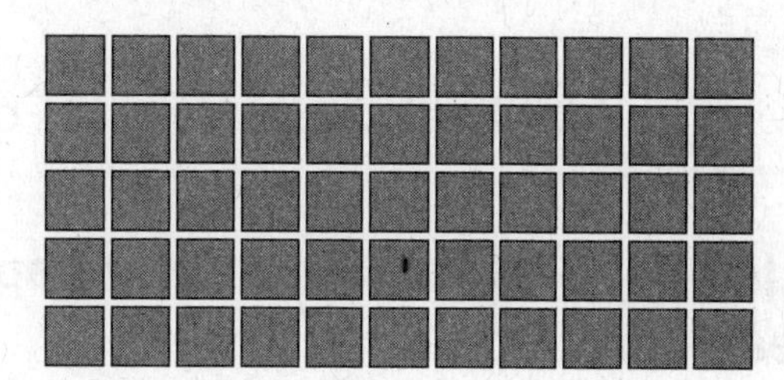

$5 \times 11 =$

____

____

3.

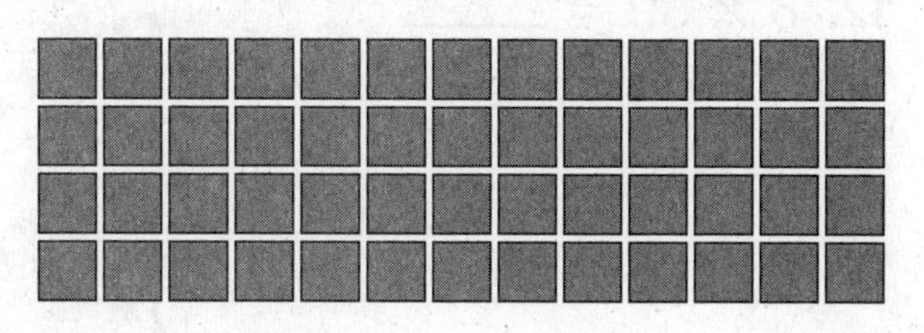

$4 \times 13 =$

____

____

4.

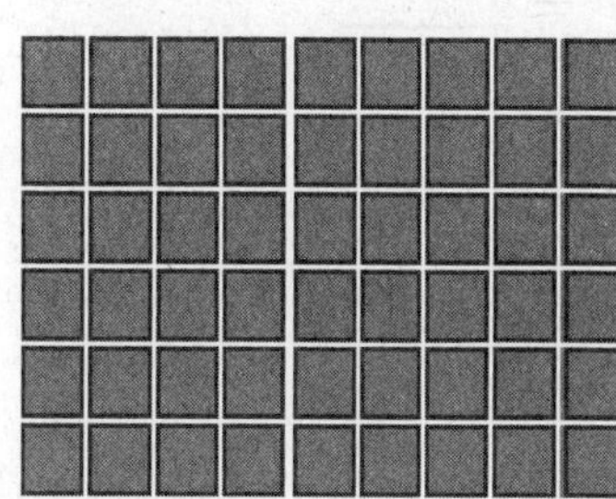

$6 \times 11 =$

____

____

Name___________________________________

Arrays with Tens and Ones

Find the product.

1. $2 \times 16 =$ ______

2. $4 \times 13 =$ ______

3. $3 \times 22 =$ ______

4. $5 \times 14 =$ ______

5. $6 \times 15 =$ ______

6. 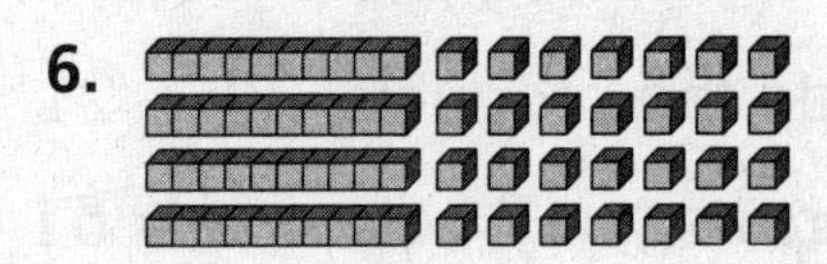

$4 \times 17 =$ ______

Use base-ten blocks or grid paper to find each product.

7. $5 \times 25 =$ ______

8. $4 \times 18 =$ ______

9. $4 \times 22 =$ ______

10. $3 \times 19 =$ ______

11. $4 \times 27 =$ ______

12. $8 \times 39 =$ ______

13. $6 \times 38 =$ ______

14. $4 \times 12 =$ ______

15. $7 \times 31 =$ ______

16. $3 \times 24 =$ ______

17. $4 \times 29 =$ ______

18. $9 \times 15 =$ ______

19. $8 \times 16 =$ ______

20. $5 \times 35 =$ ______

Name______________________________

Model 2-Digit Multiplication

You can use base-ten blocks to model 2-digit multiplication.

Find the product. Use place value or regrouping.

$$\begin{array}{r} 27 \\ \times\ 3 \\ \hline \end{array}$$

Multiply the ones

$$\begin{array}{r} \overset{2}{2}7 \\ \times\ 3 \\ \hline 1 \end{array}$$

Regroup the ones.
21 ones = 2 tens
and 1 one.

Multiply the tens
Don't forget to add
the 2 regrouped tens.

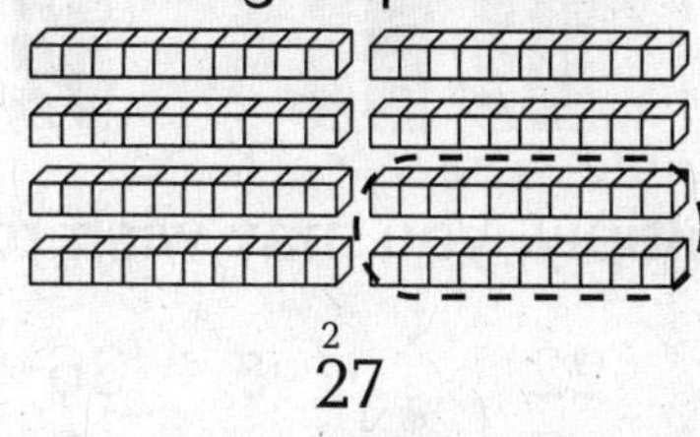

$$\begin{array}{r} \overset{2}{2}7 \\ \times\ 3 \\ \hline 81 \end{array}$$

So, $27 \times 3 = 81$.

Find the product. Use place value or regrouping.

1. $\begin{array}{r} 39 \\ \times\ 2 \\ \hline \end{array}$

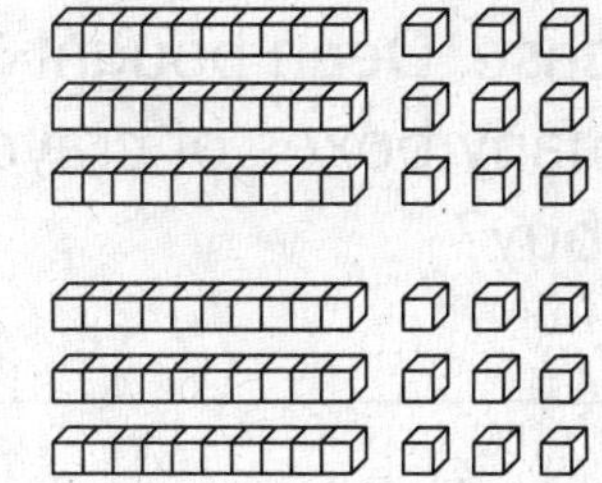

2. $\begin{array}{r} 17 \\ \times\ 3 \\ \hline \end{array}$

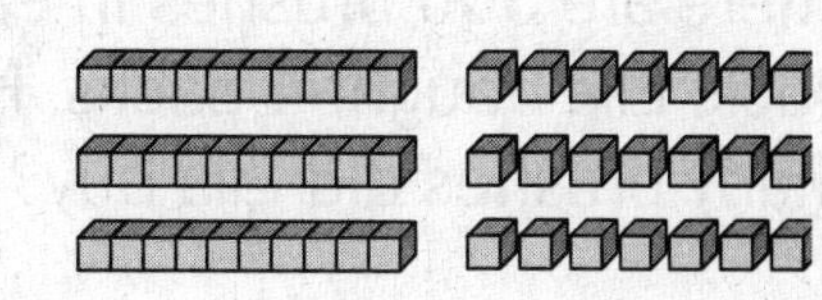

3. $\begin{array}{r} 14 \\ \times\ 5 \\ \hline \end{array}$

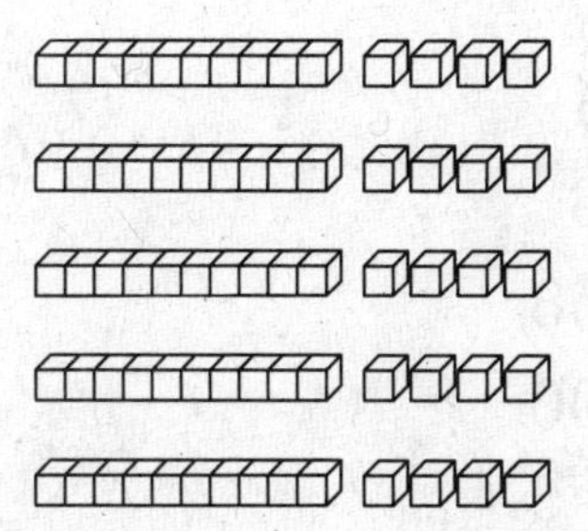

4. $\begin{array}{r} 40 \\ \times\ 2 \\ \hline \end{array}$

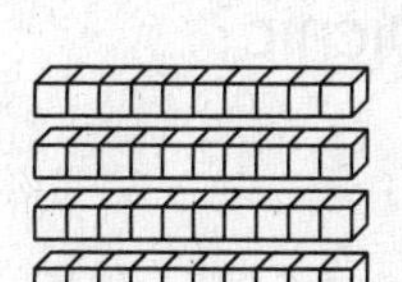

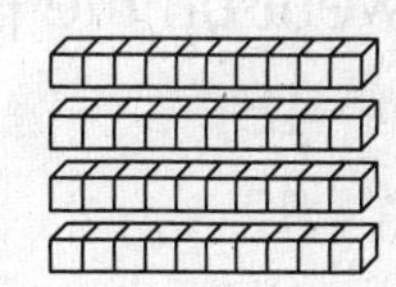

Name ______________________________ Lesson 15.3

Model Two-Digit Multiplication

Find the product. Use place value or regrouping.

1. $\begin{array}{r} 25 \\ \times\ 2 \\ \hline \end{array}$

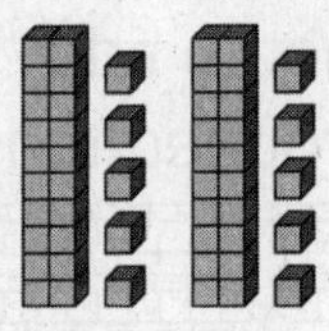

2. $\begin{array}{r} 16 \\ \times\ 4 \\ \hline \end{array}$

3. $\begin{array}{r} 34 \\ \times\ 3 \\ \hline \end{array}$

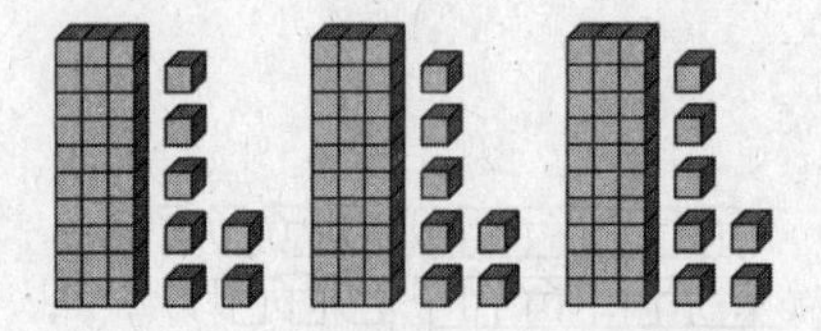

Multiply. You may wish to use base-ten blocks to help you.

4. $\begin{array}{r} 22 \\ \times\ 7 \\ \hline \end{array}$ **5.** $\begin{array}{r} 36 \\ \times\ 3 \\ \hline \end{array}$ **6.** $\begin{array}{r} 43 \\ \times\ 5 \\ \hline \end{array}$ **7.** $\begin{array}{r} 24 \\ \times\ 6 \\ \hline \end{array}$ **8.** $\begin{array}{r} 55 \\ \times\ 2 \\ \hline \end{array}$

9. $\begin{array}{r} 32 \\ \times\ 5 \\ \hline \end{array}$ **10.** $\begin{array}{r} 18 \\ \times\ 4 \\ \hline \end{array}$ **11.** $\begin{array}{r} 31 \\ \times\ 4 \\ \hline \end{array}$ **12.** $\begin{array}{r} 16 \\ \times\ 4 \\ \hline \end{array}$ **13.** $\begin{array}{r} 12 \\ \times\ 8 \\ \hline \end{array}$

Problem Solving and Test Prep

14. There are 300 brushes in each pack. Ella bought 4 packs. How many brushes did Ella buy?

____

15. There are 20 boxes of crayons in each case. Dean bought 3 cases. How many boxes of crayons did Dean buy?

____

16. Carter's class went on a picnic. There were 13 students in each of 4 groups. How many students went on the picnic?

A 48

B 52

C 56

D 60

17. Eddie reads 2 hours a day. How many hours does Eddie read in 12 weeks?

A 14

B 28

C 168

D 100

Name ______________________________ **Week 22**

Spiral Review

For 1–5, find the missing factor.

1. $\square \times 7 = 7$

2. $8 \times \square = 0$

3. $12 \times \square = 12$

4. $3 \times \square = 0$

5. $\square \times 25 = 0$

For 9–12, find the missing number in each pattern.

9. 11, 15, _____, 23, 27, 31

10. 27, 32, 37, _____, 47, 52

11. 100, 91, _____, 73, 64, 55

12. 200, _____, 198, 197, 196, 195

For 6–8, Find the volume of the solid figures.

Figure	Volume
6.	_____ cubic units
7.	_____ cubic units
8.	_____ cubic units

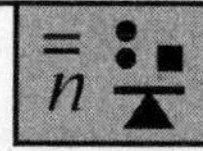

For 13–17, compare using >, <, or =.

13. $17 - 9 \bigcirc 2 + 7$

14. $8 + 9 \bigcirc 20 - 3$

15. $11 - 3 \bigcirc 10 - 4$

16. $2 + 0 \bigcirc 9 - 7$

17. $6 - 1 \bigcirc 12 - 6$

Name____

Estimate Products

When you estimate a product, you round a number to the greatest place value.

Estimate the product. Round to the greatest place value.

7×826

STEP 1

Find the greatest place value.

You can use a place value chart.

Hundreds	Tens	Ones
8	2	6

The greatest place value is 8.

8 is in the hundreds place.

So, round to the nearest hundred.

STEP 2

Round 826 to the nearest hundred.

Look at the tens place.

826

Is 2 greater than, less than, or equal to 5? Less than.

So, you round down to the nearest hundred:

800.

STEP 3

Multiply your rounded factor.

$$\begin{array}{r} 800 \\ \times\ 7 \\ \hline \end{array}$$

Remember your basic facts!

$7 \times 8 = 56$

$7 \times 80 = 560$

$7 \times 800 = 5,600$

So, the estimated product of 7×862 is 5,600

Estimate each product.

1. $\begin{array}{r} 378 \\ \times\ 5 \\ \hline \end{array}$

2. $\begin{array}{r} 692 \\ \times\ 8 \\ \hline \end{array}$

3. $\begin{array}{r} 39 \\ \times\ 6 \\ \hline \end{array}$

4. $\begin{array}{r} 912 \\ \times\ 7 \\ \hline \end{array}$

5. $\begin{array}{r} 71 \\ \times\ 4 \\ \hline \end{array}$

6. $\begin{array}{r} 219 \\ \times\ 9 \\ \hline \end{array}$

Name____

Estimate Products

Estimate each product.

1. $29 \times 6 =$ _____
2. $42 \times 9 =$ _____
3. $7 \times 31 =$ _____
4. $3 \times 56 =$ _____
5. $31 \times 5 =$ _____
6. $8 \times 72 =$ _____
7. $4 \times 27 =$ _____
8. $62 \times 3 =$ _____

9. 61×4
10. 38×5
11. 71×4
12. 316×7
13. 297×6
14. 82×8
15. 361×4
16. 72×3
17. 157×9
18. 228×3

Problem Solving and Test Prep

USE DATA For 19–20, use the table.

Beach Shops	
Name	**Number of Shops**
Bike Rentals	5
Board Rentals	15
Boat Rentals	22
Game Rentals	47

19. Each boat shop rents 33 boats a day. About how many boats are rented in each week by all of the boat shops combined?

____

20. Each game shop rents 9 volleyball nets each week. About how many volleyball nets are rented each week by all of the game shops combined?

____

21. Felix has 4 boxes of 18 stamps each and 3 boxes of 17 stamps each. About how many stamps does Felix have in all?
   - **A** 240
   - **B** 340
   - **C** 140
   - **D** 40

22. Kitty and her friends built 3 sand castles. They used 125 shells to decorate each castle. About how many shells did they use in all?
   - **A** 300
   - **B** 700
   - **C** 500
   - **D** 600

Name____ **Lesson 15.5**

Multiply 2-Digit Numbers

Sometimes you need to regroup to find the product when you multiply.

Find the product.

$$\begin{array}{r} 28 \\ \times\ 3 \\ \hline \end{array}$$

STEP 1

Multiply the ones.

$$\begin{array}{r} 28 \\ \times\ 3 \\ \hline 24 \end{array}$$ ones

STEP 2

Regroup.

24 ones = 2 tens and 4 ones.

$$\begin{array}{r} {}^{2}\ \\ 28 \\ \times\ 3 \\ \hline 4 \end{array}$$

STEP 3

Multiply the tens.

$3 \times 20 = 60$

Add the regrouped tens.

$60 + 20 = 80$

$$\begin{array}{r} {}^{2}\ \\ 28 \\ \times\ 3 \\ \hline 84 \end{array}$$

So, $28 \times 3 = 84$.

Find each product.

1. $\begin{array}{r} 71 \\ \times\ 8 \\ \hline \end{array}$

2. $\begin{array}{r} 24 \\ \times\ 4 \\ \hline \end{array}$

3. $\begin{array}{r} 76 \\ \times\ 6 \\ \hline \end{array}$

4. $\begin{array}{r} 37 \\ \times\ 8 \\ \hline \end{array}$

5. $\begin{array}{r} 94 \\ \times\ 3 \\ \hline \end{array}$

6. $\begin{array}{r} 82 \\ \times\ 3 \\ \hline \end{array}$

Name________________________________

Multiply 2-Digit Numbers

Find each product.

1. $\begin{array}{r} 23 \\ \times\ 4 \\ \hline \end{array}$

2. $\begin{array}{r} 78 \\ \times\ 6 \\ \hline \end{array}$

3. $\begin{array}{r} 77 \\ \times\ 6 \\ \hline \end{array}$

4. $\begin{array}{r} 15 \\ \times\ 9 \\ \hline \end{array}$

5. $\begin{array}{r} 34 \\ \times\ 7 \\ \hline \end{array}$

6. $\begin{array}{r} 39 \\ \times\ 7 \\ \hline \end{array}$

7. $\begin{array}{r} 92 \\ \times\ 3 \\ \hline \end{array}$

8. $\begin{array}{r} 41 \\ \times\ 7 \\ \hline \end{array}$

9. $\begin{array}{r} 84 \\ \times\ 2 \\ \hline \end{array}$

10. $\begin{array}{r} 67 \\ \times\ 3 \\ \hline \end{array}$

11. $95 \times 8 =$ _____
12. $57 \times 6 =$ _____
13. $4 \times 99 =$ _____
14. $6 \times 73 =$ _____

Problem Solving and Test Prep

USE DATA For 15–16, use the graph.

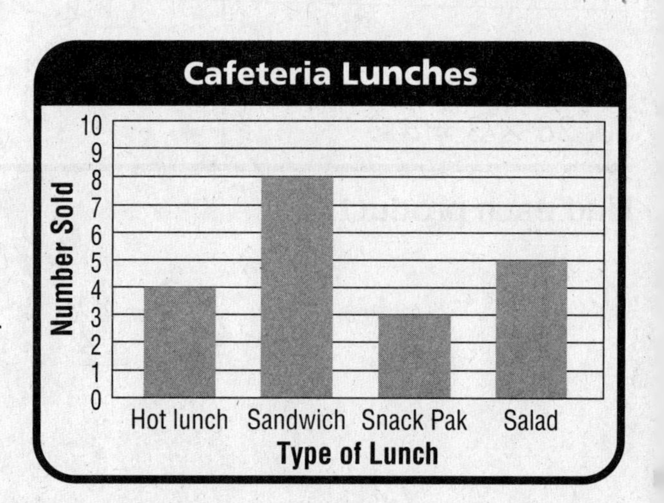

15. What number times 3, minus 16, equals the number of lunches sold in all?

____

16. What number times 10 equals the number of lunches sold in all?

____

17. Vincent uses 50 inches of wood to make a frame. How many inches of wood will Vincent need to make 9 frames?

A 475 inches
B 540 inches
C 450 inches
D 480 inches

18. Colleen listened to three CDs. Each CD is 63 minutes long. How many minutes did it take for Colleen to listen to all three CDs?

A 146 minutes
B 169 minutes
C 189 minutes
D 378 minutes

Name______________________________ Lesson 15.6

Problem Solving Workshop Strategy: Solve a Simpler Problem

The recycling center collected 67 bags of aluminum cans in June and 92 bags of aluminum cans in July. If each bag weighed about 3 pounds, how many pounds of aluminum cans did the center collect in the two months?

Read to Understand

1. Write the question as a fill-in-the blank sentence.

____

Plan

2. How can using a simpler problem help you solve the problem?

____

Solve

3. Solve the problem. Describe how you solved a simpler problem first.

____

____

4. Write your answer in a complete sentence.

____

____

Check

5. How can you check that your answer is correct?

____

____

Solve a simpler problem to solve.

6. There are 8 shelves. Each shelf holds 76 boxes of cereal. What is the maximum number of cereal boxes these shelves can hold in all?

____

7. Asha loads flowers onto 6 trucks. Each truck holds 85 bunches of flowers. How many bunches of flowers does Asha load if she fills up all 6 trucks?

____

Name________________________________ Lesson 15.6

Problem Solving Workshop Strategy: Solve a Simpler Problem

Problem Solving Strategy Practice.

1. The music club gives a concert to raise money for new sheet music. They receive $0.75 for each ticket sold. The club sold 99 tickets. How much money did they raise?

   ____

2. Brett uses toy houses to build his model village. The toy houses come in packages of 65. If Brett buys 4 packages, how many toy houses will he have?

   ____

Mixed Strategy Practice

USE DATA For 3–4, use the table.

3. Some third grade classes collected canned goods for a food drive. The table shows the foods one of the classes collected. If 4 third grade classes each collected the same number of cans of each type of food, what is the total number of cans of peas and corn these 4 classes collected?

Food	Number of cans
Peas	35
Corn	27
Chicken	15
Potatoes	22
Soup	13

   ____

4. If the 4 third grade classes each collected the same number of each type of food, how many more cans of potatoes did they collect than cans of chicken?

   ____

5. Heath volunteers at a library 3 days a week every summer. He reshelves 97 or more books each day he volunteers. What is the least number of books Heath reshelves each week?

   ____

6. A brown, a black, a white, and a gray dog are in line at a training class. The black dog is not last. The white dog in is front of the brown dog. The brown dog is second. Draw a picture to show the order of the dogs.

Name________________________________ Lesson 16.1

Algebra: Multiples of 10, 100, and 1,000

When you multiply by 10, 100, and 1,000, you use basic fact patterns to increase the number of place values you have by adding zeros.

Use a basic fact and patterns to find each product.

$2 \times 5 =$ ____ $20 \times 5 =$ ____ $200 \times 5 =$ ____

Multiply 1s.	Multiply 10s.	Multiply 100s.
$\begin{array}{r} 2 \\ \times 5 \\ \hline 10 \end{array}$	$\begin{array}{r} 20 \\ \times 5 \\ \hline 100 \end{array}$	$\begin{array}{r} 200 \\ \times 5 \\ \hline 1{,}000 \end{array}$
basic fact	**multiple of** 10	**multiple of** 100

Notice that each time a zero was added to the factor 2 that the poduct gained a zero as well.

So: $2 \times 5 = 10$, $20 \times 5 = 100$, $200 \times 5 = 1{,}000$.

Use a basic fact and patterns to find each product.

1. $8 \times 6 =$ ____
$8 \times 60 =$ ____
$8 \times 600 =$ ____
$8 \times 6{,}000 =$ ____

2. $5 \times 3 =$ ____
$5 \times 30 =$ ____
$5 \times 300 =$ ____
$5 \times 3{,}000 =$ ____

3. $8 \times 8 =$ ____
$8 \times 80 =$ ____
$8 \times 800 =$ ____
$8 \times 8{,}000 =$ ____

4. $7 \times 3 =$ ____
$7 \times 30 =$ ____
$7 \times 300 =$ ____
$7 \times 3{,}000 =$ ____

5. $4 \times 5 =$ ____
$4 \times 50 =$ ____
$4 \times 500 =$ ____
$4 \times 5{,}000 =$ ____

6. $2 \times 7 =$ ____
$2 \times 70 =$ ____
$2 \times 700 =$ ____
$2 \times 7{,}000 =$ ____

Name____ Lesson 16.1

Algebra: Multiples of 10, 100, and 1,000

Use a basic fact and patterns to find each product.

1. $5 \times 3 =$ _____

$50 \times 3 =$ _____

$500 \times 3 =$ _____

$5{,}000 \times 3 =$ _____

2. $2 \times 7 =$ _____

$2 \times 70 =$ _____

$2 \times 700 =$ _____

$2 \times 7{,}000 =$ _____

3. $9 \times 8 =$ _____

$9 \times 80 =$ _____

$9 \times 800 =$ _____

$9 \times 8{,}000 =$ _____

Find the product.

4. $500 \times 7 =$ _____ **5.** $4 \times 50 =$ _____ **6.** _____ $= 3 \times 2{,}000$

7. $400 \times 8 =$ _____ **8.** _____ $= 6 \times 9{,}000$ **9.** $4 \times 60 =$ _____

10. _____ $= 1 \times 600$ **11.** _____ $= 700 \times 3$ **12.** $2{,}000 \times 5 =$ _____

Find the missing factor.

13. $\square \times 1{,}600 = 6{,}400$ **14.** $5 \times \square = 25{,}000$ **15.** $\square \times 4{,}000 = 8{,}000$

Problem Solving and Test Prep

16. Victoria has 10 pencil cases. There are 25 pencils in each case. How many pencils does Victoria have in all?

____

17. Eddie has 4 shelves full of books. There are 29 books on each shelf. How many books does Eddie have in all?

____

18. Which has a product of 6,000?

A $12 \times 3{,}000$

B $3{,}000 \times 2$

C 6×100

D $6{,}000 \times 1{,}018$

19. Which has a product of 5,500?

A 10×500

B 10×50

C 550×10

D $5{,}550 \times 1$

Spiral Review

For 1–5, round each number to the nearest thousand.

1. 6,581 ________
2. 1,157 ________
3. 8,502 ________
4. 8,205 ________
5. 506 ________

For 10–12, use the pattern in the table to answer the questions.

number of stop signs	1	2	3	4		6
number of sides	8	16	24		40	

10. How many sides do 4 stop signs have?

____

11. How many stop signs have 40 sides in all?

____

12. How many sides do 6 stop signs have?

____

For 6–9, find the perimeter of each figure.

6. 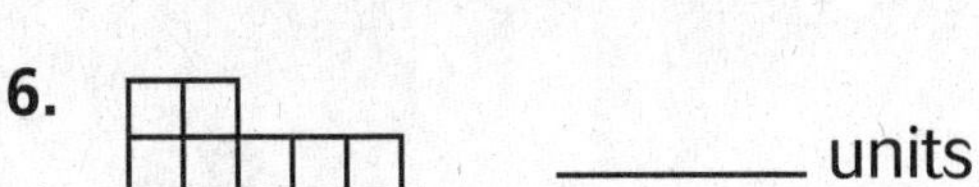 ________ units

7.  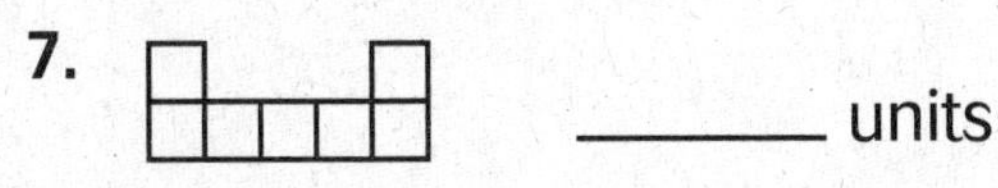 ________ units

8. ________ units

9. 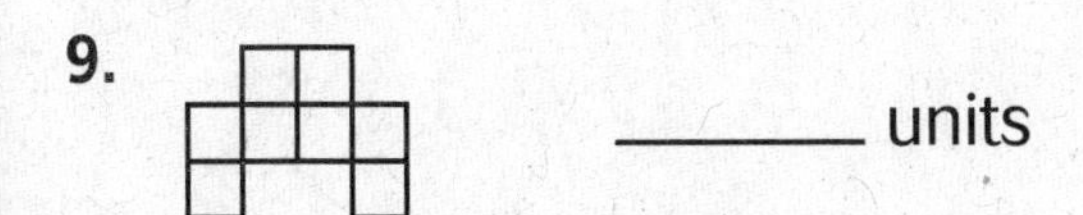 ________ units

For 13–15, complete the pattern.

13. 39, 41, 43, 45, ☐, 49, ☐

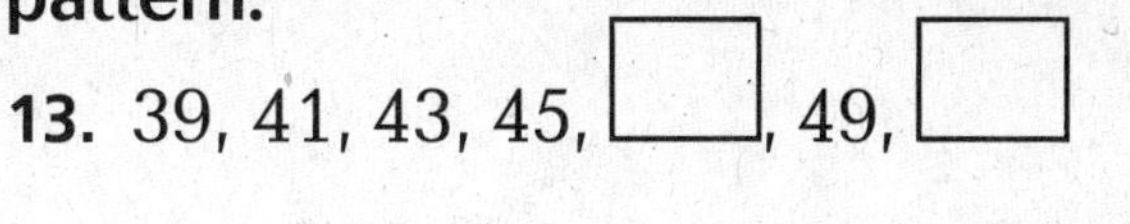

____

14. 74, 70, 66, ☐, 58, 54, ☐

____

15. 79, 76, 73, ☐, 67, 64, ☐

____

Name________________________________ **Lesson 16.2**

Problem Solving Workshop Strategy: Find a Pattern

The first 2,000 people who visit the flower show are given a bouquet of flowers. Each bouquet is made up of 3 flowers. How many flowers are given away altogether?

Read to Understand

1. Write the question as a fill-in-the-blank sentence.

____

____

Plan

2. How can finding a pattern help you solve the problem?

____

Solve

3. What basic fact can help you solve?

____

4. Write your answer as a complete sentence.

____

____

Check

5. How can you check your answer?

____

Find a pattern to solve.

6. The delivery trucks each hold 6,000 books. Only 1 truck can unload its books each hour. How many books are unloaded during 3 hours?

____

7. Asha the pirate has 5 boats. Each boat has 10 oars. Each oar requires 2 workers to row it. How many boats, oars, and workers does Asha the pirate have in all?

____

____

Name ____________________

Lesson 16.2

Problem Solving Strategy Practice: Find a Pattern

Find a pattern to solve.

1. There are 9 large flower displays at the Flower and Garden Festival. Each display has about 10,000 blooms. How many blooms are there in all?

____

2. The first 3,000 children who visit the Garden Center get a free gardening kit. Each kit has 4 seed packs. How many seed packs are given away altogether?

____

Mixed Strategy Practice

USE DATA For 3–5, use the pictures.

Garden Barn circular	
	Gloves 6 pairs for $20
PETUNIA SEEDS	Petunia seeds 100 for $9
	Rose bushes 10 for $100
	Tomato plants 50 for $100

3. Kyra's mom told her to buy 500 petunia seeds. How many packs of petunia seeds should Kyra buy? Should she estimate or buy the exact amount?

____

4. Suen Win bought 50 rose bushes and 50 tomato plants. How much money did she spend in all?

____

5. Nestor bought 6 pairs of gloves, 10 rose bushes, and 50 tomato plants. How much money did he spend in all?

____

Name____________________

Multiply 3-Digit Numbers

Find the product.

$$\begin{array}{r} 537 \\ \times\ 5 \\ \hline \end{array}$$

You can use regrouping to find the product.

STEP 1

Multiply the ones.

$$\begin{array}{r} {}^{3} \\ 537 \\ \times 5 \\ \hline 5 \end{array}$$

$7 \times 5 = 35$
Regroup 35 ones as 3 tens and 5 ones.

STEP 2

Multiply the tens. Don't forget to add the 3 regrouped tens

$$\begin{array}{r} {}^{1\,3} \\ 537 \\ \times 5 \\ \hline 85 \end{array}$$

$3 \times 5 = 15$, $15 + 3 = 18$
Regroup 18 tens as 8 tens and 1 hundred.

STEP 3

Multiply the hundreds. Don't foget to add the 1 regrouped hundred.

$$\begin{array}{r} {}^{1\,3} \\ 537 \\ \times 5 \\ \hline 2{,}685 \end{array}$$

$5 \times 5 = 25$, $25 + 1 = 26$

So, $537 \times 5 = 2{,}685$.

Find each product.

1. $\begin{array}{r} 132 \\ \times\ 3 \\ \hline \end{array}$
2. $\begin{array}{r} 318 \\ \times\ 4 \\ \hline \end{array}$
3. $\begin{array}{r} 142 \\ \times\ 2 \\ \hline \end{array}$
4. $\begin{array}{r} 333 \\ \times\ 3 \\ \hline \end{array}$
5. $\begin{array}{r} 411 \\ \times\ 3 \\ \hline \end{array}$
6. $\begin{array}{r} 931 \\ \times\ 5 \\ \hline \end{array}$
7. $\begin{array}{r} 642 \\ \times\ 7 \\ \hline \end{array}$
8. $\begin{array}{r} 497 \\ \times\ 6 \\ \hline \end{array}$
9. $\begin{array}{r} 563 \\ \times\ 8 \\ \hline \end{array}$
10. $\begin{array}{r} 217 \\ \times\ 9 \\ \hline \end{array}$
11. $\begin{array}{r} 298 \\ \times 3 \\ \hline \end{array}$
12. $\begin{array}{r} 422 \\ \times 4 \\ \hline \end{array}$
13. $\begin{array}{r} 607 \\ \times 8 \\ \hline \end{array}$
14. $\begin{array}{r} 101 \\ \times 9 \\ \hline \end{array}$
15. $\begin{array}{r} 759 \\ \times 2 \\ \hline \end{array}$

Name______________________________ Lesson 16.3

Multiply 3-Digit Numbers

Find each product.

1. 832 × 2

2. 196 × 4

3. 312 × 3

4. 375 × 4

5. 456 × 1

6. 432 × 3

7. 821 × 3

8. 139 × 3

9. 472 × 4

10. 424 × 2

11. $5 \times 304 =$ ______

12. $3 \times 153 =$ ______

13. $6 \times 413 =$ ______

Problem Solving and Test Prep

USE DATA For 14–15, use the table.

14. A calorie is a unit of energy food gives you. The chart shows the number of calories used in one hour for different sports. Asha played Ping Pong for 4 hours one weekend. How many calories did Asha burn?

____

Calories Burned	
Sport	**Calories Burned per Hour***
Soccer	206**
Softball	147**
Ping Pong	118
Tennis	206**
Water Polo	295

*for 65 pound student
**calories rounded to nearest ones

15. David plays soccer 4 times a week for 1 hour each time. How many calories does David burn in 6 weeks? ______________________

16. The store has 715 snack packs of ginger snaps. There are 8 ginger snaps in each snack pack. How many ginger snaps does the store have in all?

A 4,720
B 5,720
C 7,542
D 568

17. There are 15 stickers on a page in a sticker book. There are 501 pages in the book. How many stickers are in the book in all?

A 7,115 stickers
B 7,000 stickers
C 7,051 stickers
D 7,515 stickers

Name________________________________

Lesson 16.4

Multiply 4-Digit Numbers

Find the product.

$$\begin{array}{r} 2{,}173 \\ \times\ 4 \\ \hline \end{array}$$

You can use place value to find the product.

Step 1

Multiply the ones.

$$\begin{array}{r} 2{,}173 \\ \times 4 \\ \hline \end{array}$$

$4 \times 3 = 12$

$$\begin{array}{r} 2{,}173 \\ \times\ 4 \\ \hline 12 \end{array}$$

Step 2

Multiply the tens.

$$\begin{array}{r} 2{,}173 \\ \times 4 \\ \hline \end{array}$$

Since the 7 is in the tens place, its value is 70.

$70 \times 4 = 280$

$$\begin{array}{r} 2{,}173 \\ \times\ 4 \\ \hline 12 \\ 280 \end{array}$$

Step 3

Multiply the hundreds.

$$\begin{array}{r} 2{,}173 \\ \times 4 \\ \hline \end{array}$$

Since the 1 is in the hundreds place, its value is 100.

$100 \times 4 = 400$

$$\begin{array}{r} 2{,}173 \\ \times\ 4 \\ \hline 12 \\ 280 \\ 400 \end{array}$$

Step 4

Multiply the thousands.

$$\begin{array}{r} 2{,}173 \\ \times 4 \\ \hline \end{array}$$

Since the 2 is in the thousands place, its value is 2,000.

$2{,}000 \times 4 = 8{,}000$

$$\begin{array}{r} 2{,}713 \\ \times\ 4 \\ \hline 12 \\ 280 \\ 400 \\ 8{,}000 \end{array}$$

Step 5: Add the products. $12 + 280 + 400 + 8{,}000 = 8{,}692$.

So, $2{,}173 \times 4 = 8{,}692$.

Find each product.

1. $\begin{array}{r} 5{,}381 \\ \times\quad 2 \\ \hline \end{array}$

2. $\begin{array}{r} 1{,}358 \\ \times\quad 4 \\ \hline \end{array}$

3. $\begin{array}{r} 2{,}542 \\ \times\quad 6 \\ \hline \end{array}$

4. $\begin{array}{r} 5{,}173 \\ \times\quad 9 \\ \hline \end{array}$

5. $\begin{array}{r} 4{,}296 \\ \times\quad 2 \\ \hline \end{array}$

6. $\begin{array}{r} 3{,}753 \\ \times\quad 3 \\ \hline \end{array}$

Name ________________________

Multiply 4-Digit Numbers

Find each product.

1. $\begin{array}{r} 1,334 \\ \times \quad 2 \\ \hline \end{array}$

2. $\begin{array}{r} 3,968 \\ \times \quad 4 \\ \hline \end{array}$

3. $\begin{array}{r} 2,142 \\ \times \quad 6 \\ \hline \end{array}$

4. $\begin{array}{r} 5,372 \\ \times \quad 3 \\ \hline \end{array}$

5. $\begin{array}{r} 2,432 \\ \times \quad 9 \\ \hline \end{array}$

6. $\begin{array}{r} 4,215 \\ \times \quad 2 \\ \hline \end{array}$

7. $\begin{array}{r} 6,139 \\ \times \quad 5 \\ \hline \end{array}$

8. $\begin{array}{r} 3,472 \\ \times \quad 4 \\ \hline \end{array}$

9. $5 \times 3,042 =$ ________

10. $3 \times 7,153 =$ ________

11. $6 \times 3,413 =$ ________

Problem Solving and Test Prep

USE DATA For 12–13 use the table.

12. If every student in third grade was to donate 3 cans of food for a food drive, how many cans of food would be donated in all?

____

Grade	Number of Students
First	206
Second	147
Third	118
Fourth	236
Fifth	295

13. If each first, second, third, fourth and fifth grader saw 2 movies during the year, then how many movies would the students have seen in all?

____

14. David wrote a 5-page story. Each page has 2,170 letters. How many letters are in his whole story?

A 1,700

B 13,020

C 8,860

D 10,850

15. The Santiagos rode bikes from San Francisco to Pueblo, Colorado on the National Bike Route. Then they rode back. The distance one way is 1,579 miles. How many miles did the Santiagos ride in all?

A 3,048 miles

B 3,058 miles

C 3,148 miles

D 3,158 miles

Name________________________________ Lesson 16.5

Multiply Money Amounts

You can multiply money the same way you multiply whole numbers. But when you multiply money amounts you must remember to write a dollar sign and place the decimal point correctly in the product.

Find the product. $\begin{array}{r} \$12.32 \\ \times \quad 2 \\ \hline \end{array}$

Write the problem using whole numbers.

$$\begin{array}{r} 1{,}232 \\ \times \quad 2 \\ \hline \end{array}$$

Multiply the ones, tens, hundreds, and thousands.

$$\begin{array}{r} 1{,}232 \\ \times \quad 2 \\ \hline 2{,}464 \end{array}$$

Place the dollar sign and decimal point in the correct places. Just move them down.

$$\begin{array}{r} 1{,}232 \\ \times \quad 2 \\ \hline 2{,}464 \end{array} \qquad \begin{array}{r} \$12.32 \\ \times \quad 2 \\ \hline \$24.64 \end{array}$$

So, $12.32 × 2 = $24.64.

Find each product.

1. $\begin{array}{r} \$13.63 \\ \times \quad 4 \\ \hline \end{array}$
2. $\begin{array}{r} \$25.71 \\ \times \quad 5 \\ \hline \end{array}$
3. $\begin{array}{r} \$0.97 \\ \times \quad 3 \\ \hline \end{array}$
4. $\begin{array}{r} \$1.65 \\ \times \quad 6 \\ \hline \end{array}$
5. $\begin{array}{r} \$3.95 \\ \times \quad 5 \\ \hline \end{array}$
6. $\begin{array}{r} \$43.26 \\ \times \quad 4 \\ \hline \end{array}$
7. $\begin{array}{r} \$50.08 \\ \times \quad 2 \\ \hline \end{array}$
8. $\begin{array}{r} \$17.98 \\ \times \quad 8 \\ \hline \end{array}$
9. $\begin{array}{r} \$63.19 \\ \times \quad 3 \\ \hline \end{array}$
10. $\begin{array}{r} \$0.98 \\ \times \quad 6 \\ \hline \end{array}$

Name________________________________

Multiply Money Amounts

Find the product.

1. $3.82 × 2

2. $9.16 × 4

3. $2.21 × 3

4. $5.73 × 4

5. $7.46 × 1

6. $2.43 × 3

7. $1.28 × 3

8. $3.96 × 3

9. $7.42 × 4

10. $8.43 × 2

Problem Solving and Test Prep

USE DATA For 11–12, use the table.

11. Mike and his family visited Muir Woods. They bought 6 adult passes and 1 Golden Age pass. How much money did they spend in all?

____

12. Mr. and Mrs. Smith visit Muir Woods twice a month. Is it cheaper for them to buy 24 Adult passes or buy 2 Annual passes?

____

Muir Woods' Prices

Age	Pass Price
Adult (16+)	$3
Children	Free
Annual	$50
Golden Age (65+)	$10

13. Concert tickets cost $22.50. Rhonda needs to buy 1 for herself and 3 for her friends. How much money does Rhonda need?

A $67.50
B $90.00
C $88.50
D $90.50

14. Spring water sells for $1.95 a bottle. Which is the cost for 8 bottles?

A $12.60
B $14.30
C $14.60
D $15.60

Name ______________________________ **Week 24**

Spiral Review

For 1–5, write the fact family for each set of numbers.

1. 8, 5, 40

________ ________

________ ________

2. 6, 8, 48

________ ________

________ ________

3. 7, 3, 21

________ ________

________ ________

4. 9, 4, 36

________ ________

________ ________

5. 9, 9, 81

________ ________

For 10–11, a class takes a survey about their heroes. Write the results as tally marks.

10. 15 students chose Martin Luther King Jr.

____

11. 9 students chose Eleanor Roosevelt.

____

12. Look at the table at the right. How many more students chose Mother Teresa than chose Thomas Edison?

Who is Your Hero?	
Person	**Students Choice**
Thomas Edison	𝍸 ||
Abraham Lincoln	||||
Mother Teresa	𝍸 |||

____

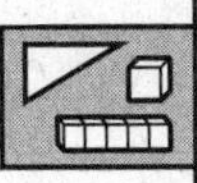

For 6–9, draw a line to match each figure to its name.

6. 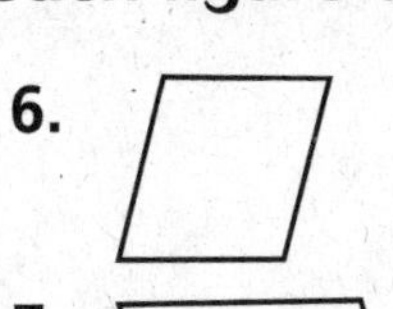

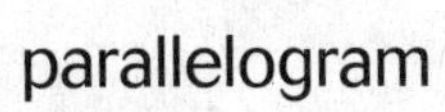

parallelogram

7. 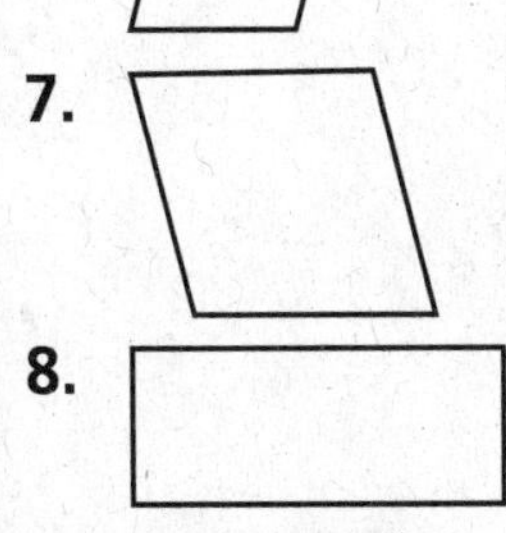

trapezoid

8.

rhombus

9.

rectangle

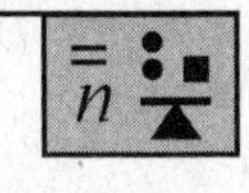

For 13–15, write the related multiplication fact.

13. $10 \times 7 = 70$

____

14. $8 \times 9 = 72$

____

15. $3 \times 6 = 18$

____

Name______________________________

Model Division

You can divide 2-digit numbers without remainders using base-ten blocks. Use the model to find the quotient. 36 ÷ 3

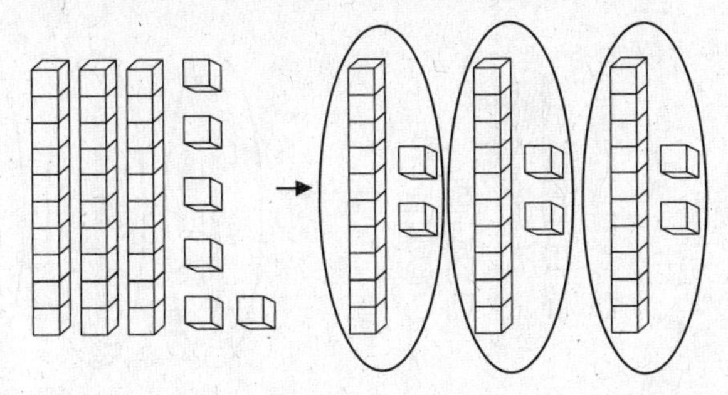

36 blocks   3 groups of 12 blocks each

You can solve 36 ÷ 3 using long division as shown below.

$$\begin{array}{r} 12 \\ 3\overline{)36} \\ -3 \\ \hline 6 \\ -6 \\ \hline 0 \end{array}$$

So, 36 ÷ 3 = 12.

Use the model to find the quotient.

1.

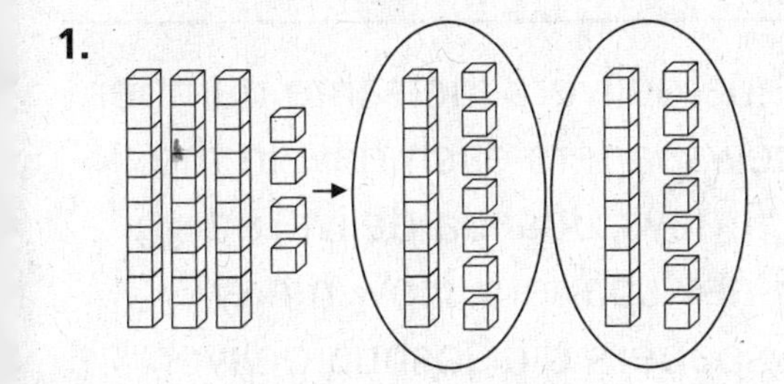

34 ÷ 2

____

2.

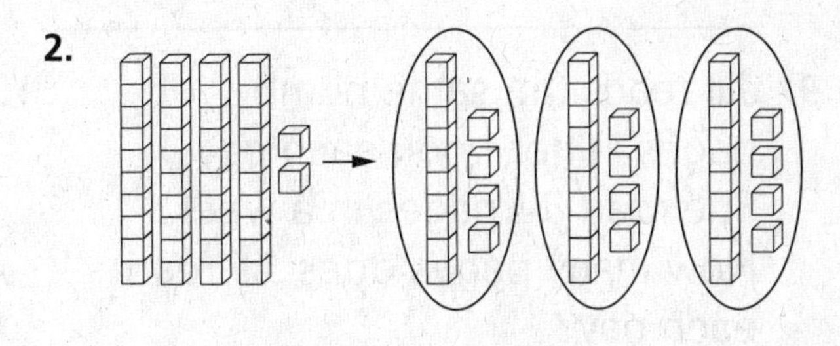

42 ÷ 3

____

NS 2.5 Solve division problems in which a multi-digit number is evenly divided by a one-digit number (135 ÷ 5 = ___)

Name ________________________________

Model Division

Use the model to find the quotient.

1.

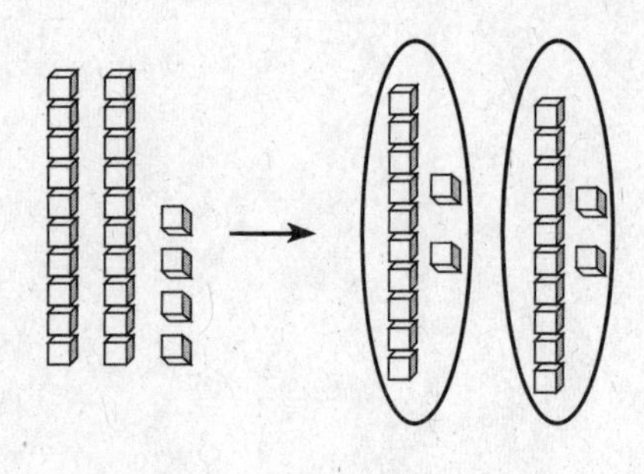

$24 \div 2 =$ ____

2.

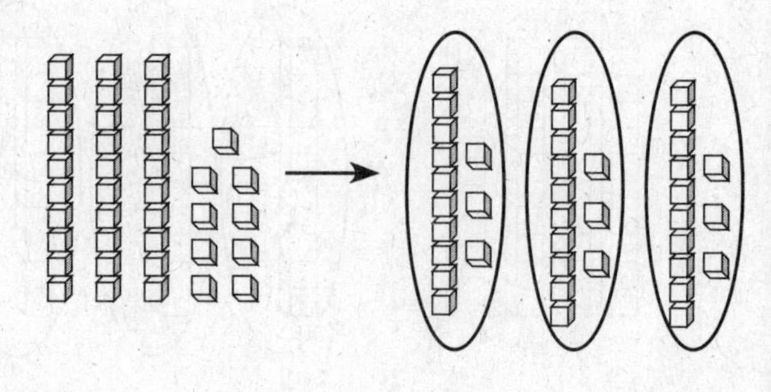

$39 \div 3 =$ ____

Divide. Use base-ten blocks to help.

3. $62 \div 2 =$ ____ **4.** $78 \div 3 =$ ____ **5.** $65 \div 5 =$ ____ **6.** $56 \div 4 =$ ____

Problem Solving and Test Prep

7. Mike uses treats to train his dog. He gives his dog the same number of treats each week for 3 weeks. If Mike gives the dog 48 treats in all, how many treats does he give his dog each week?

____

8. Isabel sends her grandmother the same number of e-mails each week. If she sends 28 e-mails in 4 weeks, how many e-mails does Isabel send her grandmother each week?

____

9. Jill reads the same number of pages in her book each day. She read 84 pages in a week. How many pages does Jill read each day?

A 7

B 12

C 14

D 21

10. Joshua delivers the same number of newspapers each day. In the last 5 days, Joshua delivered 165 newspapers. How many newspapers did Joshua deliver each day?

A 16

B 23

C 32

D 33

Name________________________________ Lesson 17.2

Algebra: Division Patterns

You can use basic division facts and patterns to divide multiples of 10, 100, and 1,000.

Use a basic fact and patterns to find each quotient below.

$30 \div 6 = \blacksquare$ Think of basic facts: $5 \times 6 = 30$.

$300 \div 6 = \blacksquare$ 1 zero is added so add 1 zero to the quotient.

$3{,}000 \div 6 = \blacksquare$ 2 zeros are added so add 2 zeros to the quotient.

So, $30 \div 6 = 5$.

So, $300 \div 6 = 50$.

So, $3{,}000 \div 6 = 500$.

Use a basic fact and patterns to find each quotient.

1. $18 \div 2 =$ ___

$180 \div 2 =$ ___

$1{,}800 \div 2 =$ ___

2. $21 \div 3 =$ ___

$210 \div 3 =$ ___

$2{,}100 \div 3 =$ ___

3. $80 \div 8 =$ ___

$800 \div 8 =$ ___

$8{,}000 \div 8 =$ ___

4. $42 \div 6 =$ ___

$420 \div 6 =$ ___

$4{,}200 \div 6 =$ ___

5. $32 \div 4 =$ ___

$320 \div 4 =$ ___

$3{,}200 \div 4 =$ ___

6. $27 \div 3 =$ ___

$270 \div 3 =$ ___

$2{,}700 \div 3 =$ ___

7. $35 \div 5 =$ ___

$350 \div 5 =$ ___

$3{,}500 \div 5 =$ ___

8. $45 \div 9 =$ ___

$450 \div 9 =$ ___

$4{,}500 \div 9 =$ ___

9. $28 \div 7 =$ ___

$280 \div 7 =$ ___

$2{,}800 \div 7 =$ ___

Name________________________________

Algebra: Division Patterns

Use a basic fact and patterns to find each quotient.

1. $16 \div 2 =$ ______ **2.** $15 \div 3 =$ ______ **3.** $40 \div 5 =$ ______

$160 \div 2 =$ ______ $150 \div 3 =$ ______ $400 \div 5 =$ ______

$1,600 \div 2 =$ ______ $1,500 \div 3 =$ ______ $4,000 \div 5 =$ ______

4. $27 \div 3 =$ ______ **5.** $42 \div 6 =$ ______ **6.** $48 \div 8 =$ ______

$270 \div 3 =$ ______ $420 \div 6 =$ ______ $480 \div 8 =$ ______

$2,700 \div 3 =$ ______ $4,200 \div 6 =$ ______ $4,800 \div 8 =$ ______

Use a basic fact and patterns to find each quotient.

7. $180 \div 3 =$ ______ **8.** ______ $= 400 \div 2$ **9.** $350 \div 7 =$ ______

10. $210 \div 7 =$ ______ **11.** ______ $= 2,500 \div 5$ **12.** $320 \div 4 =$ ______

Problem Solving and Test Prep

13. David took 400 photos while on vacation. He put the same number of photos in each of 5 albums. How many photos did David put in each album?

____

14. Laura counted 120 raisins in one box. She divided these raisins evenly into each of 3 bags. How many raisins did Laura put in each bag?

____

15. Which is the quotient of $240 \div 4$?

A 4

B 6

C 40

D 60

16. Which basic division fact can be used to find $6,000 \div 3$?

A $6 \div 2 = 3$

B $6 \div 3 = 2$

C $60 \div 2 = 30$

D $60 \div 3 = 20$

Name____

Lesson 17.3

Estimate Quotients

Compatible numbers are numbers that are easy to divide.
When a problem does not need an exact answer, you can estimate by using compatible numbers.

Estimate. Tell the compatible numbers you used.

472 ÷ 7

First, decide what number is close to 47 (472) and divisible by 7?
49 is close to 47 and divisible by 7.
So, 472 is changed to 490.

Then, use basic division facts: 49 ÷ 7 = 7.

49 ÷ 7 = 7 Use basic division facts.
Use the pattern of 1 zero in the dividend and quotient 490 ÷ 7 = 70.
So, 472 ÷ 7 can be estimated by the quotient of 490 ÷ 7, or 70.

Estimate. Tell the compatible numbers you used for each.

1. 30 ÷ 4 **2.** 34 ÷ 8 **3.** 146 ÷ 3 **4.** 454 ÷ 5 **5.** 694 ÷ 7

_____ ÷ _____ _____ ÷ _____ _____ ÷ _____ _____ ÷ _____ _____ ÷ _____

6. 117 ÷ 3 **7.** 80 ÷ 9 **8.** 223 ÷ 7 **9.** 492 ÷ 10 **10.** 169 ÷ 4

_____ ÷ _____ _____ ÷ _____ _____ ÷ _____ _____ ÷ _____ _____ ÷ _____

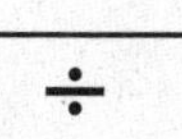

NS 2.5 Solve division problems in which a multi-digit number is evenly divided by a one-digit number (135 ÷ 5 = ___)

Name ______________________

Estimate Quotients

Estimate. Tell the compatible numbers you used for each.

1. $58 \div 6 =$ ____ ______________
2. $35 \div 3 =$ ____ ______________
3. $122 \div 4 =$ ____ ______________
4. $151 \div 3 =$ ____ ______________
5. $304 \div 3 =$ ____ ______________
6. $73 \div 6 =$ ____ ______________
7. $212 \div 3 =$ ____ ______________
8. $185 \div 9 =$ ____ ______________
9. $5\overline{)34}$ ______________
10. $4\overline{)42}$ ______________
11. $8\overline{)157}$ ______________
12. $9\overline{)444}$ ______________

Problem Solving and Test Prep

13. A group of 50 students have signed up for a ski trip. The students will be divided into 4 groups. About how many students will be in each group?

____

14. Elisabeth had a 122-ounce container of milk. She wants to divide the milk equally into 6 containers. About how many ounces of milk will be in each container?

____

15. Kenny needs to estimate $44 \div 9$. Which expression shows the best choice of compatible numbers for Kenny to use?
   - **A** $40 \div 9$
   - **B** $40 \div 10$
   - **C** $45 \div 9$
   - **D** $45 \div 10$

16. Nell needs to estimate $116 \div 6$. Which expression shows the best choice of compatible numbers for Nell to use?
   - **A** $112 \div 6$
   - **B** $115 \div 5$
   - **C** $120 \div 5$
   - **D** $120 \div 6$

Name______________________________ **Lesson 17.4**

Divide 2- and 3- Digit Numbers

You can divide 2- and 3-digit numbers without remainders.

Divide. Use multiplication to check your answer.

678 ÷ 3

Look at he first digit, 6.

Divide, $6 \div 3 = 2$.

$$\begin{array}{r} 2 \\ 3\overline{)678} \\ -6 \\ \hline 0 \end{array}$$

Bring down the second digit, 7.

Divide, $7 \div 3 = 2$ with 1 left over.

$$\begin{array}{r} 22 \\ 3\overline{)678} \\ -6\downarrow \\ \hline 07 \\ -6 \\ \hline 1 \end{array}$$

Bring down the third digit, 8.

Because 1 was left over from the tens place, divide 18 by 3.

$18 \div 3 = 6$

$$\begin{array}{r} 226 \\ 3\overline{)678} \\ -6 \\ \hline 07 \\ -6\downarrow \\ \hline 18 \\ -18 \\ \hline 0 \end{array}$$

To check your answer, multiply the quotient by the divisor.

$$\begin{array}{r} 226 \\ \times\ 3 \\ \hline 678 \end{array}$$

678 ⟵ The product equals the dividend, so the quotient is correct.

So, $678 \div 3 = 226$.

Divide. Use multiplication to check your answer.

1. $6\overline{)324}$ ____________
2. $4\overline{)128}$ ____________
3. $5\overline{)360}$ ____________
4. $2\overline{)824}$ ____________
5. $224 \div 4$ ____________
6. $843 \div 3$ ____________
7. $970 \div 5$ ____________
8. $432 \div 6$ ____________

NS 2.5 Solve division problems in which a multi-digit number is evenly divided by a one-digit number (135 ÷ 5 = ___)

Name________________________________ Lesson 17.4

Divide 2- and 3-Digit Numbers

Divide. Use multiplication to check your answer.

1. $2\overline{)456}$ **2.** $3\overline{)213}$ **3.** $5\overline{)675}$ **4.** $8\overline{)848}$ **5.** $6\overline{)120}$

6. $108 \div 9 =$ ____ **7.** $248 \div 2 =$ ____

8. $624 \div 6 =$ ____ **9.** $580 \div 4 =$ ____

Problem Solving and Test Prep

10. Joe's elementary school is going to an art show. There are 8 buses for the 240 students going to the show. How many students will be on each bus if the students are placed on the bus in equal numbers?

____

11. There are 8 parents on a field trip with 88 students. Each parent will guide an equal number of students. How many students will be with each parent?

____

12. Which shows the quotient of $207 \div 9$?

A 23
B 24
C 26
D 34

13. Which shows the quotient of $98 \div 7$?

A 9
B 12
C 14
D 24

Spiral Review

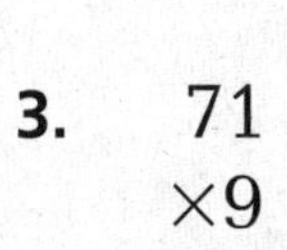

For 1–5, find the product.

1. $\begin{array}{r} 26 \\ \times 8 \\ \hline \end{array}$

2. $\begin{array}{r} 54 \\ \times 6 \\ \hline \end{array}$

3. $\begin{array}{r} 71 \\ \times 9 \\ \hline \end{array}$

4. $\begin{array}{r} 603 \\ \times 3 \\ \hline \end{array}$

5. $\begin{array}{r} 481 \\ \times 2 \\ \hline \end{array}$

For 10–12, answer the questions using the information in the line plot.

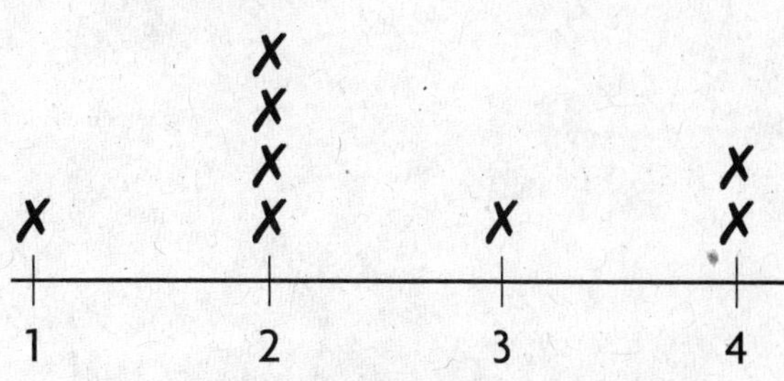

How Many Books Did You Read this Month?

10. How many students read 4 books? ____________

11. What is the range of the data?

____

12. What is the mode of the data?

____

For 6–9, draw lines to match each description with the correct quadrilateral.

6. four sides the same length, four right angles — rhombus

7. two sides parallel, two sides not parallel — square

8. two pairs of sides the same length, four right angles — rectangle

9. four sides the same length, angles not the same measure — trapezoid

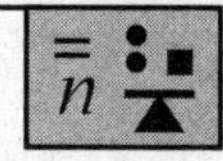

For 13–15, fill in the blank.

13. 2 feet= ______ inches

14. 6 yards= ______ feet

15. 36 inches= ______ feet

Name____ Lesson 17.5

Problem Solving Workshop Strategy: Solve a Simpler Problem

Millie and her family are having family portraits taken. The studio cost is $37. This cost includes a $25 sitting fee and a $2 per person charge. How many people are in Millie's family?

Read to Understand

1. Write the question as a fill-in-the-blank sentence.

____

Plan

2. What are the important details in the problem? _______________

____

Solve

3. Subtract the sitting fee from the studio cost.

____

4. Divide the difference from question 3 by the per person charge.

____

5. Write your answer to the problem in a complete sentence.

____

Check

6. How can you check your answer using multiplication? _______________

____

Use the "Solve a Simpler Problem" strategy to help solve.

7. Hanna takes a singing class. Her class meets 3 times a week for 3 weeks. The overall cost of the class is $110. This cost includes a $20 audition fee. How much does Hanna pay for each class?

____

8. Peter, a child, and his family are going to a musical. The admission charge for the family is $38. The cost is $6 for a child ticket and $10 for an adult ticket. If 2 adults went with Peter, how many children are in his family?

____

NS 2.5 Solve division problems in which a multidigit number is evenly divided by a one digit number (135 ÷ 5 = ___).

Name________________________________ Lesson 17.5

Problem Solving Workshop Strategy: Solve a Simpler Problem

Problem Solving Strategy Practice

1. Jason has rowing lessons 2 times a week. The boat marina charges him $180 every 4 weeks. This fee includes a $20 fee for use of the boat. How much does each lesson cost? ____________

2. Trudy takes a tumbling class. Her class meets 3 times a week for 3 weeks. The cost of the class is $310. The cost includes a $40 insurance charge. How much does Trudy pay for each class?

____

3. David went bowling and spent $20. David paid $3 to rent bowling shoes and another $7 on food. How many games did David bowl if each game costs $2? ____________

4. Peter's family paid $38 to go to a movie. Each child's ticket costs $6. Each adult's ticket costs $10. If Peter's mother and father went to the movie, how many children went with them including Peter?

____

Mixed Strategy Practice

5. Mia is taking piano lessons. She goes to lessons 2 times a week. The cost for 4 weeks of lessons is $186. This cost includes a $10 fee for music tapes. How much money does Mia pay for each lesson?

____

6. The school library has 172 books on the third grade reading list. Exactly 104 of these books have been checked out. How many books on the reading list are still in the library? Show your work.

____

USE DATA For 7–8, use the table.

7. Which tree received the most votes?

____

8. How many votes did birch receive?

____

Favorite Type of Tree

Type	Number of Votes
Apple	10
Birch	3
Maple	4
Oak	6

Name____ **Lesson 17.6**

Divide Money Amounts

You can divide money amounts much the same way you divide whole numbers.

Divide. $1.28 ÷ 4

The first digit, 1, cannot be divided evenly by 4.
The first two digits make 12, which can be divided evenly by 4.

$$\begin{array}{r} 3 \\ 4\overline{)\$1.28} \\ -12 \\ \hline 0 \end{array}$$

Look at the last digit, 8. Can 8 be divided by 4?
Yes, 8 ÷ 4 = 2.

$$\begin{array}{r} 32 \\ 4\overline{)\$1.28} \\ -12\downarrow \\ \hline 08 \\ -8 \\ \hline 0 \end{array}$$

Don't forget to include the dollar sign and the decimal point in your quotient.

So, $1.28 ÷ 4 = $0.32.

Divide.

1. $2\overline{)\$9.32}$ **2.** $3\overline{)\$5.46}$ **3.** $5\overline{)\$2.95}$ **4.** $6\overline{)\$7.56}$

_______ _______ _______ _______

5. $5.52 ÷ 4 **6.** $8.01 ÷ 3 **7.** $7.95 ÷ 5 **8.** $0.84 ÷ 6

_______ _______ _______ _______

NS 3.3 Solve problems involving addition, subtraction, multiplication, and division of money amounts in decimal notation and multiply and divide money amounts in decimal notation using whole-number multiplers and divisors.

Name________________________________ Lesson 17.6

Divide Money Amounts

Divide.

1. $3\overline{)\$9.72}$ **2.** $5\overline{)\$3.45}$ **3.** $6\overline{)\$6.54}$ **4.** $2\overline{)\$3.22}$ **5.** $4\overline{)\$9.04}$

6. $\$0.91 \div 7 =$ ______________ **7.** $\$5.34 \div 2 =$ ______________

8. $\$2.28 \div 6 =$ ______________ **9.** $\$6.05 \div 5 =$ ______________

Problem Solving and Test Prep

10. Al, Luisa, and Jane held a car wash. They earned $9.30. If they share the money equally, how much money should each person receive?

____

11. Kyle pulled weeds for 3 hours. He was paid $9.75 in all. How much money did Kyle earn per hour?

____

12. Jordan is paid $7.50 to walk 6 dogs. Jordan receives an equal amount of money for each dog. How much money does Jordan receive for walking each dog?

A $1.20
B $1.25
C $1.35
D $1.50

13. Rowan was paid $8.40 for picking 4 bushels of apples. He receives an equal amount of money for each bushel picked. How much money was Rowan paid for each bushel he picked?

A $0.21
B $1.20
C $2.10
D $21.00

Name________________________________

Model Division with Remainders

When you divide two numbers evenly, the result is called the quotient. If the numbers do not divide evenly, the amount left over is called the **remainder**.

You can use counters to model division with remainders.

Use counters to find the quotient and remainder.
15 ÷ 2.

Use 15 counters.
Make 2 circles.

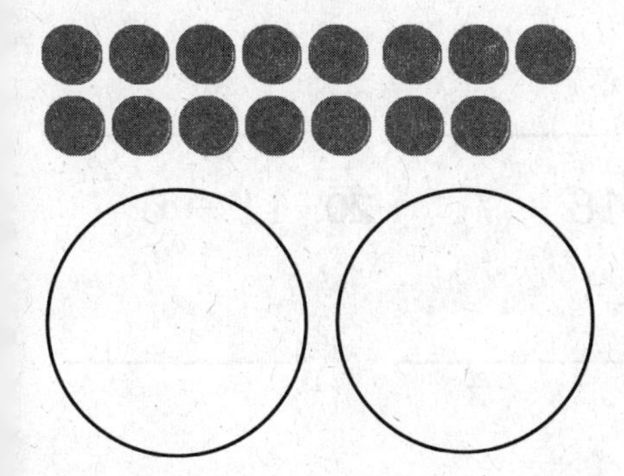

Divide the counters evenly into the 2 circles. 7 counters fit evenly into each circle.
The 1 counter left over is the remainder.

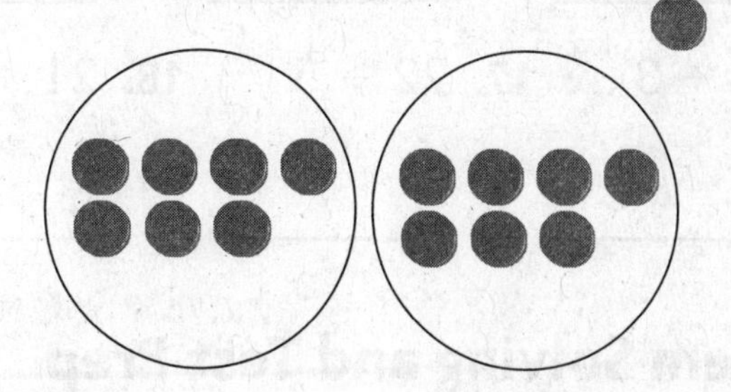

So, 15 ÷ 2 = 7 r1.

Use counters to find the quotient and remainder.

1. 13 ÷ 2 ________

2. 16 ÷ 3 ________

3. 14 ÷ 5 ________

4. 23 ÷ 4 ________

5. 25 ÷ 4 ________

6. 28 ÷ 9 ________

7. 17 ÷ 5 ________

8. 21 ÷ 6 ________

NS 2.0 Students calculate and solve problems involving addition, subtraction, multiplication, and division

Name____

Model Division with Remainders

Use counters to find the quotient and remainder.

1. $13 \div 2$ **2.** $15 \div 4$ **3.** $20 \div 3$ **4.** $17 \div 4$ **5.** $25 \div 3$

______ ______ ______ ______ ______

6. $17 \div 6$ **7.** $11 \div 2$ **8.** $14 \div 3$ **9.** $19 \div 6$ **10.** $16 \div 7$

______ ______ ______ ______ ______

11. $16 \div 5$ **12.** $15 \div 6$ **13.** $10 \div 4$ **14.** $11 \div 5$ **15.** $19 \div 4$

______ ______ ______ ______ ______

16. $19 \div 3$ **17.** $22 \div 5$ **18.** $21 \div 4$ **19.** $18 \div 7$ **20.** $11 \div 6$

______ ______ ______ ______ ______

Problem Solving and Test Prep

21. Karen collected 23 shark teeth on her vacation. She wants to make 3 necklaces with the same number of shark teeth on each necklace. Can Karen divide the 23 teeth equally into 3 groups?

____

22. Ronald bought 22 stickers. He wants to share the same number of stickers with 4 friends. How many stickers will he have left over for himself?

____

23. Bill's coin collection has 17 Canadian pennies. Bill wants to divide the Canadian pennies evenly into 3 boxes. How many Canadian pennies will be left over?

A 1

B 2

C 3

D 4

24. Lindsay made 12 muffins. She wants to share the muffins evenly with each of the 4 people in her family. How many muffins will be left over?

A 1

B 2

C 3

D 0

Name____

Model Part of a Whole

A fraction is a number that names part of a whole or part of a group.

Write a fraction in numbers and in words to name the shaded part.

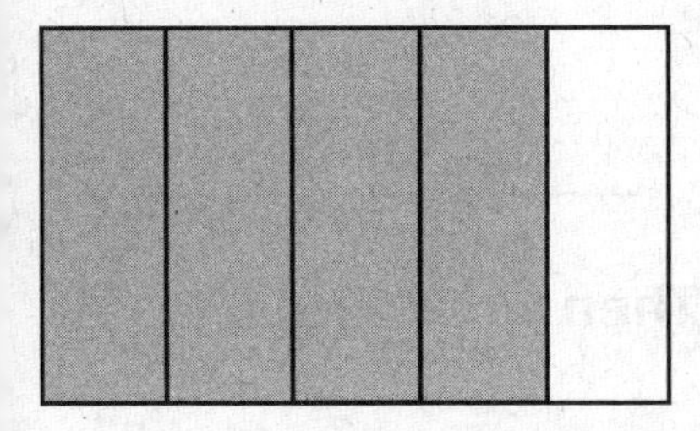

The rectangle is divided into 5 equal sections. 4 out of the 5 sections are shaded.

The numerator tells how many parts are being counted. There are 4 parts shaded, so 4 will be the numerator.

The denominator tells how many equal parts are in the whole. There are 5 equal parts in the rectangle, so 5 will be the denominator.

Write: $\frac{4}{5}$

Read: four fifths

So, the shaded part is $\frac{4}{5}$, or four fifths.

Write a fraction in numbers and in words to name the shaded part.

1. 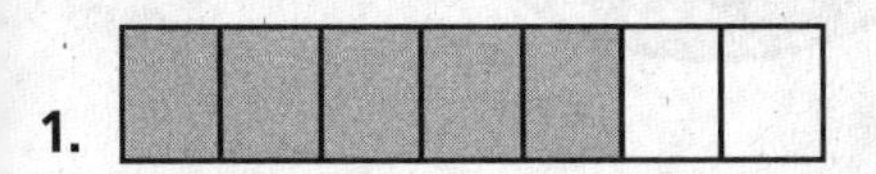

2. 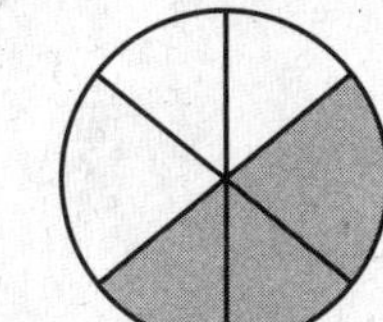

3. 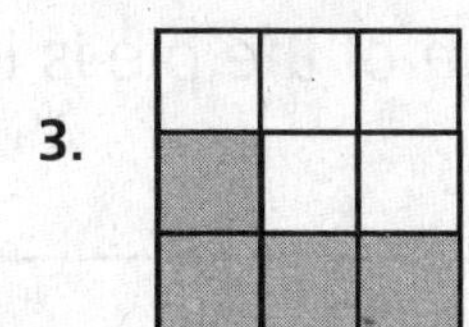

______________ ______________ ______________

Use fraction piece to make a model of each. Then write the fraction by using numbers.

4. seven out of nine

5. three out of six

6. five eighths

______________ ______________ ______________

Name____________________

Lesson 18.1

Model Part of a Whole

Write a fraction in numbers and in words that names the shaded part.

1. 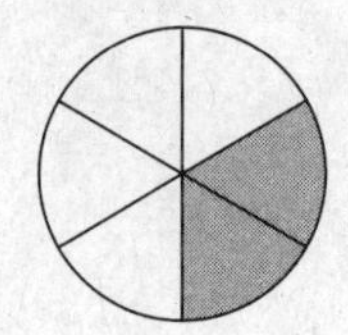

2. 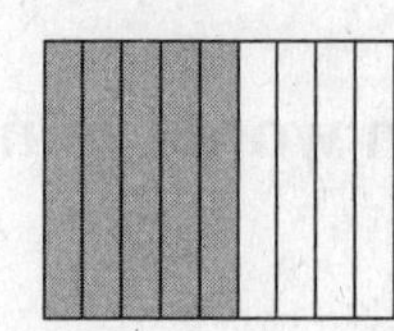

3. 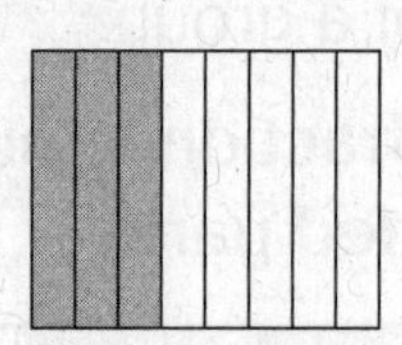

____________ ____________ ____________

Use fraction circle pieces to make a model of each. Then write the fraction by using numbers.

4. two fifths

5. seven tenths

6. five out of eight

____________ ____________ ____________

Problem Solving and Test Prep

7. Sam cut the apple pie into 6 equal slices. He ate one slice of pie. How much of the pie is left?

____

8. Sam gave Jenny 2 slices of a six slice pie with slices of equal size. How much of the pie is left?

____

9. Which fraction of the shape is not shaded?

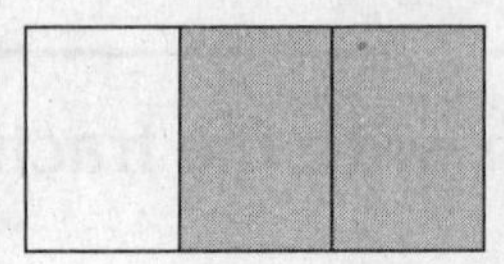

A $\frac{1}{2}$

B $\frac{1}{3}$

C $\frac{2}{3}$

D $\frac{1}{4}$

10. Which fraction of the shape is shaded?

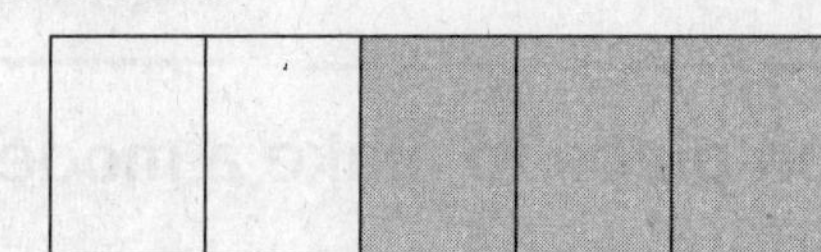

A $\frac{2}{5}$

B $\frac{3}{3}$

C $\frac{1}{3}$

D $\frac{3}{5}$

Spiral Review

For 1–5, find each quotient.

1. $63 \div 7 =$ ____

2. $42 \div 6 =$ ____

3. $56 \div 7 =$ ____

4. $48 \div 8 =$ ____

5. $35 \div 7 =$ ____

For 6–9, use the corner of a piece of paper to decide whether each angle is a *right angle, greater than* a right angle, or *less than* a right angle.

6. ______________

7. ______________

8. ______________

9. ______________

For 10–12, use the line plot to answer the questions.

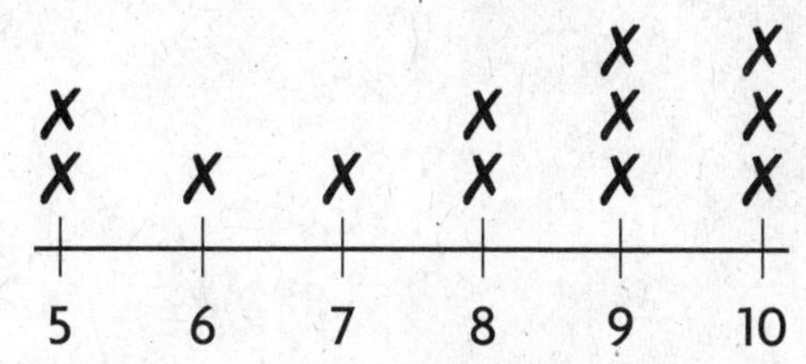

Number of Questions Answered Incorrectly

10. What was the least amount of incorrect answers? ____

11. How many people answered 7 questions incorrectly? ____

12. What is the greatest amount of incorrect answers? ____

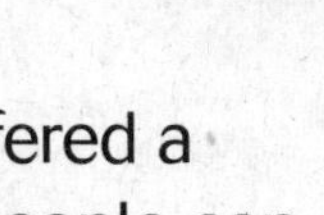

For 13–15, use the rule to answer the questions.

Rule: The movie theater offered a special on Thursday. Two people can see a movie for $7.00.

13. How much would it cost for 4 people to see the movie?

____

14. How much would it cost for 10 people to see the movie?

____

15. How much would it cost for 6 people to see the movie?

____

Name ________________________________ **Lesson 18.2**

Model Part of a Group

Write a fraction that names the gray part of the group below.

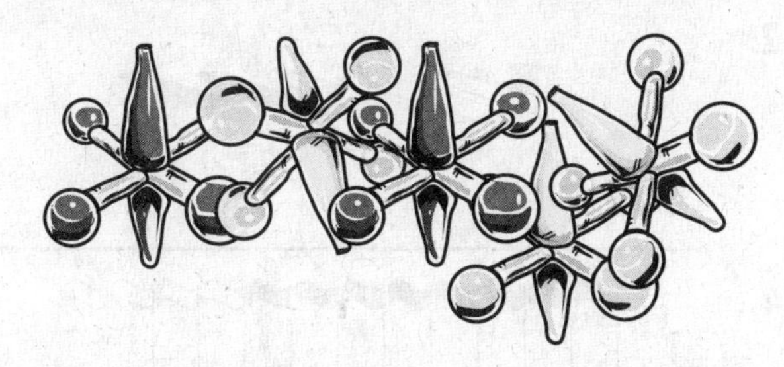

There are 5 jacks in this gorup.
Of these 5 jacks, 2 are shaded gray.

Write a fraction to show 2 out
Of these 5 jacks, 2 are shaded gray.

gray jacks → 2 ← numerator
total jacks → 5 ← denominator

So, $\frac{2}{5}$ of the jacks are shaded gray.

Write a fraction that names the part of each group that is shaded.

1.

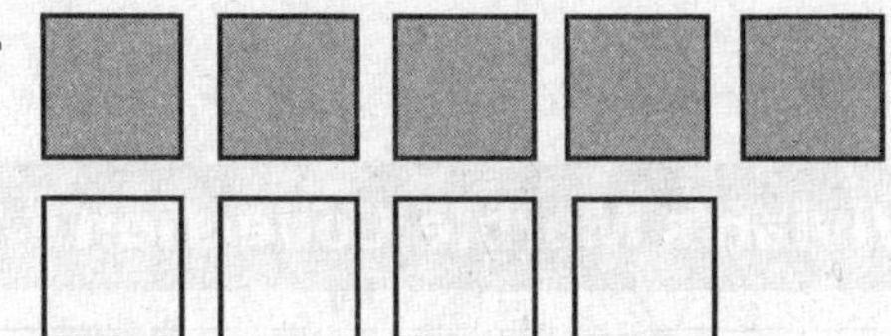

____

2.

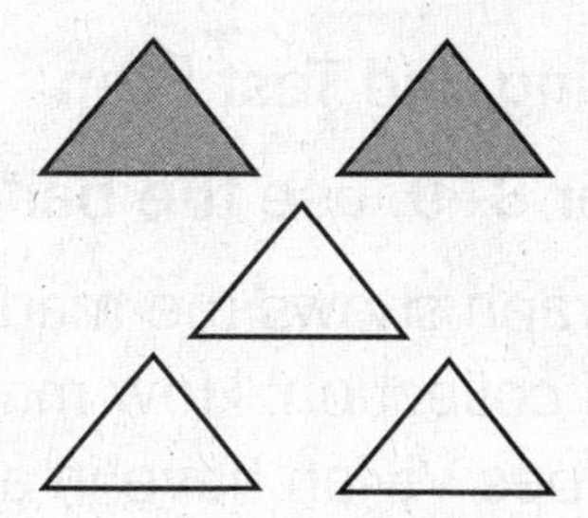

____

Draw each. Then write a fraction that names the shaded part.

3. Draw 12 circles.
Shade 4 circles.

____

4. Draw 6 squares.
Shade 1 square.

____

5. Draw 9 triangles.
Shade 8 triangles.

____

Name________________________________ Lesson 18.2

Model Part of a Group

Write a fraction that names the black part of each group.

1. 

____

2. 

____

3. 

____

4. 

____

Draw each. Then write the fraction that names the shaded part.

5. Draw 5 squares. Shade 2 squares.

____

6. Draw 8 circles. Shade 5 circles.

____

7. Draw 4 diamonds. Shade 3 diamonds.

____

Problem Solving and Test Prep

USE DATA For 8–9, use the bar graph.

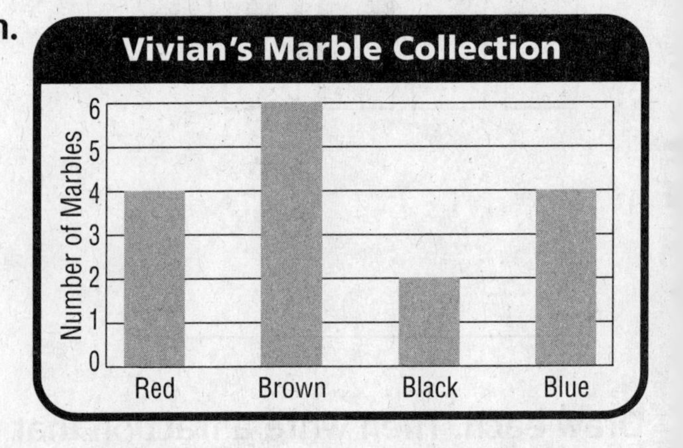

8. The bar graph shows the marbles in Vivian's collection. How many marbles does Vivian have in all?

____

9. What fraction of the marbles are brown?

____

10. Which fraction of the animals are rabbits?

A $\frac{1}{2}$

B $\frac{1}{8}$

C $\frac{4}{8}$

D $\frac{2}{8}$

11. Jack has 10 toy trucks. Of these, $\frac{1}{5}$ are red. How many trucks are red?

A 2

B 5

C 1

D 10

Name____

Equivalent Fractions

Two or more fractions that name the same amount are called **equivalent fractions.**

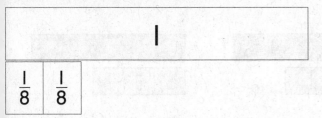

Use $\frac{1}{4}$ fraction bars to show the same about as the two $\frac{1}{8}$ bars.

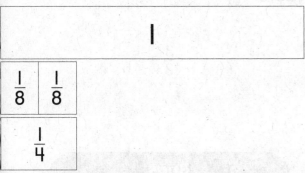

One $\frac{1}{4}$ fraction bar is the same amount as two $\frac{1}{8}$ fraction bars.

So, $\frac{1}{4}$ is equivalent to $\frac{2}{8}$.

Find an equivalent fraction. Use fraction bars.

1.

1		
$\frac{1}{6}$	$\frac{1}{6}$	$\frac{1}{6}$

____

2.

1			
$\frac{1}{12}$	$\frac{1}{12}$	$\frac{1}{12}$	$\frac{1}{12}$

____

3.

1			
$\frac{1}{5}$	$\frac{1}{5}$	$\frac{1}{5}$	$\frac{1}{5}$

____

MG 3.1 Compare fractions represented by drawings or concrete materials to show equivalency and to add and subtract simple fractions in context (e.g. $\frac{1}{2}$ of a pizza is the same amount as $\frac{2}{4}$ of another pizza that is the same size; that $\frac{3}{8}$ is larger than $\frac{1}{4}$).

Equivalent Fractions

Find an equivalent fraction. Use fraction bars.

1.

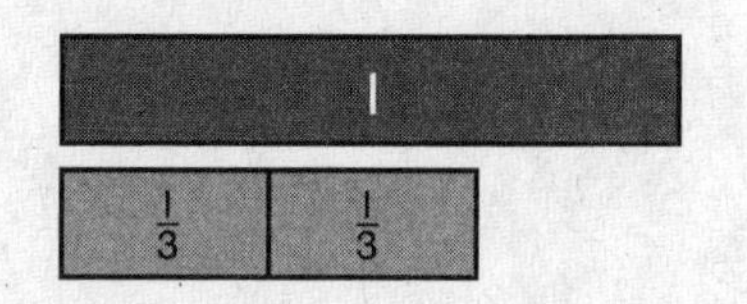

____

2.

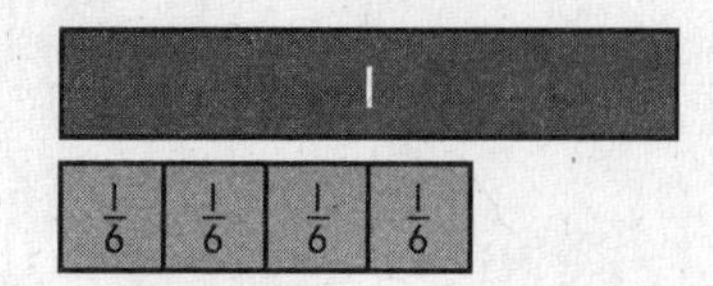

____

3.

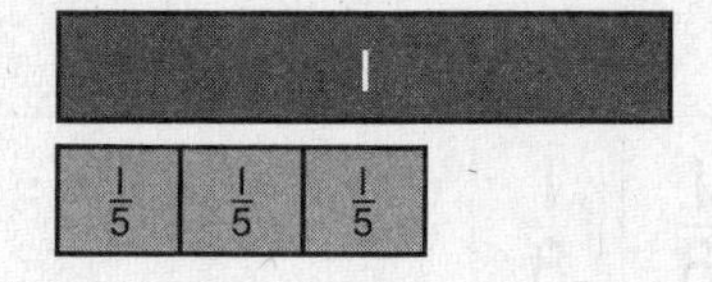

____

Find the missing numerator. Use fraction bars.

4. $\frac{2}{8} = \frac{\square}{4}$

5. $\frac{1}{2} = \frac{\square}{10}$

6. $\frac{3}{3} = \frac{\square}{6}$

Problem Solving and Test Prep

USE DATA For 7, use the table.

7. The bar graph shows the weights of three different kinds of make-believe bugs. How many beetles would it take to equal the weight of one dragonfly?

____

Bugs

Type	Weight
Beetle	$\frac{1}{8}$ gram
Grasshopper	$\frac{1}{2}$ gram
Dragonfly	$\frac{3}{4}$ gram

8. Erin has 8 fish. Of these, 3 are blue. What fraction of her fish are blue?

A $\frac{1}{2}$

B $\frac{5}{8}$

C $\frac{3}{8}$

D $\frac{2}{4}$

9. What is the missing numerator?

$\frac{4}{12} = \frac{}{3}$

A 2

B 4

C 1

D 6

Name____

Compare and Order Fractions

You can use fraction bars and number lines to compare fractions.

Use the number line below to compare. Write <, >, or =.

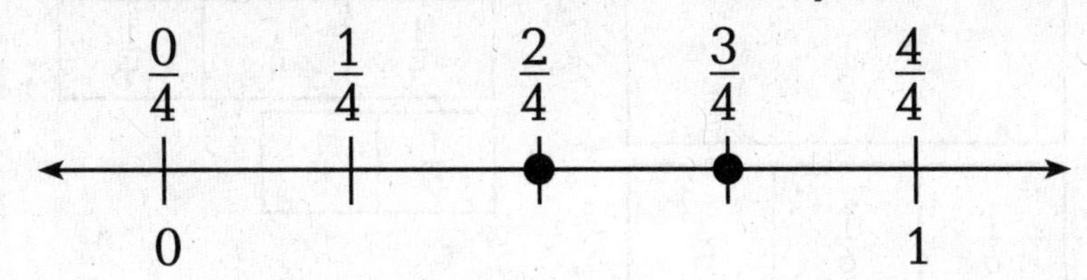

As you move to the right on a number line, the fractions become greater. Locate $\frac{2}{4}$ and $\frac{3}{4}$ on the number line. Since $\frac{3}{4}$ is farther to the right on the number line than $\frac{2}{4}$, $\frac{3}{4}$ is greater.

So, $\frac{2}{4} < \frac{3}{4}$.

Compare. Write <, >, or = for each circle.

1. 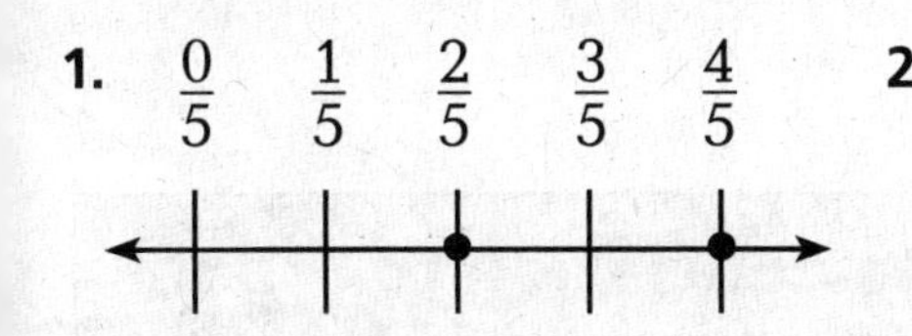

$\frac{2}{5}$ ◯ $\frac{4}{5}$

2. 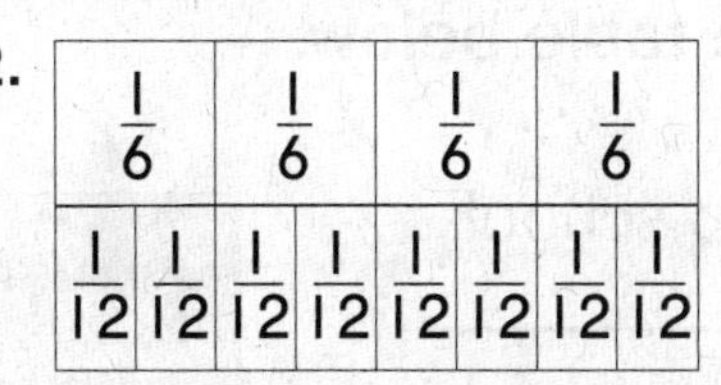

$\frac{4}{6}$ ◯ $\frac{8}{12}$

3. 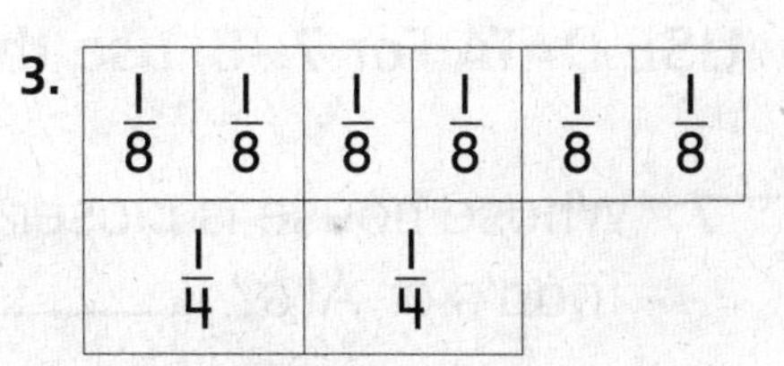

$\frac{6}{8}$ ◯ $\frac{2}{4}$

4. $\frac{0}{6}$ $\frac{1}{6}$ $\frac{2}{6}$ $\frac{3}{6}$ $\frac{4}{6}$ $\frac{5}{6}$ $\frac{6}{6}$

$\frac{0}{6}$ ◯ $\frac{6}{6}$

5. $\frac{1}{4}$ $\frac{1}{4}$ $\frac{1}{4}$

$\frac{1}{8}$ $\frac{1}{8}$ $\frac{1}{8}$ $\frac{1}{8}$ $\frac{1}{8}$ $\frac{1}{8}$

$\frac{3}{4}$ ◯ $\frac{6}{8}$

6. 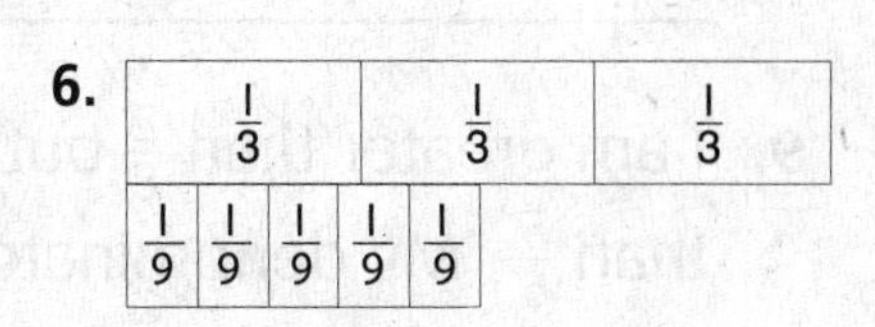

$\frac{3}{3}$ ◯ $\frac{5}{9}$

NS 3.1 Compare fractions represented by drawings or concrete materials to show equivalency and to add and subtract simple fractions in context (e.g. 1/2 of a pizza is the same as 2/4 of another pizza that is the same size; show that 3/8 is larger than 1/4).

Name________________________________

Compare and Order Fractions

Compare. Write <, >, or = for each ○.

1.

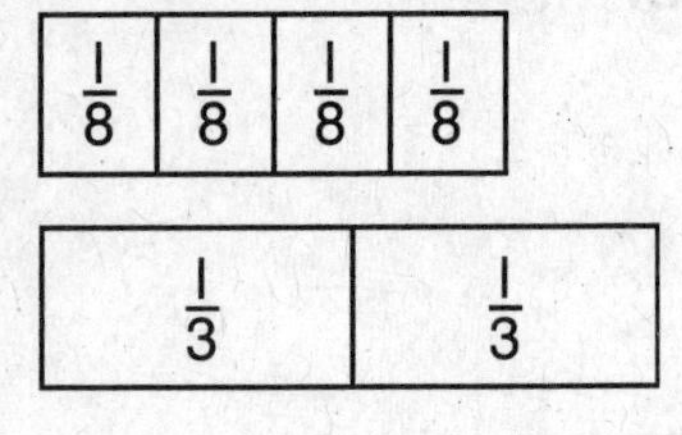

$\frac{4}{8}$ ○ $\frac{2}{3}$

2.

$\frac{1}{2}$

$\frac{1}{6}$ | $\frac{1}{6}$ | $\frac{1}{6}$

$\frac{1}{2}$ ○ $\frac{3}{6}$

3.

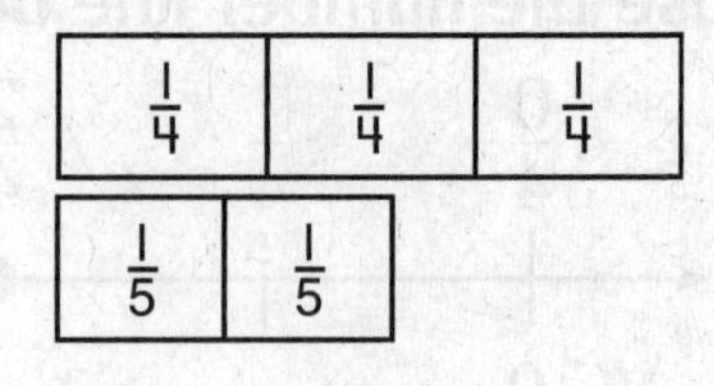

$\frac{3}{4}$ ○ $\frac{2}{5}$

4. $\frac{3}{8}$ ○ $\frac{1}{4}$

5. $\frac{2}{3}$ ○ $\frac{5}{6}$

6. $\frac{4}{8}$ ○ $\frac{3}{6}$

Problem Solving and Test Prep

USE DATA For 7–8, use the table below.

7. Whose house is closer to school, Todd's or Al's? ____________

8. Dan walked from his house to school. Then Dan walked to Todd's house. Which distance is farther?

____

Houses in Relation to School

Whose House?	Distance from School
Al's	$\frac{3}{6}$ mile
Dan's	$\frac{2}{5}$ mile
Todd's	$\frac{3}{4}$ mile

9. I am greater than $\frac{2}{8}$ but less than $\frac{5}{6}$. My denominator is 2. Which fraction am I?

A $\frac{1}{2}$ **C** $\frac{3}{8}$

B $\frac{0}{2}$ **D** $\frac{2}{2}$

10. Which fraction is greater than $\frac{3}{5}$?

A $\frac{3}{6}$ **C** $\frac{7}{8}$

B $\frac{1}{4}$ **D** $\frac{6}{10}$

Name____ Lesson 18.5

Problem Solving Workshop Strategy: Make a Model

Lewis made a wax candle at the carnival. He made $\frac{2}{8}$ of it blue wax, $\frac{1}{2}$ of it green, and $\frac{1}{4}$ of it yellow. Which color did he use the most?

Read to Understand

1. What information is given?

____

____

Plan

2. How can making a model help you solve the problem?

____

Solve

3. Solve the problem. Describe the strategy used.

____

____

____

____

____

4. Which color did Lewis use the most? _______________

Check

5. What other strategy could you use to solve the problem?

____

Make a model to solve.

6. Tim spent $\frac{1}{3}$ of his money on food at the carnival, $\frac{1}{6}$ of his money on rides, and $\frac{3}{6}$ on games. What did Tim spend the most money on?

____

7. Alyssa painted a picture that was $\frac{3}{6}$ red, $\frac{1}{6}$ green, and $\frac{1}{3}$ blue. Which color did Alyssa use the most?

____

MG 3.1 Compare fractions represented by drawings or concrete materials to show equivalency and to add and subtract simple fractions in context (e.g., $\frac{1}{2}$ of a pizza is the same amount as $\frac{2}{4}$ of another pizza that is the same size; show that $\frac{3}{8}$ is larger than $\frac{1}{4}$).

Name____

Problem Solving Workshop Strategy: Make a Model

Choose a strategy. Then solve.

1. Lisa and Michelle played a ring toss game at the carnival. Lisa tossed $\frac{2}{3}$ of her 12 rings around the bottle. Michelle tossed $\frac{5}{6}$ of her 12 rings around the bottle. Who tossed more rings around the bottle?

____

2. Chris and his friends ordered ice cream sundaes at the food stand. Chris ate $\frac{1}{3}$ of his sundae. Hayden ate $\frac{3}{5}$ of his sundae, and Jacob ate $\frac{2}{8}$ of his sundae. Who ate the greatest amount of their sundae?

____

Mixed Strategy Practice

USE DATA For 3–4, use the table.

Balloon Pop Game	
Name of Player	**Fraction of Balloons Popped**
Taylor	$\frac{5}{8}$
Sean	$\frac{3}{4}$
Roseanne	$\frac{1}{2}$

3. For the Balloon Pop game, players have 8 chances to pop balloons. Who popped the greatest number of balloons?

____

4. Who popped the fewest number of balloons?

____

5. Eileen read $\frac{7}{10}$ of the book for school. Shelly read $\frac{4}{5}$ of the same book and Kara read $\frac{1}{2}$ of this book. Who read the greatest amount of the book?

____

6. Richard's team took turns working the scoreboard at the game. Each player worked the scoreboard for $\frac{1}{6}$ of an hour. The game lasted for 2 hours. How many players are on the team?

____

Spiral Review

For 1–3, use the Theater Prices to solve the problems.

Theater Prices	
Adult Ticket $10	Child Ticket $5
Popcorn $4	Pickle $3
Candy $4	Drink $3

1. Toni wants to buy two adult tickets and two drinks. He has $25. Does he have enough money? _____
2. The school movie club collected $50 to go to the theater. They have 6 students going. With the money left over, how many adults can go with them to the theater?

   ____
3. Natalie's grandmother gave her $20 to go to the movies. She buys a child ticket, a drink, a pickle, and popcorn. How much change will she get back? _____

For 4–7, write the number of edges for each figure.

	Figure	Number of Edges
4.	 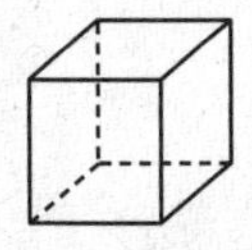	_____
5.	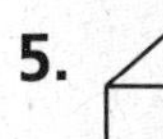	_____
6.	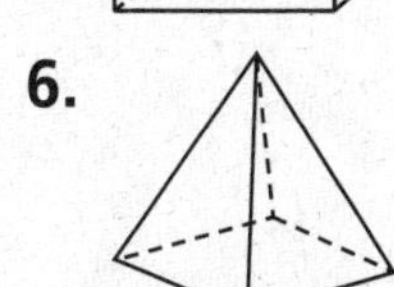	_____
7.	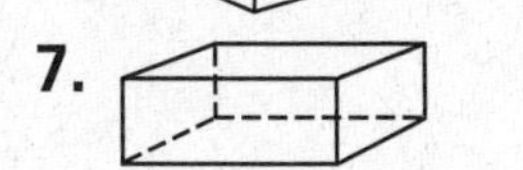	_____

For 8–11, use the table to complete the pattern.

rows	1	2	3		5
chairs	4	8		16	

8. How many chairs are in 3 rows?

   ____
9. How many rows have 16 chairs in all?

   ____
10. How many chairs are in 5 rows?

    ____
11. Complete the statement: For every row, there are _____ chairs.

For 12–14, complete the number sentence.

12. 2 hr = __________ min
13. 3 yr = __________ mo
14. 21 days = _______ wks

Name________________________________ **Lesson 19.1**

Add Like Fractions

Fractions that have the same denominator are called **like fractions.** You can add like fractions using fraction bars.

Use fraction bars to find the sum.

$\frac{3}{6} + \frac{2}{6} = \square$

The denominators are the same.
Use $\frac{1}{6}$ fraction bars to add.

$\frac{1}{6}$	$\frac{1}{6}$	$\frac{1}{6}$

$\frac{1}{6}$	$\frac{1}{6}$

Count the number of $\frac{1}{6}$ fraction bars to find the sum.

$\frac{1}{6} + \frac{1}{6} + \frac{1}{6} + \frac{1}{6} + \frac{1}{6} = \frac{5}{6}$

So, $\frac{3}{6} + \frac{2}{6} = \frac{5}{6}$.

Use fraction bars to find each sum.

1.

$\frac{1}{6}$	$\frac{1}{6}$	$\frac{1}{6}$	$\frac{1}{6}$

$\frac{1}{6}$

$\frac{4}{6} + \frac{1}{6} =$ ______

2.

$\frac{1}{12}$	$\frac{1}{12}$	$\frac{1}{12}$

$\frac{1}{12}$	$\frac{1}{12}$	$\frac{1}{12}$	$\frac{1}{12}$	$\frac{1}{12}$

$\frac{3}{12} + \frac{5}{12} =$ ______

NS 3.1 Compare fractions represented by drawings or concrete materials to show equivalency and to add and subtract simple fractions in context (e.g., $\frac{1}{2}$ of a pizza is the same amount as $\frac{2}{4}$ of another pizza that is the same size; show that $\frac{3}{8}$ is larger than $\frac{1}{4}$).

Name ______________________________ Lesson 19.1

Add Like Fractions

Use fraction bars to find each sum.

1. | $\frac{1}{6}$ | $\frac{1}{6}$ | $\frac{1}{6}$ | | $\frac{1}{6}$ | $\frac{1}{6}$ |

$\frac{3}{6} + \frac{2}{6} =$ ______

2. | $\frac{1}{8}$ | $\frac{1}{8}$ | $\frac{1}{8}$ | $\frac{1}{8}$ | | $\frac{1}{8}$ |

$\frac{4}{8} + \frac{1}{8} =$ ______

3. | $\frac{1}{9}$ | $\frac{1}{9}$ | $\frac{1}{9}$ | $\frac{1}{9}$ | $\frac{1}{9}$ | | $\frac{1}{9}$ | $\frac{1}{9}$ |

$\frac{5}{9} + \frac{2}{9} =$ ______

Find each sum.

4. $\frac{3}{10} + \frac{4}{10} =$ ______

5. $\frac{1}{4} + \frac{1}{4} =$ ______

6. $\frac{2}{12} + \frac{6}{12} =$ ______

Problem Solving and Test Prep

7. Hannah made a bracelet with $\frac{2}{10}$ yard of pink ribbon and $\frac{4}{10}$ yard of green ribbon. How much ribbon did Hannah use in all?

____

8. Mike bought 2 jars of peanuts. Each jar contains $\frac{2}{6}$ cup of peanuts. How many peanuts does Mike have in all?

____

9. Which is the sum?

$\frac{4}{8} + \frac{3}{8} =$ ______

A $\frac{1}{2}$

B $\frac{7}{8}$

C $\frac{5}{8}$

D $\frac{5}{4}$

10. Which is the sum?

$\frac{4}{12} + \frac{4}{12} =$ ______

A $\frac{2}{6}$

B $\frac{8}{12}$

C $\frac{6}{18}$

D $\frac{6}{12}$

Name____

Add Like Fractions

A fraction is in **simplest form** when it is modeled with the largest fraction bar or bars possible.

Find the sum. Write the answer in simplest form.

$\frac{3}{5} + \frac{2}{5} = \square$

Use fraction bars to find the sum.

$\frac{1}{5}$	$\frac{1}{5}$	$\frac{1}{5}$	$\frac{1}{5}$	$\frac{1}{5}$

$\frac{3}{5} + \frac{2}{5} = \frac{5}{5}$ The sum is $\frac{5}{5}$.

Now use fraction bars to write $\frac{5}{5}$ in simplest form.

$\frac{1}{5}$	$\frac{1}{5}$	$\frac{1}{5}$	$\frac{1}{5}$	$\frac{1}{5}$
1				

So, the simplest form of $\frac{5}{5}$ is 1.

Find the sum. Write each answer in simplest form.

1.

$\frac{1}{8}$	$\frac{1}{8}$	$\frac{1}{8}$	$\frac{1}{8}$	$\frac{1}{8}$	$\frac{1}{8}$
$\frac{1}{4}$		$\frac{1}{4}$		$\frac{1}{4}$	

$\frac{2}{8} + \frac{4}{8} = \square$ or $\square$

____

2.

$\frac{1}{6}$	$\frac{1}{6}$	$\frac{1}{6}$
$\frac{1}{2}$		

$\frac{2}{6} + \frac{1}{6} = \square$ or $\square$

____

3.

$\frac{1}{10}$	$\frac{1}{10}$	$\frac{1}{10}$	$\frac{1}{10}$
$\frac{1}{5}$		$\frac{1}{5}$	

$\frac{3}{10} + \frac{1}{10} = \square$ or $\square$

____

Name______________________________ Lesson 19.2

Add Like Fractions

Find each sum. Write the answer in simplest form.

1. $\frac{1}{4}$ $\frac{1}{4}$ | $\frac{1}{4}$

$\frac{2}{4} + \frac{1}{4} =$ ____

2. $\frac{1}{10}$ $\frac{1}{10}$ | $\frac{1}{10}$

$\frac{2}{10} + \frac{1}{10} =$ ____

3. $\frac{1}{9}$ $\frac{1}{9}$ $\frac{1}{9}$ $\frac{1}{9}$ $\frac{1}{9}$ $\frac{1}{9}$ | $\frac{1}{9}$ $\frac{1}{9}$

$\frac{6}{9} + \frac{2}{9} =$ ____

4. $\frac{5}{12} + \frac{1}{12} =$ ____

5. $\frac{2}{6} + \frac{2}{6} =$ ____

6. $\frac{1}{8} + \frac{5}{8} =$ ____

7. $\frac{4}{5} + \frac{6}{5} =$ ____

Problem Solving and Test Prep

8. Terry has 2 bags of flour. One bag weighs $\frac{2}{8}$ of a pound and the other weighs $\frac{4}{8}$ of a pound. How much do the two bags of flour weigh in all? Write your answer in simplest form.

____

9. Rachel read $\frac{3}{12}$ of her book on Sunday and $\frac{5}{12}$ on Monday. How much of her book did Rachel read in all?

____

10. Scott walked $\frac{2}{6}$ of a mile to the park. Al walked $\frac{1}{6}$ of a mile. How far did Scott and Al walk in all?

A $\frac{1}{2}$

B $\frac{1}{6}$

C $\frac{4}{6}$

D $\frac{6}{3}$

11. Which fraction is equivalent to $\frac{4}{12}$?

A $\frac{1}{6}$

B $\frac{6}{24}$

C $\frac{1}{3}$

D $\frac{6}{12}$

Name____

Subtract Like Fractions

You can use fraction bars to subtract like fractions.

Use fraction bars to find the difference.

$\frac{7}{8} - \frac{2}{8} = \square$

$\frac{1}{8}$	$\frac{1}{8}$	$\frac{1}{8}$	$\frac{1}{8}$	$\frac{1}{8}$	$\frac{1}{8}$	$\frac{1}{8}$

→

$\frac{1}{8}$	$\frac{1}{8}$

The dashed line around the fraction bars shows what is being subtracted.

Count the number of $\frac{1}{8}$ fraction bars left.

There are five $\frac{1}{8}$ fraction bars left.

So, $\frac{5}{8}$ is the difference.

Use fraction bars to find each difference.

1.

$\frac{1}{9}$	$\frac{1}{9}$	$\frac{1}{9}$	$\frac{1}{9}$	$\frac{1}{9}$	$\frac{1}{9}$	$\frac{1}{9}$

$\frac{7}{9} - \frac{5}{9} = \square$

____

2.

$\frac{1}{5}$	$\frac{1}{5}$	$\frac{1}{5}$

$\frac{3}{5} - \frac{2}{5} = \square$

____

3.

$\frac{1}{12}$	$\frac{1}{12}$	$\frac{1}{12}$	$\frac{1}{12}$	$\frac{1}{12}$	$\frac{1}{12}$	$\frac{1}{12}$	$\frac{1}{12}$	$\frac{1}{12}$

$\frac{9}{12} - \frac{3}{12} = \square$

____

NS 3.1 Compare fractions represented by drawings or concrete materials to show equivalency and to add and subtract simple fractions in context (e.g., $\frac{1}{2}$ of a pizza is the same amount as $\frac{2}{4}$ of another pizza that is the same size; show that $\frac{3}{8}$ is larger than $\frac{1}{4}$).

Name________________________________

Subtract Like Fractions

Use fraction bars to find each difference.

1.

$\frac{1}{9}$	$\frac{1}{9}$	$\frac{1}{9}$	$\frac{1}{9}$	$\frac{1}{9}$	$\frac{1}{9}$	$\frac{1}{9}$
$\frac{1}{9}$	$\frac{1}{9}$					

$\frac{7}{9} - \frac{2}{9} =$ ____

2.

$\frac{1}{8}$	$\frac{1}{8}$	$\frac{1}{8}$	$\frac{1}{8}$	$\frac{1}{8}$	$\frac{1}{8}$
$\frac{1}{8}$					

$\frac{6}{8} - \frac{1}{8} =$ ____

3.

$\frac{1}{6}$	$\frac{1}{6}$	$\frac{1}{6}$	$\frac{1}{6}$	$\frac{1}{6}$
$\frac{1}{6}$	$\frac{1}{6}$	$\frac{1}{6}$	$\frac{1}{6}$	

$\frac{5}{6} - \frac{4}{6} =$ ____

Find each difference.

4. $\frac{10}{12} - \frac{5}{12} =$ ____

5. $\frac{5}{8} - \frac{2}{8} =$ ____

6. $\frac{2}{5} - \frac{1}{5} =$ ____

7. $\frac{3}{5} - \frac{1}{5} =$ ____

Problem Solving and Test Prep

8. Jake ate $\frac{2}{10}$ of a pizza. Mark ate $\frac{4}{10}$ of the same pizza. How much more of the pizza did Mark eat than Jake?

____

9. Jane played soccer for $\frac{1}{6}$ of an hour on Monday. She played for $\frac{4}{6}$ of an hour on Tuesday. How much longer did Jane play soccer on Tuesday than on Monday?

____

10. Which is the difference?

$\frac{6}{8} - \frac{4}{8} =$ ____

A $\frac{1}{8}$

B $\frac{2}{8}$

C $\frac{2}{4}$

D $\frac{10}{8}$

11. Which is the difference?

$\frac{7}{10} - \frac{3}{10} =$ ____

A $\frac{10}{10}$

B $\frac{3}{10}$

C $\frac{4}{5}$

D $\frac{4}{10}$

Name___________________________ **Lesson 19.4**

Subtract Like Fractions

You can subtract like fractions and write the difference in simplest form.

Compare. Find the difference. Write the answer in simplest form.

$\frac{6}{6} - \frac{2}{6} = \square$

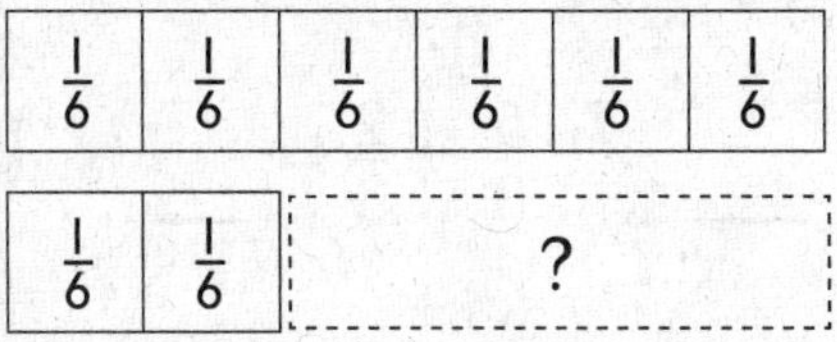

Compare the bars to find the difference.

The difference is $\frac{4}{6}$.

Find the largest fraction bar that is equivalent to $\frac{4}{6}$.

So, $\frac{4}{6}$ written in simplest form is $\frac{2}{3}$.

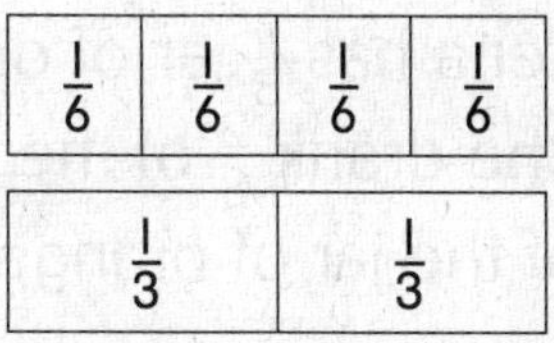

Compare. Find each difference. Write the answer in simplest form.

1.

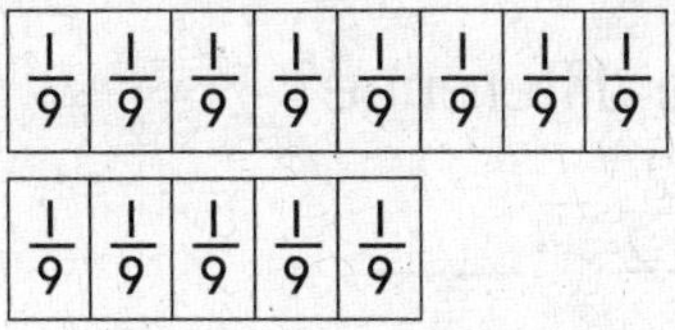

$\frac{8}{9} - \frac{5}{9} = \square$

____

2.

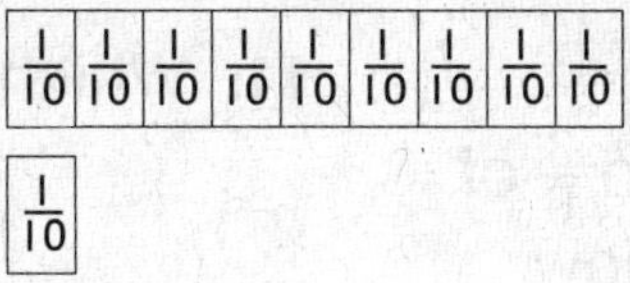

$\frac{9}{10} - \frac{1}{10} = \square$

____

3.

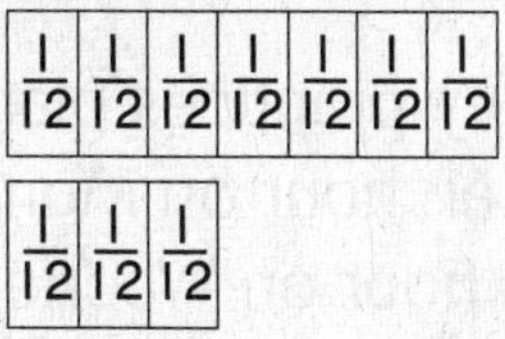

$\frac{7}{12} - \frac{3}{12} = \square$

____

4.

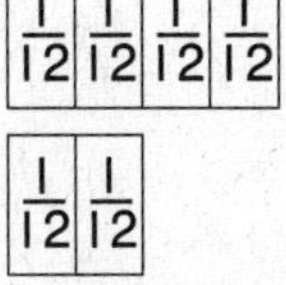

$\frac{4}{12} - \frac{2}{12} = \square$

____

5.

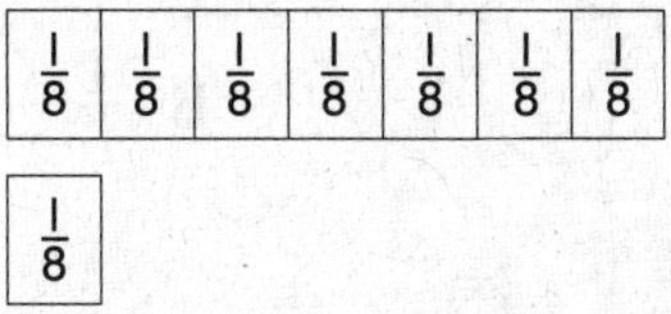

$\frac{7}{8} - \frac{1}{8} = \square$

____

6.

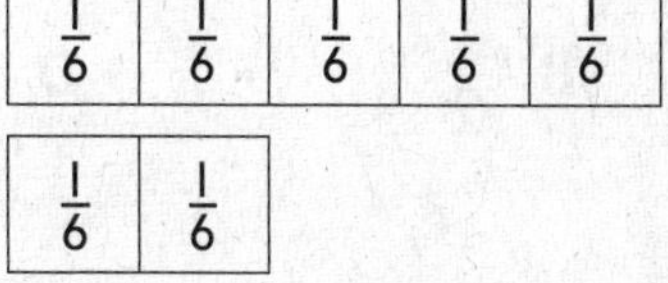

$\frac{5}{6} - \frac{2}{6} = \square$

____

Name______________________________

Subtract Like Fractions

Compare. Find each difference. Write the answer in simplest form.

1.

$\frac{1}{10}$	$\frac{1}{10}$	$\frac{1}{10}$	$\frac{1}{10}$	$\frac{1}{10}$	$\frac{1}{10}$	$\frac{1}{10}$
$\frac{1}{10}$	$\frac{1}{10}$	?				

$\frac{7}{10} - \frac{2}{10} =$ ____

2.

$\frac{1}{8}$	$\frac{1}{8}$	$\frac{1}{8}$	$\frac{1}{8}$	$\frac{1}{8}$	$\frac{1}{8}$
$\frac{1}{8}$	$\frac{1}{8}$	$\frac{1}{8}$	$\frac{1}{8}$	?	

$\frac{6}{8} - \frac{4}{8} =$ ____

3.

$\frac{1}{6}$	$\frac{1}{6}$	$\frac{1}{6}$	$\frac{1}{6}$	$\frac{1}{6}$
$\frac{1}{6}$	$\frac{1}{6}$	$\frac{1}{6}$	?	

$\frac{5}{6} - \frac{3}{6} =$ ____

4. $\frac{8}{12} - \frac{5}{12} =$ ____

5. $\frac{6}{8} - \frac{2}{8} =$ ____

6. $\frac{7}{10} - \frac{1}{10} =$ ____

7. $\frac{5}{3} - \frac{2}{3} =$ ____

Problem Solving and Test Prep

8. Lena has $\frac{6}{8}$ jar of orange juice. She drank $\frac{2}{8}$ of the jar. How much of the jar of orange juice is left?

____

9. Mark walked $\frac{3}{4}$ mile to school. Kevin walked $\frac{1}{4}$ mile to school. How much farther did Mark walk than Kevin walked?

____

10. Riley practices piano for $\frac{2}{6}$ of an hour on Monday and $\frac{3}{6}$ of an hour on Wednesday. How much more time did Riley practice on Wednesday than on Monday?

A $\frac{1}{6}$

B $\frac{1}{2}$

C $\frac{5}{6}$

D $\frac{6}{6}$

11. Which is the difference?

$\frac{8}{12} - \frac{2}{12} =$ ____

A $\frac{1}{3}$

B $\frac{1}{6}$

C $\frac{10}{12}$

D $\frac{1}{2}$

Spiral Review

For 1–5, find the sum.

1. $\frac{3}{8} + \frac{2}{8} =$

2. $\frac{1}{4} + \frac{1}{4} =$

3. $\frac{2}{9} + \frac{4}{9} =$

4. $\frac{1}{5} + \frac{3}{5} =$

5. $\frac{5}{10} + \frac{4}{10} =$

For 6–7, combine the given figures to make a new figure. Draw an outline of the new figure.

6.     

7.   

For 8–10, use the pattern in the table to answer the questions.

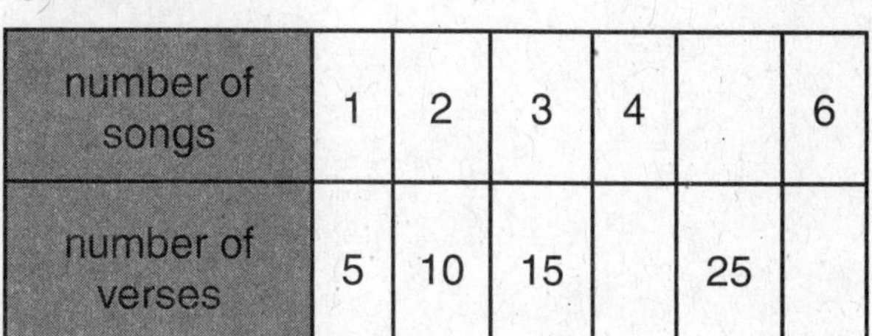

number of songs	1	2	3	4		6
number of verses	5	10	15		25	

8. How many verses are in 4 songs?

____

9. How many songs have 25 verses in all?

____

10. How many verses are in 6 songs?

____

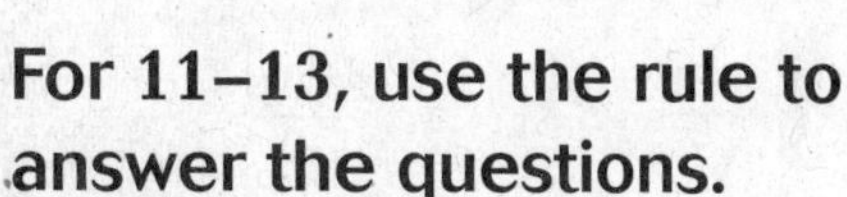

For 11–13, use the rule to answer the questions.

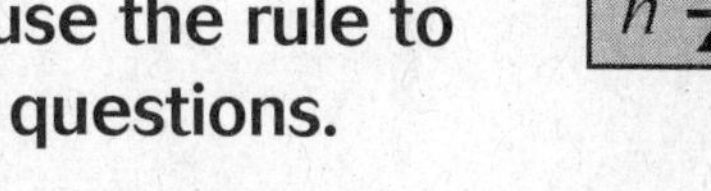

Rule: The electronics shop has a sale. Customers can purchase 3 batteries for $5.

11. How much would it cost to purchase 9 batteries?

____

12. How much would it cost to purchase 15 batteries?

____

13. How many batteries can you purchase with $30.00?

____

Name____

Problem Solving Workshop Skill: Too Much/Too Little Information

There are 8 books on a shelf. Three of the books are short stories, some are coloring books, and the rest are sports books. What fraction of the books are NOT short stories?

1. What are you asked to find?

____

2. What information do you need to solve the problem?

____

____

3. Is there too much or too little information in the problem? Explain.

____

____

4. Can you solve the problem? If so, solve.

____

____

Tell whether there is too much or too little information. Solve if there is enough information.

5. Jim has 12 fish in his fish tank. He also has a dog. He separates the fish into 2 equal groups. One of the groups has all red fish and one has all goldfish. What fraction of the fish are not red?

____

6. Allen bought turkey, ham and cheddar cheese at the deli counter. He bought $\frac{2}{4}$ pound of turkey and $\frac{3}{4}$ pound of ham. How many pounds of food did Allen buy at the deli counter?

____

MR 1.1 Analyze problems by identifying relationships, distinguishing relevant from irrelevant information, sequencing and prioritizing information, and observing patterns.

Name ____________________

Problem Solving Workshop Skill: Too Much/Too Little Information

Problem Solving Skill Practice

Tell if there is too much or too little information. Solve if there is enough information.

1. Erin, Jackie, and John walk home from school together. Erin's house is $\frac{2}{6}$ mile from school. Jackie lives $\frac{4}{6}$ mile from school and John lives $\frac{3}{6}$ mile from school. How many miles do Erin and John live from school in all?

____

2. Mrs. Williams brought a cake to class. Christine ate $\frac{1}{8}$ of the cake. Kellie ate $\frac{2}{8}$ of the cake and Tom ate $\frac{2}{8}$ of the cake. How much of the cake did Edgar eat?

____

Mixed Applications

3. Fiona wrote $\frac{3}{10}$ of her paper on Thursday. She wrote $\frac{2}{10}$ of her paper on Friday and $\frac{4}{10}$ of her paper on Saturday. On which day did she write the most amount of her paper out of these 3 days?

____

4. Derek has $20.00. He bought 3 books at the book store. Each book cost $6.00. How much money does he have left? Which operations will you use to solve?

____

USE DATA For 5–6, use the table.

5. Who spent the least time working on their homework?

____

6. How much time did Jamie spend on her homework in minutes?

____

Time Spent on Homework	
Student	**Time**
Lara	$\frac{1}{4}$ hour
Jamie	$\frac{2}{3}$ hour
Todd	$\frac{1}{2}$ hour

Name________________________________ **Lesson 20.1**

Model Tenths

A **decimal** is a number with one or more digits to the right of a decimal point. It shows part of a whole. A *fraction* also shows part of a whole.

Look at decimal model A. It is a whole. Count the parts of the whole model. There are 10 equal parts. When a whole has 10 equal parts, each part is called a *tenth*.

A. 

Write the fraction and decimal for the shaded part.

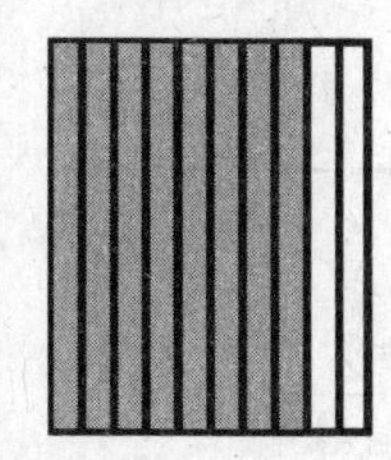

The model is a whole with 10 equal parts. Eight of the parts are shaded.

Fraction	Decimal
Write: $\frac{8}{10}$	**Write:** 0.8
Read: eight tenths	Read: eight tenths

So, $\frac{8}{10}$ and 0.8 name the shaded part.

Write the fraction and decimal for the shaded part

1. 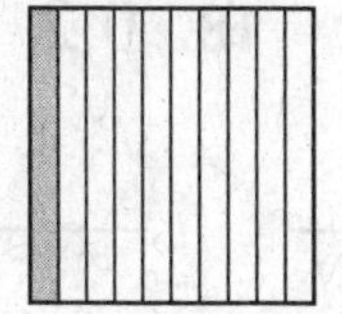

____

____

2. 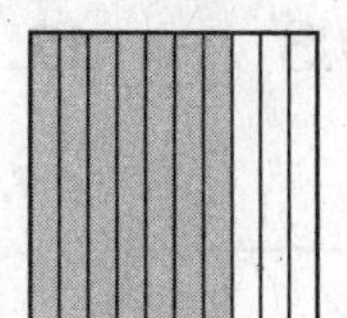

____

____

3. 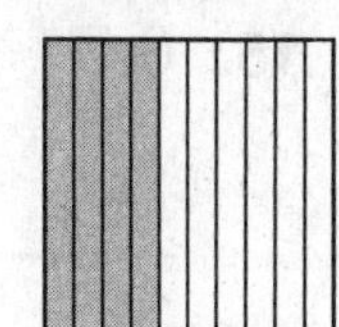

____

____

4. 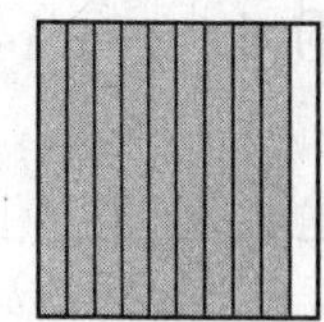

____

____

Shade the blocks to show the tenths. Show the tenths as fractions and decimals.

5. nine tenths

____

____

6. seven tenths

____

____

7. one tenth

____

____

8. six tenths

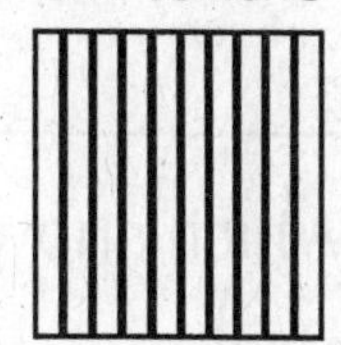

____

____

Name ____________________

Model Tenths

Write the fraction and decimal for the shaded part.

1. 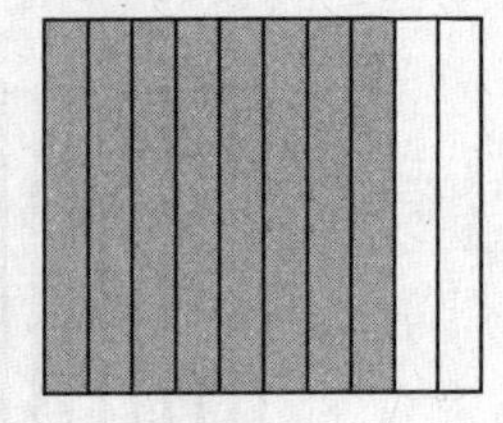
____

2. 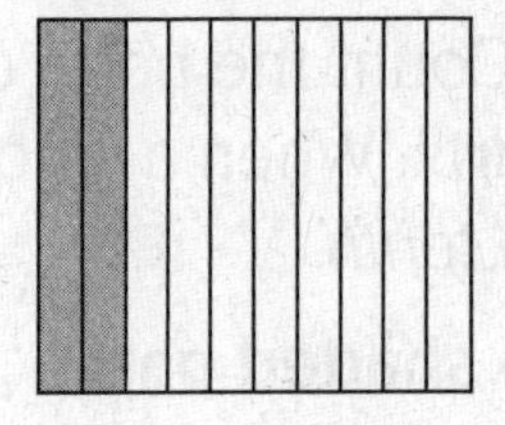
____

3. 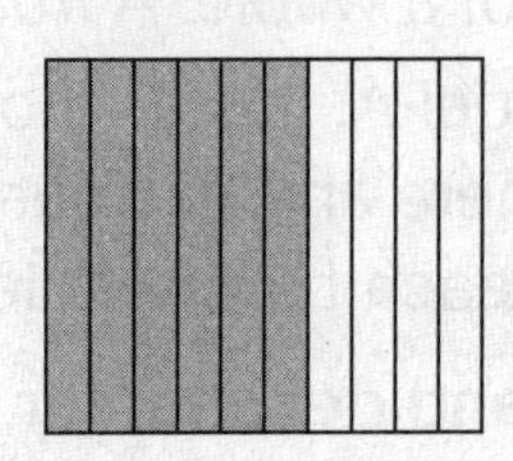
____

4. 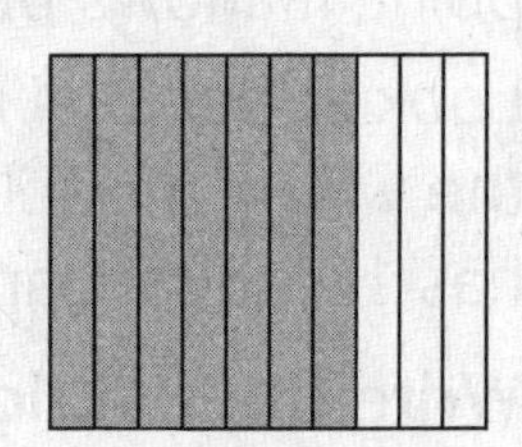
____

Write each fraction as a decimal.

5. $\frac{7}{10}$ __________
6. $\frac{3}{10}$ __________
7. $\frac{8}{10}$ __________
8. $\frac{1}{10}$ __________
9. $\frac{2}{10}$ __________

Write each decimal as a fraction.

10.

ONES	.	TENTHS
0	.	9

____

11.

ONES	.	TENTHS
0	.	6

____

12.

ONES	.	TENTHS
0	.	2

____

13. 0.4 __________
14. 0.1 __________
15. 0.7 __________
16. 0.3 __________
17. 0.9 __________
18. 0.2 __________

Problem Solving and Test Prep

19. There are ten balls in the gym. Six balls are red. Four balls are blue. Write a decimal to show what part of the balls are blue.

____

20. Thomas played basketball. He shot the ball ten times. The ball went in the basket six times. Write a fraction to show how many baskets Thomas made.

____

21. Which shows the fraction for 0.8?

A $\frac{6}{10}$
B $\frac{3}{10}$
C $\frac{8}{10}$
D $\frac{1}{10}$

22. Which shows the decimal for $\frac{7}{10}$?

A 0.7
B 0.6
C 0.8
D 0.2

Name____ **Lesson 20.2**

Model Hundredths

Decimal models can be used to show amounts written as fractions or decimals.

Decimal Model

100 equal parts.

1 part shaded.

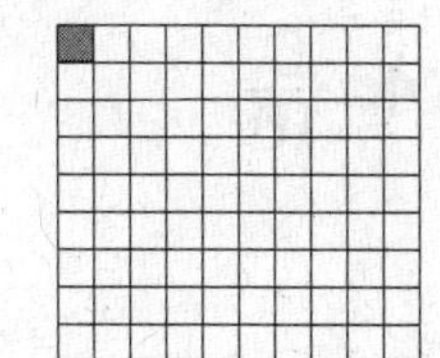

Fraction: $\frac{1}{100}$

Decimal: 0.01

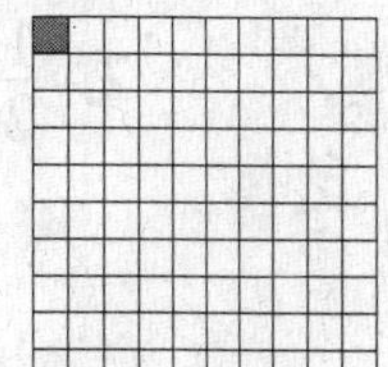

Use a decimal model to show the amount below. Then write the fraction as a decimal.

$\frac{60}{100}$

Shade in 60 hundredths of a decimal model.

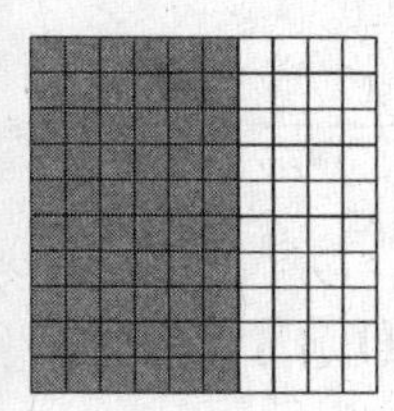

Sixty hundredths written as a decimal is 0.60.

So, $\frac{60}{100}$ is the same as 0.60.

Use decimal models to show each amount. Then write each fraction as a decimal.

1. $\frac{66}{100}$

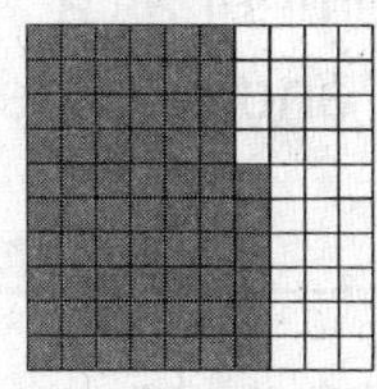

____

2. $\frac{20}{100}$

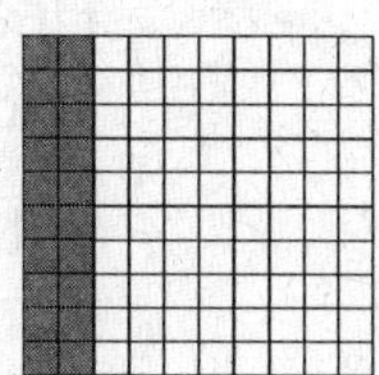

____

3. $\frac{42}{100}$

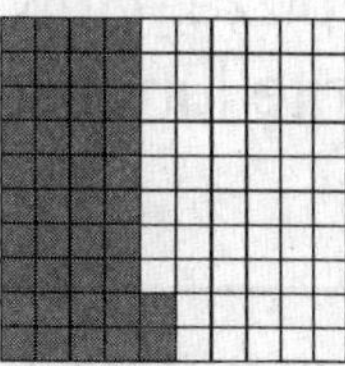

____

Name____

Model Hundredths

Use a decimal model to show each amount. Then write each fraction as a decimal.

1. $\frac{7}{10}$ **2.** $\frac{11}{20}$ **3.** $\frac{3}{5}$ **4.** $\frac{6}{10}$ **5.** $\frac{4}{10}$

______ ______ ______ ______ ______

Write each decimal as a fraction.

6.

ONES	.	TENTHS	HUNDREDTHS
0	.	6	6

____

7.

ONES	.	TENTHS	HUNDREDTHS
0	.	7	5

____

Write each decimal as a fraction and in expanded form.

8. 0.04 **9.** 0.24 **10.** 0.70 **11.** 0.43 **12.** 0.33

______ ______ ______ ______ ______

______ ______ ______ ______ ______

Problem Solving and Test Prep

13. Nicole surveyed 100 students about their favorite lunch. Of 100 students, 0.60 favorite lunch was pizza. How many students' favorite lunch was pizza?

____

14. Heather measures her pencil in centimeters. It is 7 centimeters long. 100 centimeters = 1 meter. Write Heather's measurement as a fraction of a meter and as a decimal.

____

15. Which decimal shows four hundredths?

A 0.40
B 0.004
C 4.00
D 0.04

16. Which decimal is equal to $\frac{19}{100}$?

A 0.19
B 19.00
C 1.90
D 0.019

Name____

Decimals Greater Than One

Decimals can represent numbers greater than one. The decimal point separates the whole number from the fractional part.

Write the word form and the expanded form for the example below.

Ones	•	Tenths	Hundredths
4		9	8

Use decimal models to help solve.

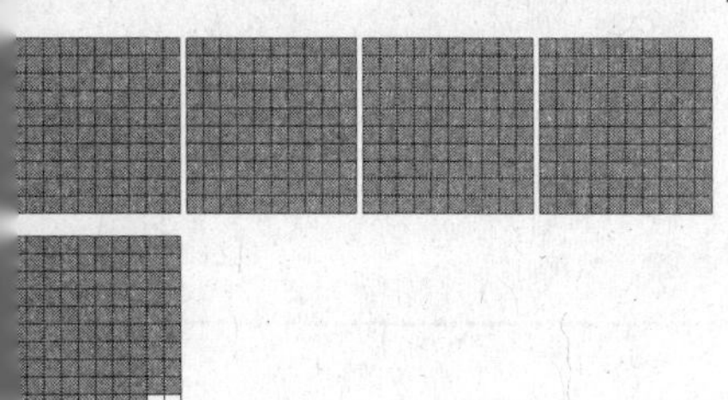

There are 4 whole models shaded and 98 hundredths of a 5th model shaded. The standard form for this number is 4.98.

So, the word form is: four and ninety-eight hundredths, and the expanded form is: 4 + 0.9 + 0.08.

Write the expanded form of the decimal. Then write the word form of the decimal.

1.

Ones	•	Tenths	Hundredths
1		7	0

____

____

2.

Ones	•	Tenths	Hundredths
2		0	8

____

____

3.

Ones	•	Tenths	Hundredths
5		2	1

____

____

4.

Ones	•	Tenths	Hundredths
3		3	3

____

____

Write the decimal for each.

5. three and seven tenths ______________

6. nine and sixty-two hundredths ______________

7. five and two tenths ______________

Name________________________________ Lesson 20.3

Decimals Greater Than One

Write the word form and expanded form for each.

1.

ONES	·	TENTHS	HUNDREDTHS
2	.	5	

____

____

2.

ONES	·	TENTHS	HUNDREDTHS
2	.	7	9

____

____

3. 1.53

____

____

4. 0.14

____

____

Write the decimal for each.

5. two and three tenths ____________

6. five and two tenths ____________

7. eleven and one hundredth ____________

Problem Solving and Test Prep

8. Ms. Peachtree is planting tulips along the front of her house. It will take 100 tulips to fill the front. She has planted 15 so far. What decimal shows how many tulips Ms. Peachtree has planted?

____

9. I am greater than 2, but less than 4. All my digits are odd. My tenths digit is 3 times my ones digit. My hundredths digit is 5. What decimal am I?

____

10. Which shows 8.3 in word form?

A eight hundred three

B eight and three hundredths

C eight and three tenths

D eight and thirty hundredths

11. Which fraction represents 0.13?

A $\frac{30}{100}$

B $\frac{13}{10}$

C $\frac{13}{100}$

D $\frac{30}{1000}$

Name ______________________________ **Week 29**

Spiral Review

For 1–5, compare the fractions using the number lines.

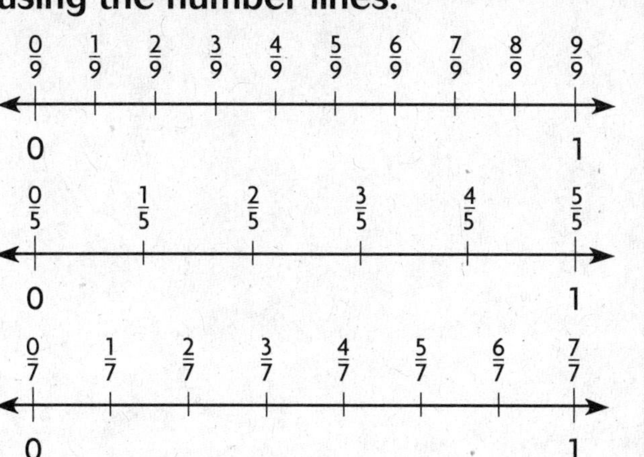

1. $\frac{4}{9}$ ◯ $\frac{2}{5}$

2. $\frac{1}{7}$ ◯ $\frac{1}{5}$

3. $\frac{8}{9}$ ◯ $\frac{6}{7}$

4. $\frac{3}{5}$ ◯ $\frac{4}{7}$

5. $\frac{2}{9}$ ◯ $\frac{3}{5}$

For 6–9, draw lines to match the object with the best unit of measure.

6. distance between two cities	yard
7. length of a sports field	inch
8. width of butterfly wings	mile
9. height of a building	foot

For 10–11, a class takes a survey about their favorite movies. Write the results as tally marks.

10. 13 students liked cartoons. ____________

11. 7 students like drama. ____________

12. Look at the table at the right. How many more students prefer puzzle videogames than prefer action video games? ______

Favorite Video Game	
Type	**Students**
Action	卌 |
Fantasy	卌
Puzzle	卌 |||

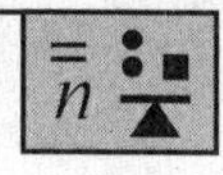

For 13–15, use the rule to answer the questions.

Rule: Two gardeners can plant 3 trees in an hour.

13. How long would it take two gardeners to plant 6 trees? ____________

14. How long would it take two gardeners to plant 15 trees? ____________

15. How long would it take four gardeners to plant 6 trees? ____________

Name______________________________________ Lesson 20.4

Relate Fractions, Decimals, and Money

You can write an amount of money less than a dollar as a fraction of a dollar.

Write the amount of money shown, then write the amount as a fraction of a dollar.

Count the value of the coins.

25 cents + 20 cents + 15 cents = 60 cents.

quarters dimes nickels

60 cents = \$0.60

Use a decimal model to help write this amount as a fraction.

The model shows the value of the coins.

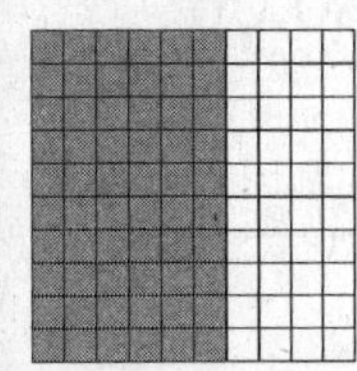

Count how many hundredths are shaded: 60 hundredths.

So, the amount money shown is \$0.60 and this amount written as a fraction of a dollar is $\frac{60}{100}$.

Write the amount as a fraction of a dollar.

1. 

____

2. 

____

Write the money amount for each fraction of a dollar.

3. $\frac{82}{100}$

\$______ . ______ ______

4. $\frac{57}{100}$

\$______ . ______ ______

Name____ Lesson 20.4

Relate Fractions, Decimals, and Money

Write the amount of money shown. Then write the amount as a fraction of a dollar.

1.

____

2.

____

Write the money amount for each fraction of a dollar.

3. $\frac{32}{100}$ ________

4. $\frac{5}{100}$ ________

5. $\frac{13}{100}$ ________

6. $\frac{89}{100}$ ________

7. $\frac{8}{100}$ ________

Write each money amount as a fraction of a dollar.

8. $0.53 ________

9. $0.28 ________

10. $0.99 ________

11. $0.15 ________

12. $0.06 ________

Write the money amount.

13. three hundredths of a dollar ______________

14. eleven hundredths of a dollar ______________

15. sixteen hundredths of a dollar ______________

Problem Solving and Test Prep

16. Jason spends $\frac{33}{100}$ of a dollar for an eraser. Write how much money he has left, of his dollar, as a decimal.

____

17. Which is greater, $0.35 or $\frac{4}{10}$ of a dollar?

____

18. A pencil costs $\frac{4}{10}$ of a dollar. Which amount equals the price of the pencil?

A $0.40 **C** $0.04

B $4.00 **D** $0.41

19. What decimal represents $\frac{79}{100}$ of a dollar?

A $0.80 **C** $79.00

B $0.77 **D** $0.79

Name____

Add and Subtract Decimals and Money

When adding money make sure to line up the decimals. Add decimals like whole numbers. Regroup if necessary. Don't forget to bring down the dollar sign and the decimal point. Include these two items in the sum.

Step 1	Step 2	Step 3
Add the pennies or hundredths.	Add the dimes or tenths.	Add the dollars or ones.
$\begin{array}{r} \$1.75 \\ +\ \$0.52 \\ \hline 7 \end{array}$	$\begin{array}{r} {}^{1} \\ \$1.75 \\ +\ \$0.52 \\ \hline 27 \end{array}$	$\begin{array}{r} {}^{1} \\ \$1.75 \\ +\ \$0.52 \\ \hline \$2.27 \end{array}$

When subtracting money make sure to line up the decimals. Subtract decimals like whole numbers. Don't forget to write the decimal point and the dollar sign in the difference.

Step 1	Step 2	Step 3
Subtract the pennies or hundredths.	Subtract the dimes or tenths.	Subtract the dollars or ones.
$\begin{array}{r} \$5.67 \\ -\ \$2.20 \\ \hline 7 \end{array}$	$\begin{array}{r} \$5.67 \\ -\ \$2.20 \\ \hline 47 \end{array}$	$\begin{array}{r} \$5.67 \\ -\ \$2.20 \\ \hline \$3.47 \end{array}$

Add.

1. $\begin{array}{r} \$4.35 \\ +\ \$1.82 \\ \hline \end{array}$ **2.** $\begin{array}{r} \$3.74 \\ +\ \$2.52 \\ \hline \end{array}$ **3.** $\begin{array}{r} \$5.73 \\ +\ \$2.62 \\ \hline \end{array}$ **4.** $\begin{array}{r} \$7.62 \\ +\ \$1.54 \\ \hline \end{array}$ **5.** $\begin{array}{r} \$4.76 \\ +\ \$4.42 \\ \hline \end{array}$

Subtract.

6. $\begin{array}{r} \$5.35 \\ -\ \$1.23 \\ \hline \end{array}$ **7.** $\begin{array}{r} \$6.74 \\ -\ \$2.23 \\ \hline \end{array}$ **8.** $\begin{array}{r} \$6.79 \\ -\ \$3.55 \\ \hline \end{array}$ **9.** $\begin{array}{r} \$7.63 \\ -\ \$4.21 \\ \hline \end{array}$ **10.** $\begin{array}{r} \$8.47 \\ -\ \$5.25 \\ \hline \end{array}$

NS 3.3 Solve problems involving addition, subtraction, multiplication, and division of money amounts in decimal notation and multiply and divide money amounts in decimal notation by using whole-number multipliers and divisors.

Name________________________________

Add and Subtract Decimals and Money

Add or Subtract.

1. $\begin{array}{r} 7.0 \\ 3.4 \\ +2.5 \\ \hline \end{array}$

2. $\begin{array}{r} 0.57 \\ -0.43 \\ \hline \end{array}$

3. $\begin{array}{r} 1.4 \\ -0.2 \\ \hline \end{array}$

4. $\begin{array}{r} \$1.46 \\ +\$2.47 \\ \hline \end{array}$

5. $\begin{array}{r} \$6.84 \\ -\$2.79 \\ \hline \end{array}$

6. $\begin{array}{r} 3.91 \\ +2.25 \\ \hline \end{array}$

7. $\begin{array}{r} 6.9 \\ 1.1 \\ +3.8 \\ \hline \end{array}$

8. $\begin{array}{r} 0.88 \\ +0.33 \\ \hline \end{array}$

9. $\begin{array}{r} \$3.26 \\ +\$1.55 \\ \hline \end{array}$

10. $\begin{array}{r} 7.67 \\ -5.82 \\ \hline \end{array}$

11. $\begin{array}{r} \$8.88 \\ +\$2.22 \\ \hline \end{array}$

12. $\begin{array}{r} 1.76 \\ -0.82 \\ \hline \end{array}$

13. $\begin{array}{r} 2.3 \\ 3.5 \\ +6.9 \\ \hline \end{array}$

14. $\begin{array}{r} 5.6 \\ 3.3 \\ +2.8 \\ \hline \end{array}$

15. $\begin{array}{r} 0.3 \\ +0.8 \\ \hline \end{array}$

Problem Solving and Test Prep

16. Randy runs a race. His time is 0.06 seconds better than Stan's time of 1 minute and 32.06 seconds. What is Randy's time?

____

17. Ruthie wants to buy a pencil for $1.30, a ruler for $0.59, and a book bag for $9.99. She has $12.00. Does Ruthie have enough money to purchase these supplies?

____

18. Jenny buys books for $8.76. She also buys pencils for $2.21. How much money does Jenny spend in all?

A $12.76

B $12.97

C $10.97

D $0.24

19. Stan races his brother Don and wins. He finishes in 1.38 minutes. He beats Don by 0.29 minutes. What is Don's time?

A 0.09 minutes

B 1.67 minutes

C 1.09 minutes

D 1.19 minutes

Name____________________ Lesson 20.6

Problem Solving Workshop Strategy: Make a Model

Beth bought popcorn and a bottle of water. The popcorn cost $2.26 and the water cost $1.56. How much money did she spend in all?

Read to Understand

1. Write the question as a fill-in-the-blank sentence.

____

Plan

2. How can making a model help you solve the problem?

____

____

Solve

3. Fill in the decimal models to solve. Explain how this helped solve.

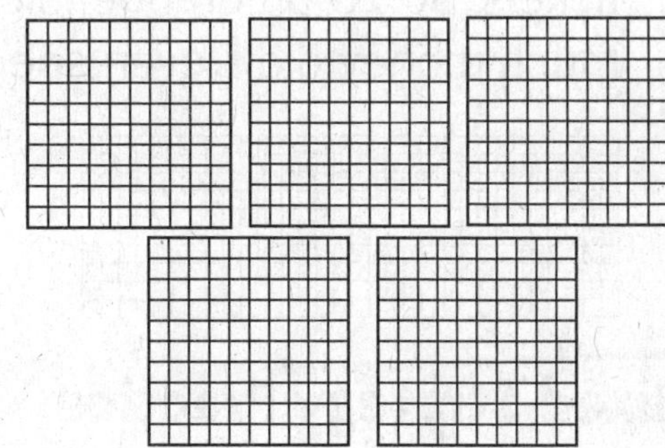

____

____

____

4. Write your answer in a complete sentence.

____

____

Check

5. Is there another strategy you could use to solve the problem?

____

Make a model to solve.

6. Brad bought a sandwich and a can of juice. The sandwich cost $3.95, and the juice cost $1.19. How much money did Brad spend in all?

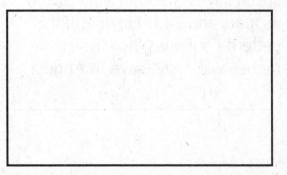

____

NS 3.3 Solve problems involving addition, subtraction, multiplication, and division of money amounts in decimal notation and multiply and divide money amounts in decimal notation by using whole-number multipliers and divisors.

Name________________________________ Lesson 20.6

Problem Solving Workshop Strategy: Make a Model

Problem Solving Strategy Practice

Shade decimal squares to solve.

1. Cindy buys a book for $7.99 and a pen for $1.89. How much money does Cindy spend in all? ______________________

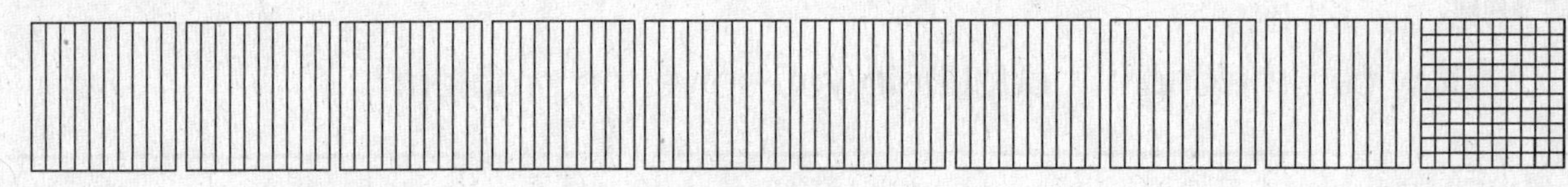

2. Devan has $3.00 to spend at the school fair. He spends $1.00 on ride tickets, $0.75 on orange juice, and $0.63 on a penny pitching game. How much money did Devan spend in all? ______________________

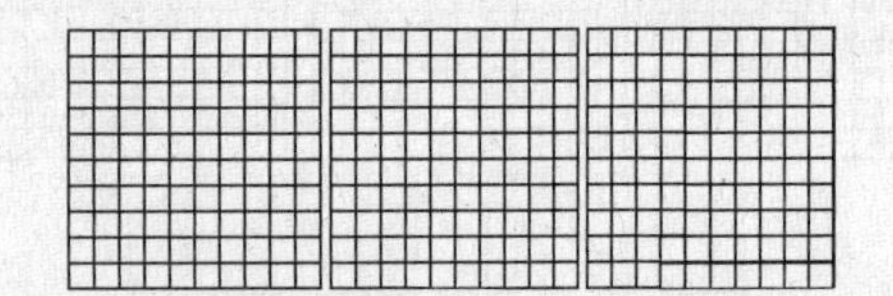

Mixed Strategy Practice

3. Use Data Jesse's mother took her to the toy store. She bought Jesse a book bag, a ball, and a puzzle. She paid with $45.00. How much money did she spend in all? Show your work.

Toy Prices	
book bag	$32.50
ball	$4.75
puzzle	$3.50

____

4. Justin saves $2.00 in May, $3.00 in June, and $4.00 in July. If this pattern continues, how much money will Justin save in August?

____

5. Ronnie earns $1.50 each week. He wants to buy a pair of sneakers for $21.00. How many weeks must he save for the sneakers?

____

Name____ Lesson 21.1

Record Outcomes

Use the data in the table below to answer the questions.

How many more yellow marbles were pulled than blue marbles?

Marble Experiment

Color	Number
red	7
blue	4
green	8
yellow	6

6 yellow marbles were pulled.

4 blue marbles were pulled.

Subtract the number of blue marbles from the number of yellow marbles. 6 – 4 = 2.

So, 2 more yellow marbles were pulled than blue marbles.

Were there more blue and green marbles pulled or more red and yellow marbles pulled?

4 blue and 8 green marbles were pulled.

Add the blue and green marbles together: 4 + 8 = 12.

7 red and 6 yellow marbles were pulled.

Add the red and yellow marbles together: 7 + 6 = 13.

Compare the sums: 13 > 12.

Since 13 is greater than 12, there were more red and yellow marbles pulled than blue and green marbles pulled.

For 1–3, use the table to the right.

Raffle Tickets Sold

Class	Number of tickets
Mrs. Ray	17
Mr. Li	12
Ms. Joy	19
Mr. Terrance	15

1. The classes sold raffle tickets. How many tickets did the 4 classes sell in all?

____

2. How many more tickets did Ms. Joy's class sell than Mr. Li's class sold?

____

3. Which two classes sold more tickets, Mrs. Ray and Mr. Terrence's classes or Mr. Li and Ms. Joy's classes?

____

SDAP 1.3 Summarize and display the results of probability experiments in a clear and organized way (e.g., use a bar graph or a line plot).

Name ______________________________ Lesson 21.1

Record Outcomes

For 1–4, use the data in the Spinner Game table.

Spinner Game	
Color	Tallies
Red	
Blue	
Yellow	
Green	

1. How many times was the spinner spun?

____

2. How many more times did the spinner land on red than on blue?

____

3. How many more times did the spinner land on blue than on green?

____

4. Which color did the spinner land on most often?

____

Problem Solving and Test Prep

USE DATA For 5–6, use the data in the Points Score table.

Points Scored	
Player	Points
Ben	10
Scott	12
Ryan	9
Jake	8
Trevor	6

5. The coach recorded the number of points each player on his team scored during the basketball game. How many points did Jake score?

____

6. How many points did the entire team score during the game?

____

7. A Favorite Juice table shows 𝍸 |||| next to apple. How many people said apple was their favorite juice?

A 9
B 6
C 4
D 15

8. Jess rolls a color cube. Her results are: 7 reds, 6 blues, and 5 yellows. Which group of tallies shows the number of reds Jess rolls?

A 𝍸 |
B 𝍸 ||
C 𝍸 |||
D 𝍸

Spiral Review

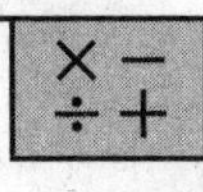

For 1–5, write the decimal for the given fraction.

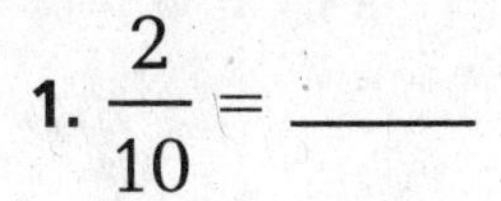

1. $\frac{2}{10}$ = ____

2. $\frac{9}{10}$ = ____

3. $\frac{1}{10}$ = ____

4. $\frac{8}{100}$ = ________

5. $\frac{57}{100}$ = ________

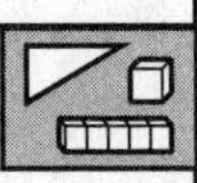

For 6–9, find the volume of the solid figures.

	Figure	Volume
6.	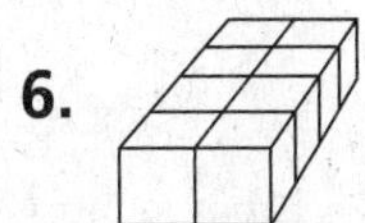	____ cubic units
7.	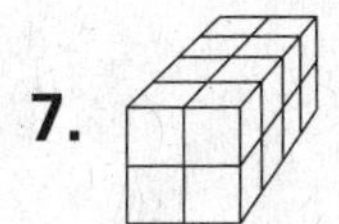	____ cubic units
8.	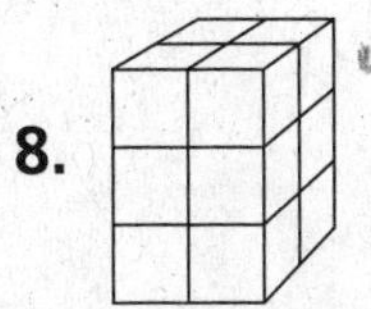	____ cubic units
9.	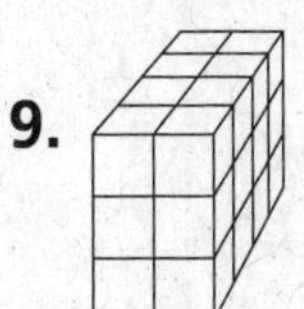	____ cubic units

For 10–12, answer the questions using the information in the line plot.

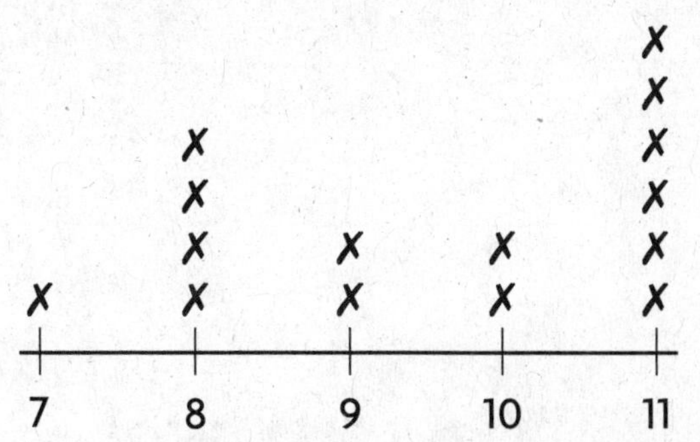

Ages of Students in the Nurse's Office

10. How many students who went to the nurse's office were 7 years old?

____

11. What is the range of the data? ____

12. What is the mode of the data? ____

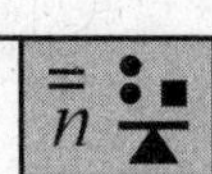

For 13–15, fill in the blank.

13. 3 feet = ____ inches

14. 2 yards = ____ feet

15. 108 inches = ____ yards

Name____ Lesson 21.2

Problem Solving Workshop Strategy: Make a Graph

Ben tossed a number cube 30 times. It landed on one 4 times, on two 6 times, on three 4 times, on four 8 times, on five 3 times, and on six 2 times. Make a bar graph to show Ben's results. Which bar is twice as long as the bar for six?

Read to Understand

1. What information is given?

____

Plan

2. How will making a bar graph help you solve this problem?

Solve

3. Solve by making a bar graph.
   - Write labels on the side and bottom.
   - Write the number for each bar.
   - Draw the bar for two.

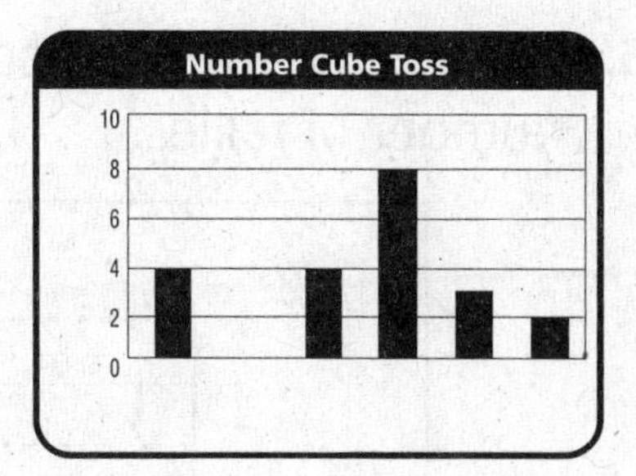

4. Which bar is twice as long as the bar for six?

Check

5. What if 3 of the rolls had been sixes instead of fours? How would the graph change?

____

____

Use the bar graph from above to solve.

6. Which bar is twice as long as the bar for five?

____

7. Which bar is four times as long as the bar for six?

____

Name____

Problem Solving Workshop Strategy: Make a Graph

Problem Solving Strategy Practice

USE DATA For 1–6, use the table below.

A group of students played a game with a spinner. They recorded the result of each spin in the tally table at the right.

Spinner Results	
Color	**Tally**
Red	𝍸 I
Green	𝍸 𝍸 I
Blue	𝍸 III
Yellow	IIII

Show these results in a bar graph.

1. Write the title at the top of the graph.
2. Write a label for the side and for the bottom.
3. Write the four colors to identify each bar.
4. Choose a scale. Write the scale numbers starting at 0.
5. Draw a bar for each color. The length of the bar is equal to the color's number of tallies.

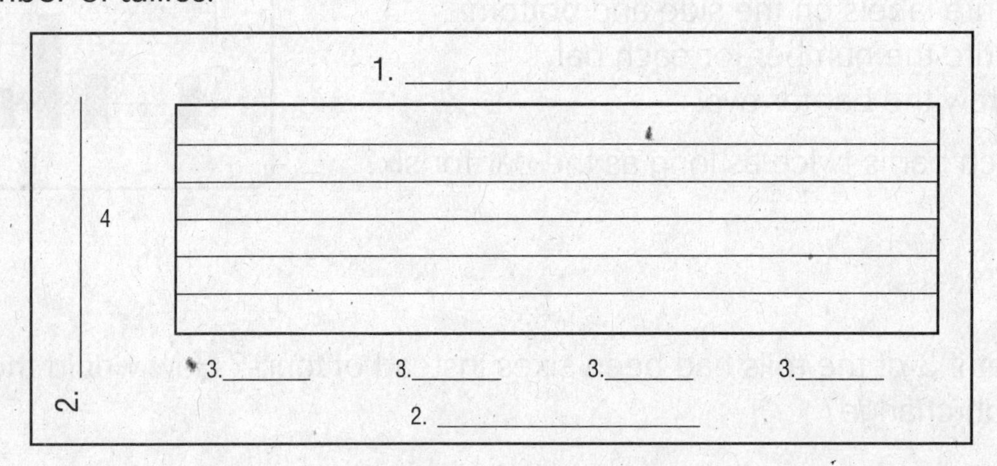

6. How many times did the spinner land on yellow and red combined?

____

Mixed Strategy Practice

7. Jerry wanted to make a different sandwich of bread and deli meat for each of his 3 friends. He has rye and pumpernickel breads. He has turkey, pastrami, and salami deli meats. What are the possible 1 bread – 1 deli meat combinations.

____

____

Name____________________

Probability: Likelihood of Events

Probability tells us how likely something is to happen. Something that is **certain** always happens. Something that is **impossible** never happens. Something that is likely happens most of the time. Something that is **unlikely** happens only some of the time.

The spinner at the right contains equal sized portions. Tell if it is likely, unlikely, certain, or impossible that the pointer will land on purple.

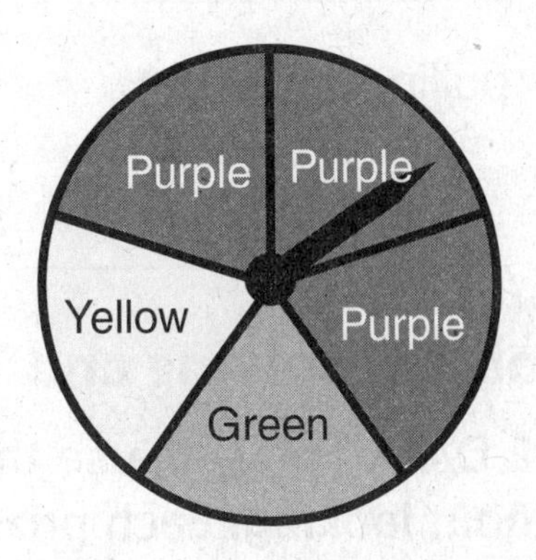

You know that it is not ***certain*** that the pointer will land on purple because it is possible that it could land on another color.

You also know that it is not ***impossible*** for the pointer to land on purple, because it is represented along with the two other colors.

Now you need to decide whether it is ***likely*** or ***unlikely*** that the pointer will land on purple.

Add up the sections: $3 + 1 + 1 = 5$

Three of the five sections are purple, and 3 is more than half of 5.

So, it is *likely* that the pointer will land on purple.

A prize bag is filled with toy cars of equal size. There are 9 black, 2 red, 5 yellow, and 3 green cars. One car is pulled from the bag without looking.

1. Which color toy car is most likely to be pulled from the bag?

____

2. Is it certain or impossible to pull a toy car from the bag?

____

3. Is it likely or unlikely that a black toy car is pulled from the bag?

____

4. Is it impossible to pull a red car from the bag?

____

Name: ______________________________

Lesson 21.3

Probability: Likelihood of Events

For 1–6 use the bag of tiles. Each tile is the same size and shape. Tell whether each event is *likely, unlikely, certain*, or *impossible*.

B is blue **G** is green
R is red **Y** is yellow

1. pulling a blue tile

____

2. pulling a red tile

____

3. pulling a white tile

____

4. pulling a yellow tile

____

5. pulling a tile

____

6. pulling a green, blue, yellow, or red tile

____

Problem Solving and Test Prep

USE DATA For 7–8, use the table. Ben pulls one prize from the bag without looking. Each prize is the same size and shape.

Prize Bag	
Prizes	**Number**
blue ball	3
red ball	5
green ball	1

7. Is it certain or impossible that Ben will pull a stuffed toy?

____

8. Is it likely or unlikely that Ben will pull a red ball?

____

9. Charles grabs a shirt from his drawer without looking. Four of his shirts are white, 1 is yellow, and 5 are blue. Which represents the likelihood that Charles grabs a yellow shirt, if all of the shirts are the same size?

   A likely **C** certain
   B unlikely **D** impossible

10. Sara is playing a game using a spinner. The spinner contains 8 sections of equal size: 1 green, 3 blue, 2 white, and 2 red. Which color is Sara least likely to spin?

   A green **C** white
   B blue **D** red

Name____

Possible Outcomes

When you choose something out of a bag there are usually a number of possible **outcomes**, or possible results. You can use knowledge about the contents of the bag to **predict** what will be pulled.

Billy says that pulling a red marble is equally likely as pulling a green marble. Does his statement make sense?

Use the bag below, which contains marbles of equal size.

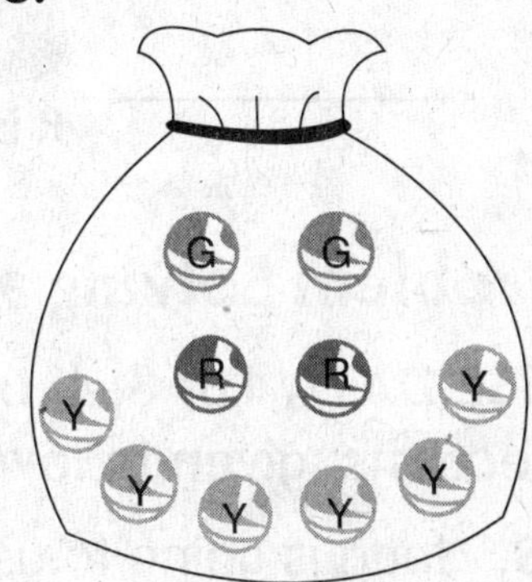

Every color marble in the bag is a possible outcome.

Possible outcomes: red, green, yellow

Now which outcomes are equally likely?

Count each number of marbles in the bag:

6 yellow, 2 green, 2 red

Two of the colors are represented by an equal number of marbles: red and green.

So, it is ***equally likely*** that a red marble or a green marble will be pulled out of the bag.

Billy's statement makes sense.

Key
G = Green
R = Red
Y = Yellow

For 1–2, list the possible outcomes for each, then tell which outcomes are equally likely.

1. List the possible outcomes for pulling one marble out of the bag below, if all the marbels are of equal size. Then tell which outcomes are equally likely.

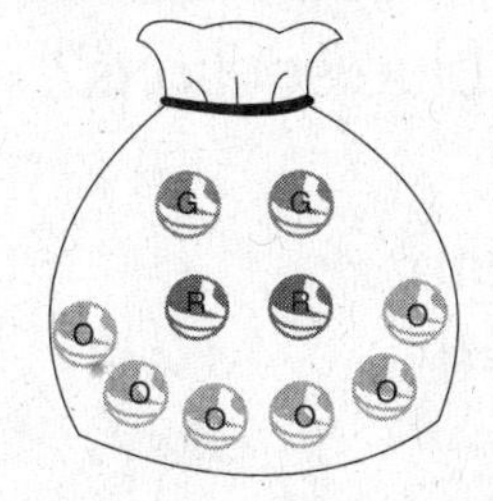

Key
G = Green
R = Red
O = Orange

____

____

____

____

2. List the possible outcomes for pulling one tile out of the bag below, if all the tilles are of equal siz and shape. Then tell which outcomes are equally likely.

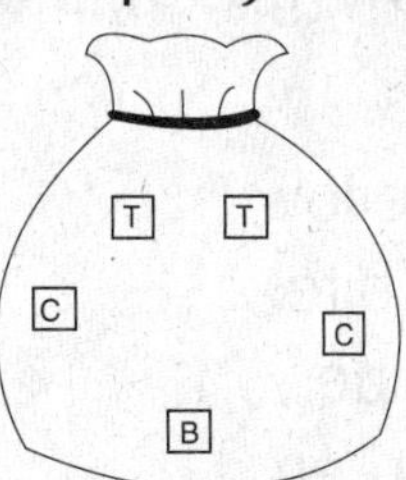

____

____

____

____

Name ____________________

Possible Outcomes

For 1–2, list the possible outcomes for each.

1. Elizabeth will pull a marble from the bag.

______

______

______

R is red **G** is green

B is blue

2. Joe will use the spinner.

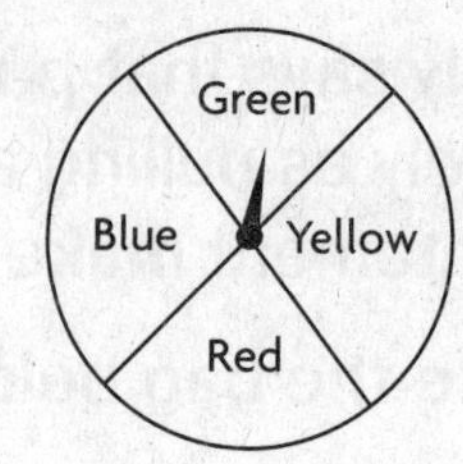

______

______

______

______

Problem Solving and Test Prep

USE DATA For 3–4, use the spinner, with equal sized sections, given below.

3. John is going to use the spinner. What are the possible outcomes?

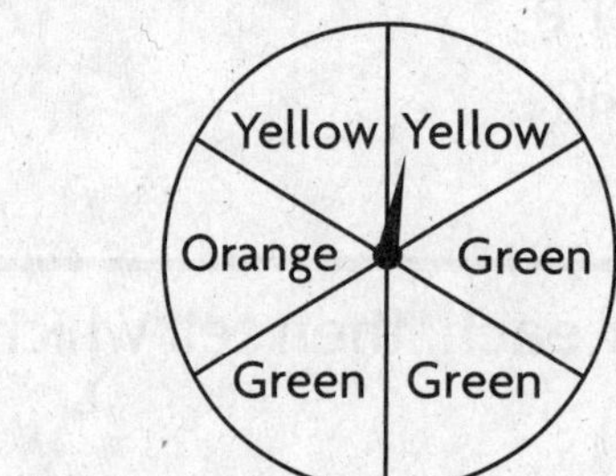

______

______

______

______

4. If John spins the spinner one time, is it equally likely that it will land on green or orange?

______

5. Which is NOT a possible outcome of spinning a spinner with the following colors one time: yellow, green, blue?

A yellow
B white
C blue
D green

6. Which are equally likely outcomes for a spinner with sections of equal size, with these colored sections : 2 yellow sections, 3 red sections, 4 white sections, 2 blue sections?

A yellow and red
B yellow and white
C yellow and blue
D red and white

Name______________________________

Experiments

An **experiment** is a way to explore something such as probability. **Probability** is the likelihood of an event happening.

Use the bar graph and the bag below, which contains tiles of equal size and shape, to answer the following question.

The bar graph below shows the results of a tile being pulled from the bag 20 times. The tile is returned to the bag after each pull.

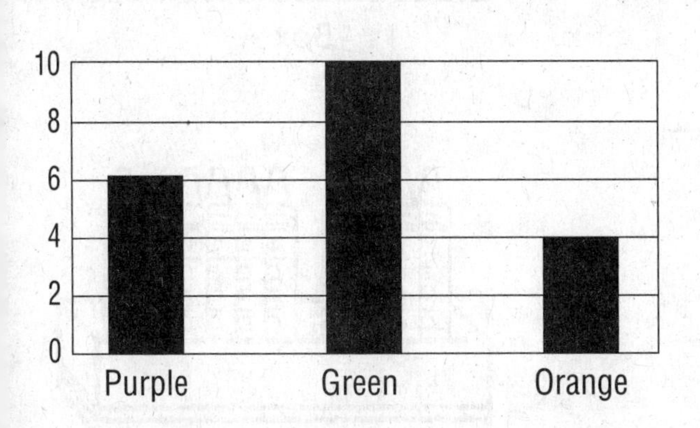

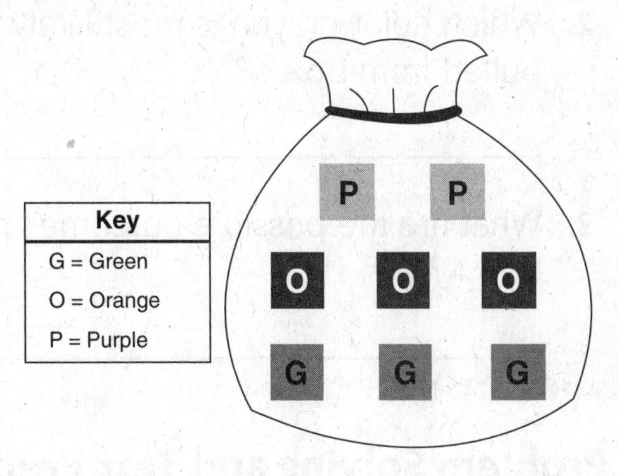

Look at the bar graph. Which tile is most likely to be pulled next?

The bar for green is the highest. So, a green tile was pulled the most often.

So, it is most *likely* that a green tile will be pulled from the bag on the next pull.

For 1-4, use the spinner at the right which contains sections of equal size.

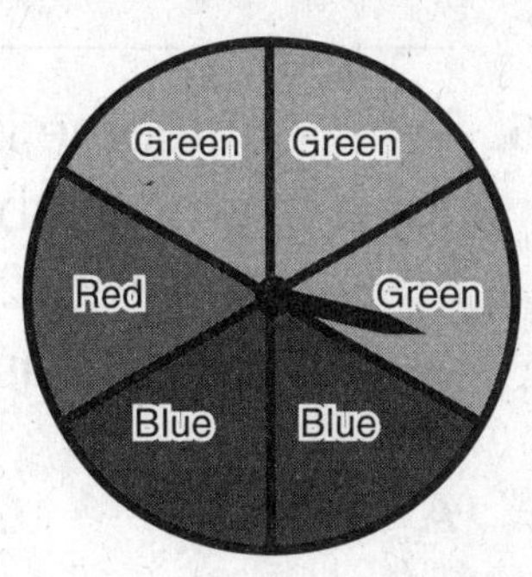

1. What are the possible outcomes? ______________ ______________
2. Which outcome is most likely? ______________
3. Which outcome is least likely? ______________
4. What is the probability of landing on blue? ______________

Name ____________________ **Lesson 21.5**

Experiments

For 1–3, use the boxes of crayons. Each crayon is the same size and shape.

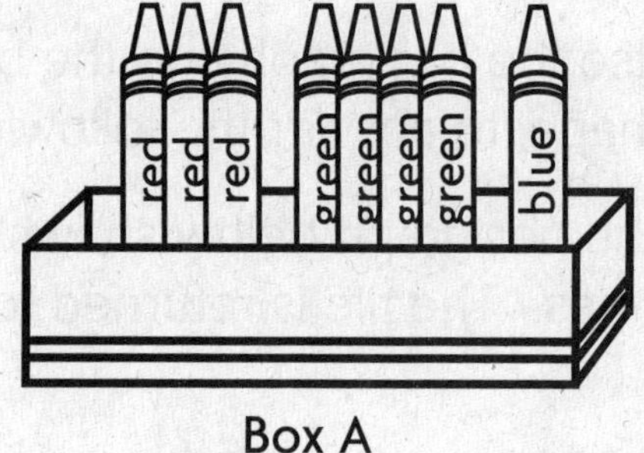

Box A

1. In Box B, which outcomes are equally likely?

____

2. Which color crayon is most likely to be pulled from Box A?

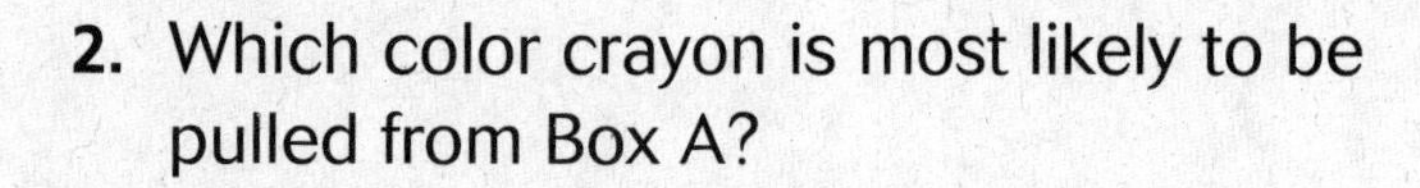

____

3. What are the possible outcomes for Box A?

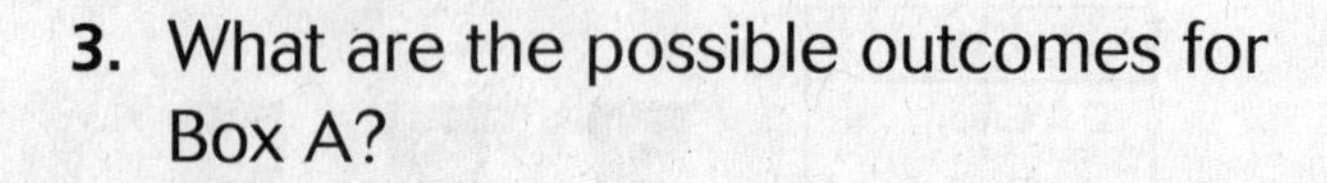

____

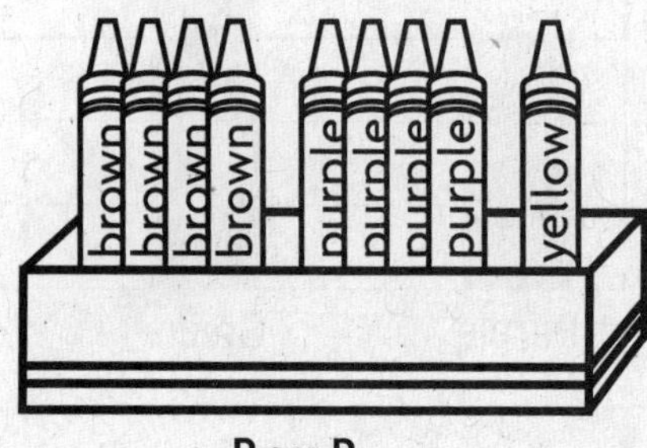

Box B

Problem Solving and Test Prep

4. A box of cookies, each of equal size and shape, contains 4 raisin, 4 oatmeal, and 6 ginger cookies. Which cookie is most likely to be pulled in one pull?

____

5. Which outcomes are equally likely for a bag of marbles, each of equal size, with 2 red, 3 green, and 2 yellow marbles?

____

6. Which outcome is least likely for a bag of marbles each of equal size, with 4 red, 2 blue, 1 green, and 3 yellow marbles?

   A red **C** green

   B blue **D** yellow

7. Which is the probability of pulling a green marble from a bag, with marbles of equal size, of 4 red, 2 blue, 1 green, and 3 yellow marbles?

   A 1 out of 10 **C** 3 out of 10

   B 2 out of 10 **D** 4 out of 10

Name______________________________

Lesson 21.6

Line Plots

A **line plot** uses an X to show each data value on a number line. A line plot can show the same data as a tally table.

What is the range of Steve's data?

Spinner Experiment	
Number Landed	Tally
1	|||
2	~~||||~~ |
3	|||
4	||
5	||||
6	||

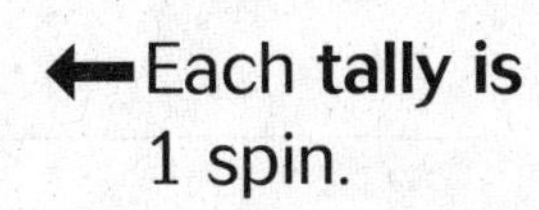

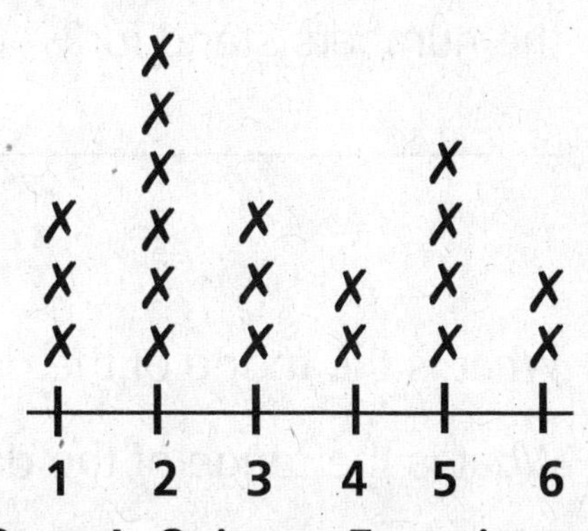

Steve's Spinner Experiment

The tally table has 20 tally marks in all.

A **range** is the difference between the greatest number and the least number in a set of data.

The line plot has 20 X's in all.

In this case 6 is the greatest number while 1 is the least number.

So 6 – 1 = 5, 5 is the range.

For 1-6, use the line plot below.

1. How many times did Jose roll a 3? _____
2. What is the range of Jose's data? _____
3. What is the mode of Jose's data? _____
4. Which number did Jose roll the least number of times? __________
5. How many more times did Jose roll a 1 than a 3? __________
6. How many more times did Jose roll a 6 than a 4? __________

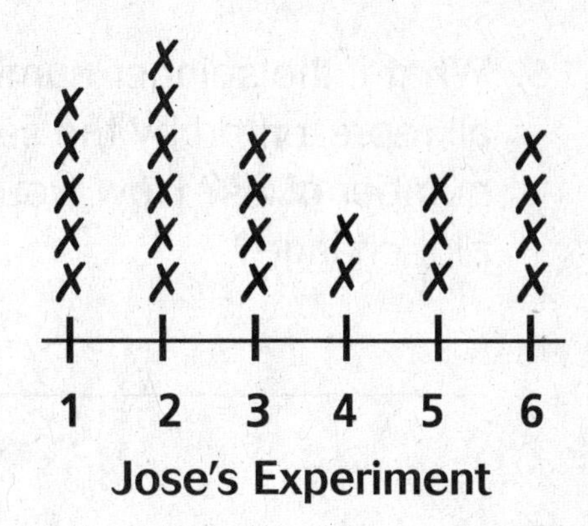

Jose's Experiment

Name____ **Lesson 21.6**

Line Plots

For 1–3, use the Swimming Lanes line plot.

1. The *X*s on the line plot stand for the number of swimmers. What do the numbers stand for?

____

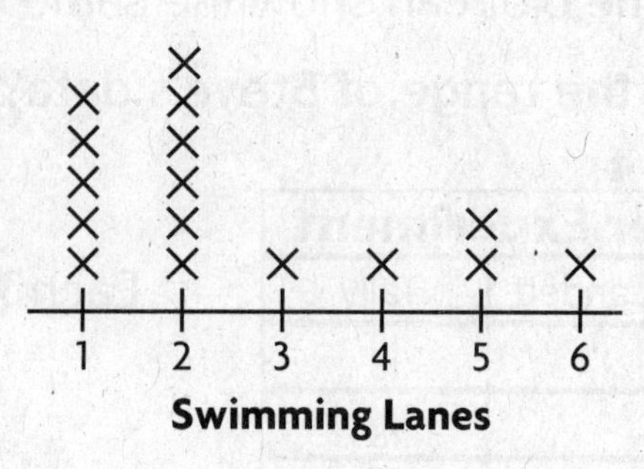

2. What is the mode of the data? ______________________________

3. What is the range of the data? ______________________________

Problem Solving and Test Prep

For 4–7, use the Spinner Experiment line plot below.

4. Did more spins land on 1 and 2 combined, or on 4 and 5 combined? Explain.

____

5. **What** if the spinner numbers were all represented by the same number of Xs? How would the line plot change?

____

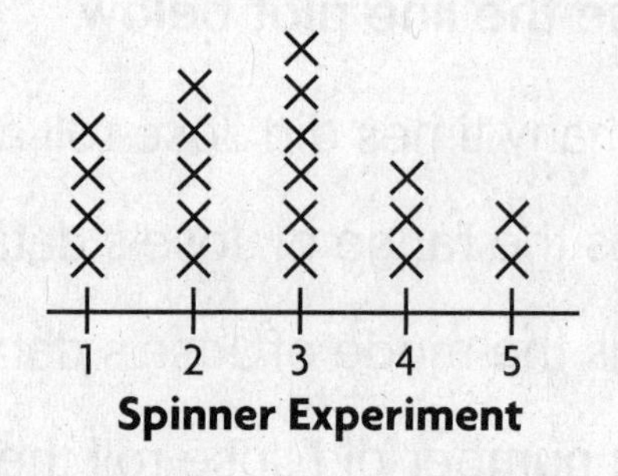

6. How many more spins landed on 3 than on 1?

A 1 **C** 3

B 2 **D** 6

7. How many spins landed on 3 and 4 in all?

A 3 **C** 6

B 4 **D** 9

Spiral Review

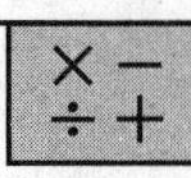

For 1–5, solve the problems.

1. $\$16.54 - \$15.99$

2. $\$40.52 + \$54.84$

3. $\$18.19 + \$94.45 + \$3.91$

4. $\$0.78 \times 6$

5. $\$89.10 \times 9$

For 10–12, use the line plot to answer the questions.

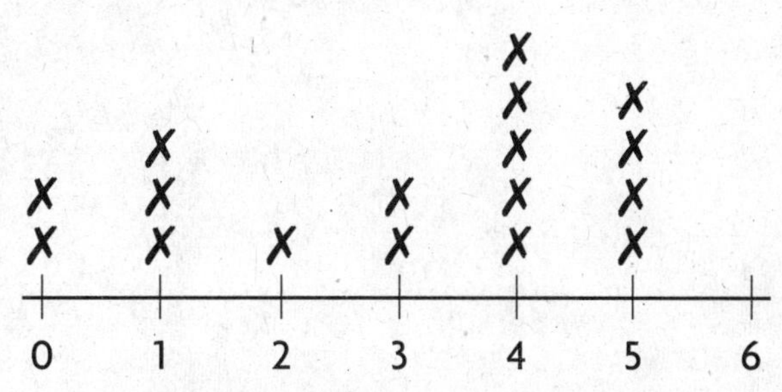

10. How many more players scored 4 home runs than scored 0 home runs? ____

11. How many players scored 5 home runs? ________

12. How many players scored 0 home runs? ________

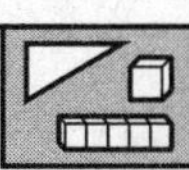

For 6–9, draw lines to match the figure with its name.

6.  pentagon

7. 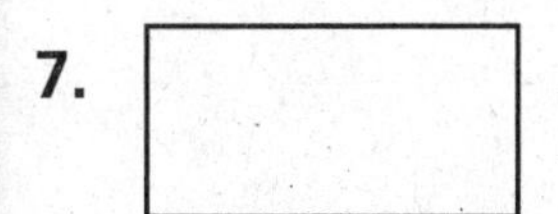 rectangle

8. 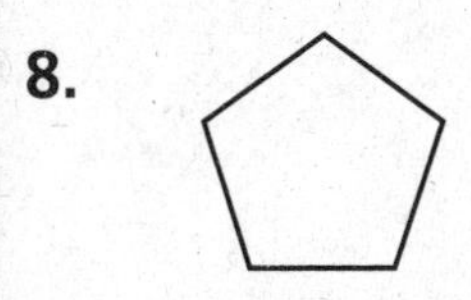 octagon

9. 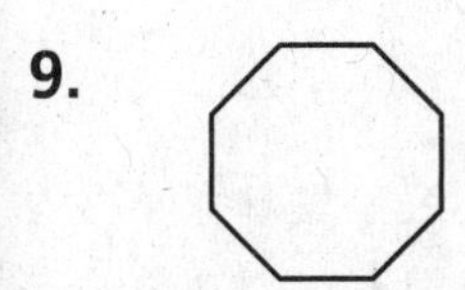 hexagon

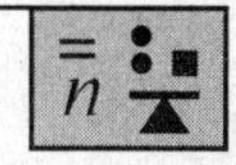

For 13–15, use the rule to answer the questions.

Rule: The bowling alley lets 2 people bowl 4 games for $5.

13. How much money would it cost for 4 people to bowl 4 games?

____

14. How much money would it cost for 2 people to bowl 8 games?

____

15. How many people could bowl 4 games for $20?

____

Name______________________________ Lesson 21.7

Predict Future Events

The results of an experiment can be shown in a tally table. You can use the results of a probability experiment to make predictions. A *prediction* is a good guess of what might happen in the future.

The tally table below shows the results of spinning the pointer on a spinner which contains sections of equal size, 22 times. Predict the color on the next spin.

Spinner Results	
Color	**Tallies**
Purple	𝍸 𝍸 ||||
Green	|||
Red	𝍸

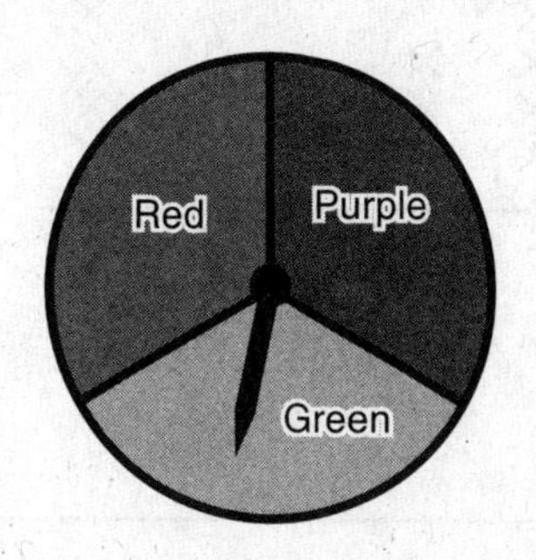

Which color did the spinner land on the most out of 22 spins?

Notice that the spinner landed on purple the most, 14 out of 22 times.

So, a good prediction for the next spin (or future event) is that the spinner will land on purple again.

Use data For 1–3, use the line plot below.

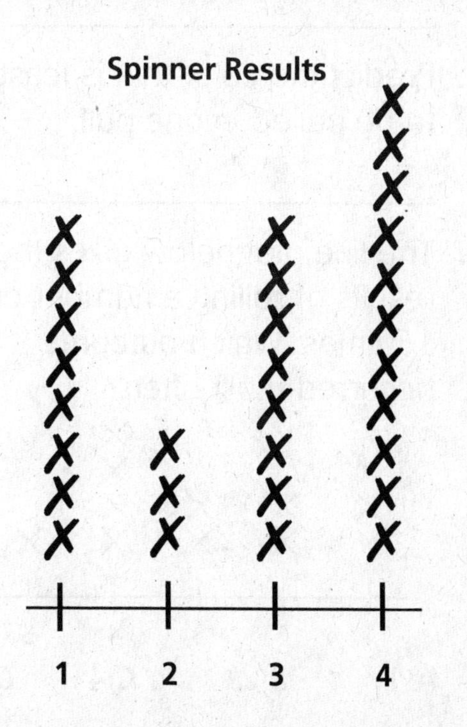

1. The line plot at the right shows the results of spinning the pointer on a spinner that has sections of equal size, 30 times. Predict the number on the next spin.

____

2. Predict the numbers that are equally likely to be the result of spinning.

____

3. Predict which number is least likely to be the result of the next spin.

____

Name ____________________

Predict Future Events

Use the line plot and tally table.

1. The line plot below shows the results of a number cube that was rolled 20 times. Predict the number on the next roll. Explain.

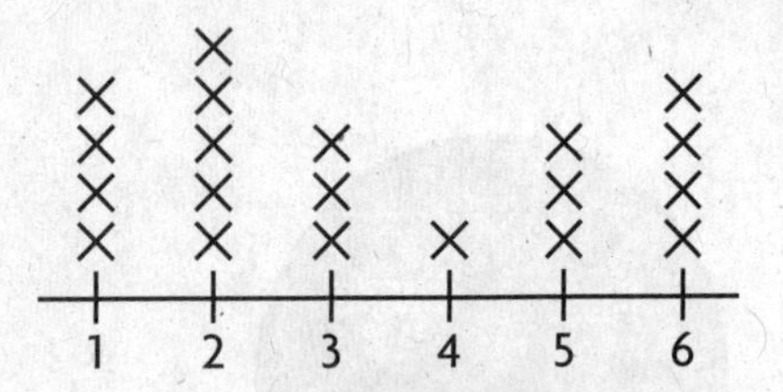

____

____

2. The tally table below shows the results of spinning the pointer on a spinner with sections of unequal size, 25 times. Predict the color on the next spin. Explain.

Spinner Results	
Color	**Tallies**
Red	卌 II
Green	卌 I
Blue	卌 卌 II

____

____

Problem Solving and Test Prep

USE DATA For 3–4, use the graph.

3. Predict the colors that are equally likely to be pulled in one pull.

____

4. Predict the color that is least likely to be pulled in one pull.

____

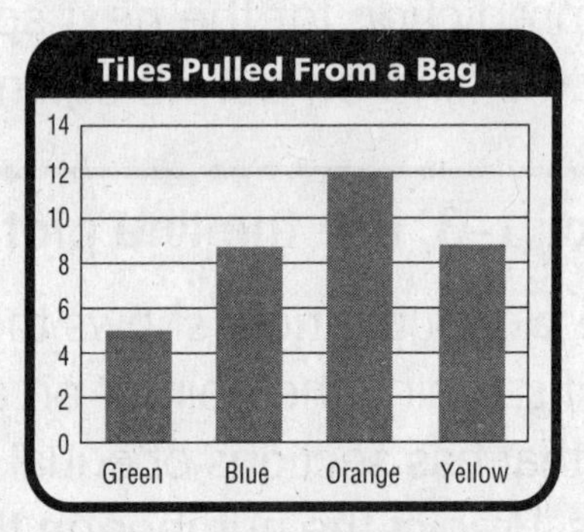

5. The line plot below gives the results of rolling a number cube 15 times. Which outcome occurred most often?

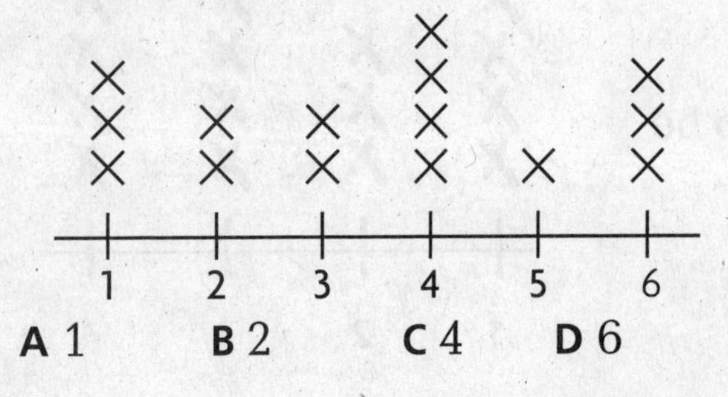

A 1 **B** 2 **C** 4 **D** 6

6. The tally table below shows the results for tossing a coin 30 times. Which is the best prediction for the next coin toss?

Side	Tallies
Heads	卌 卌 卌
Tails	卌 卌 卌

A Heads **C** Neither

B Tails **D** Both are equally likely

Name______________________________ Lesson 21.8

Problem Solving Workshop Strategy: Make an Organized List

Lana's class is having a party. Each student can choose one pencil and one notebook. There are smiley face pencils and animal pencils. The notebooks are blue, red, or yellow. How many combinations of 1 pencil and 1 notebook are there?

Read to Understand

1. Write the question as a fill-in-the-blank sentence.

____

Plan

2. How can making an organized list help you solve this problem?

____

Solve

3. Make an organized list to show the possible combinations of pencils and notebooks.

4. How many combinations of 1 pencil and 1 notebook are there? Write your answer in a complete sentence.

____

Check

5. What is another way you can solve this problem?

____

Make an organized list to solve.

6. Tina plans to make a pasta dish. She has spaghetti, macaroni, and elbow noodles. She has marinara sauce and pesto sauce. How many possible combinations of noodles and sauce are there?

____

7. George plans to make a pizza. He has olives and onions. He has marinara sauce and alfredo sauce. How many possible combinations of topping and sauce are there?

____

SDAP 1.2: Record possible outcomes for a simple event (e.g., tossing a coin) and systematically keep track of its outcomes when the event is repeated many times.

Name ____________________

Problem Solving Workshop Strategy: Make an Organized List

Problem Solving Strategy Practice.

USE DATA For 1–2, use the table.

Meat	Cheese	Bread
roast beef	swiss	white
turkey	cheddar	wheat
ham		

1. Peter wants to make a sandwich with 1 type of meat and 1 type of bread. How many different meat and bread combinations can Peter make?

____

2. Lizzy wants to make a sandwich with 1 type of bread and 1 type of cheese. How many different bread and cheese combinations can Lizzy make? ____________________

Mixed Strategy Practice

3. Mara baked treats for a family gathering. She baked sugar cookies, oatmeal cookies, brownies, muffins, and banana bread. She wants to give each of her 20 family members the same number of muffins. She made 60 muffins. How many muffins will each family member receive?

____

4. Frank has 19 coins in his pocket. Yesterday he had 14 coins. How many coins has Frank added to his pocket since yesterday?

____

5. **Use Data** Jamal and his sister need school supplies. They each need 8 pencils, 10 markers, 10 colored pencils, and 2 folders. How many packs of each type of school supply do Jamal and his sister need to buy? Draw a diagram to help solve.

____

____

School Supplies	
Type	**Number per Pack**
pencils	12
markers	8
colored pencils	10
folders	4

Name____ **Lesson 22.1**

Compare Money Amounts

You can use $<$, $>$, or $=$ to compare amounts of money..

Compare the amounts of money below.

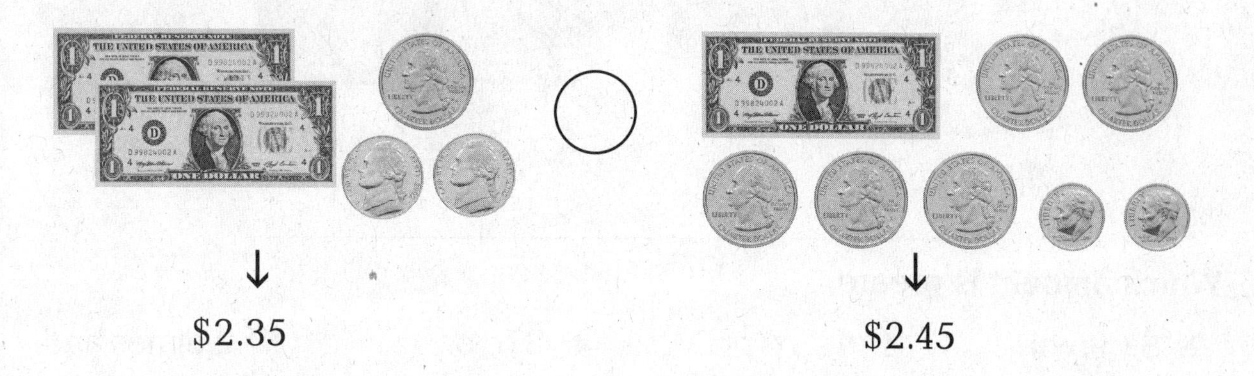

$2.35 $2.45

So, $\$2.35 < \$2.45$

Use $<$, $>$, or $=$ to compare the amounts of money.

1.   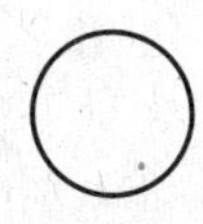    

2.   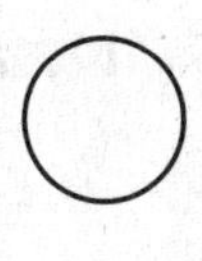 

3.  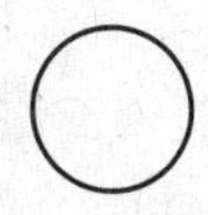 

Name____ Lesson 22.1

Compare Money Amounts

Use <, >, or = to compare the amounts of money.

1.

____

Which amount is greater?

2. $7.95 or $7.89 ____________

3. $2.10 or 9 quarters ____________

4. 87¢ or 18 nickels ____________

5. 2 dimes and 6 pennies or 1 quarter ____________

6. 2 dimes and 2 nickels or 1 quarter ____________

7. $2.25 or 5 quarters ____________

8. 35 pennies or 2 dimes, 2 nickels, and 2 pennies ____________

9. $1.12 or 11 dimes ____________

Problem Solving and Test Prep

10. Aidan has 7 quarters, 3 dimes, 3 nickels, and 4 pennies. Maria has 1 one-dollar bill, 3 quarters, 2 dimes, and 3 nickels. Who has more money, Aidan or Maria?

____

11. Matt has 5 quarters, 6 dimes, and 4 nickels. Hal has $2.51. Who has more money, Matt or Hal?

____

12. Becky has only dimes. She has more than 60¢. Which amount could Becky have?

A 75¢ **C** 81¢
B 50¢ **D** 70¢

13. Danny has only quarters and dimes. He has at least 1 quarter and 1 dime. He has more than 25¢. Which amount could Danny have?

A 45¢ **C** 64¢
B 30¢ **D** 40¢

Name____________________________________

Model Making Change

Change is the money you get back if you pay more than an item costs.

Find the amount of change. Use play money to help.

Brenda buys a comic book for $3.42. She pays with a $5 bill.

Start with the cost of the comic book. Count up to the amount Brenda paid. Use play money to help.

Cost of Comic Book       Amount Paid

$3.42 →  **$3.45** → **$3.50** → **$4.00** → **$5.00** ← Amount Paid

Now, count the value of the coins and bills Brenda received as change.

So, Brenda received $1.58 in change.

Find the amount of change. Use play money to help.

1. Ben buys a book for $2.95. He pays with a $5 bill.

____

2. Rosa buys an apple for 55¢. She pays with a $1 bill.

____

3. Oscar buys a bird treat for 79¢. He pays with a $5 bill.

____

4. Amy buys a marker for $1.99. She pays with a $10 bill.

____

5. Adam gets a haircut for $9.98. He pays with a $10 bill.

____

6. Maya buys a dozen pencils for $2.26. She pays with a $5 bill.

____

NS 3.3 Solve problems involving addition, subtraction, multiplication, and division of money amounts in decimal notation and multiply and divide money amounts in decimal notation by using whole-number multipliers and divisors.

Name____________________

Lesson 22.2

Model Making Change

Find the amount of change. Use play money to help.

1. Ali buys a collar for her dog for $4.69. She pays with a $10 bill.

____

2. Roger buys a banana for $0.49. He pays with a $1 bill.

____

3. Finn buys a bouncy ball for $0.25. He pays with a $1 bill.

____

4. Samantha buys a pom-pom for $7.42. She pays with a $10 bill.

____

Find the amount of change. List the coins and bills.

5. Pay with a $10 bill.

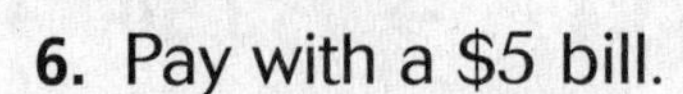

6. Pay with a $5 bill.

Problem Solving and Test Prep

7. Isaac buys sunglasses for $3.99. He pays with a $10 bill. How much change does Isaac receive? List the coins and bills.

____

8. Zoe wants to buy a clown wig that costs $5.99 and face paint that costs $1.15. She has $6.10. Is this enough money? If not, how much more does Zoe need?

____

9. Sally uses a $5 bill to buy a book that costs $3.54. How much change will Sally receive?

A $2.46
B $1.54
C $1.46
D $8.54

10. Lori wants to buy 2 CDs. They cost $10.39. She has $7.50. How much more money does Lori need?

A $2.89
B $3.11
C $17.89
D $3.89

Name______________________________

Problem Solving Workshop Strategy: Compare Strategies

Tyler has 7 quarters, 4 dimes, and 6 nickels. He wants to buy a hot dog for $1.45. How much money will Tyler have left?

Read to Understand

1. What are you asked to find?

____

____

Plan

2. How can making a table or drawing a picture help you solve the problem?

____

____

Solve

3. How much money will Tyler have left?

____

Check

4. What other strategy could you use to solve the problem? Explain.

____

Solve by comparing strategies.

5. Andy has one $1 bill and 5 quarters. He wants to buy juice for $1.25. How much money will Andy have left?

____

6. Faith has two $1 bills, 6 quarters, 3 dimes, and 2 nickels. She wants to buy a puzzle book for $2.25. How much money will Faith have left?

____

Problem Solving Strategy Workshop: Compare Strategies

Problem Solving Strategy Practice

Draw a picture or write a number sentence to solve.

1. Greg has three $1 bills and five coins. He has a total of $3.41. What type of coin does Greg have the most of?

____

2. Kyle wants to go bowling during the bowling alley's $8 dollar special. He only has coins. He has 24 quarters, 10 dimes and16 nickels. Does Kyle have enough money to go bowling?

____

Mixed Strategy Practice

3. Jenna has more money than Frank. Frank has less money than Earl. Earl has the most money. Put the students in order from greatest amount of money to least amount of money.

____

____

4. Ray has 3 quarters in his pocket. Earlier today he had $0.87, and twice as many coins in his pocket as he has now. What coins did Ray have in his pocket earlier today?

____

____

5. **USE DATA** Jenna has one $10 bill. She wants to buy three bananas, seven kiwis, and six peaches. Does Jenna have enough money? If not, how much more money does she need?

____

Fruit	Price for One
banana	$0.23
kiwi	$0.99
peach	$0.45

Spiral Review

For 1–5, compare the fractions using the number lines.

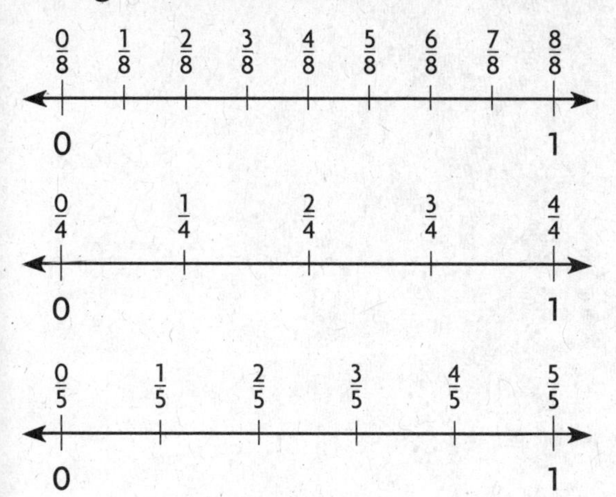

1. $\frac{4}{8} \bigcirc \frac{2}{4}$
2. $\frac{1}{4} \bigcirc \frac{1}{5}$
3. $\frac{3}{5} \bigcirc \frac{5}{8}$
4. $\frac{1}{5} \bigcirc \frac{1}{4}$
5. $\frac{2}{8} \bigcirc \frac{1}{5}$

For 9–12, tell whether each event is *likely, unlikely, certain,* or *impossible* with one pull from the bag containing marbles.

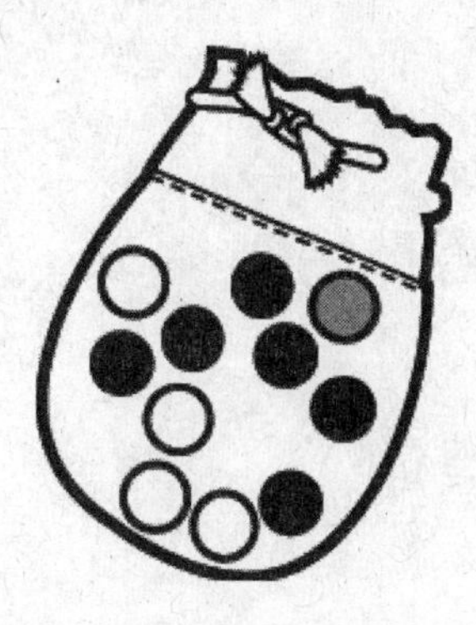

9. pulling a black marble ________
10. pulling a marble ________
11. pulling a gray marble ________
12. pulling a penny ________

For 6–9, draw the given quadrilateral.

Quadrilateral	Drawing
6. rectangle	
7. rhombus	
8. square	

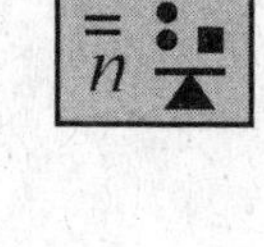

For 13–15, predict the next number in each pattern. Explain.

13. 79, 78, 77, 76, 75, ☐

____

14. 650, 625, 600, 575, ☐

____

15. 1, 3, 9, 27, ☐

____

Name______________________________ Lesson 22.4

Add Money Amounts

Adding money mounts is like adding whole numbers.
You can also estimate to see if your answer is reasonable.

Estimate. Then find the sum.

- Estimate to the nearest dollar.

$$\begin{array}{r} \$6.25 \\ +\ \$3.12 \\ \hline \end{array} \qquad \begin{array}{r} \$6.00 \\ +\ \$3.00 \\ \hline \$9.00 \end{array}$$

- Line up the decimal points and add. Write your answer in dollars and cents.

$$\begin{array}{r} \$6.25 \\ +\ \$3.12 \\ \hline \$9.37 \end{array}$$

- Since \$9.37 is close to the estimate of \$9.00, the answer is reasonable.

So, the sum is \$9.37.

Estimate. Then find the sum.

1. $\begin{array}{r} \$0.59 \\ +\ \$0.33 \\ \hline \end{array}$ ______ ______

2. $\begin{array}{r} \$2.10 \\ +\ \$1.33 \\ \hline \end{array}$ ______ ______

3. $\begin{array}{r} \$0.99 \\ +\ \$0.44 \\ \hline \end{array}$ ______ ______

4. $\begin{array}{r} \$1.89 \\ +\ \$0.53 \\ \hline \end{array}$ ______ ______

5. $\begin{array}{r} \$11.45 \\ +\ \$6.21 \\ \hline \end{array}$ ______ ______

6. $\begin{array}{r} \$20.00 \\ +\ \$5.15 \\ \hline \end{array}$ ______ ______

7. $\begin{array}{r} \$55.98 \\ +\ \$33.44 \\ \hline \end{array}$ ______ ______

8. $\begin{array}{r} \$0.99 \\ +\ \$0.02 \\ \hline \end{array}$ ______ ______

9. $\begin{array}{r} \$32.14 \\ +\ \$16.87 \\ \hline \end{array}$ ______ ______

10. $\begin{array}{r} \$35.12 \\ +\ \$11.09 \\ \hline \end{array}$ ______ ______

NS 3.3 Solve problems involving addition, subtraction, multiplication, and division of money amounts in decimal notation and multiply and divide money amounts in decimal notation by using whole-number multipliers and divisors.

Name______________________________________ Lesson 22.4

Add Money Amounts

Estimate. Then find the sum.

1. $0.52 + $0.36

2. $3.45 + $4.56

3. $18.29 + $2.21

4. $0.99 + $0.87

5. $9.82 + $2.38

6. $14.75 + $5.02

7. $7.43 + $7.68

8. $12.91 + $22.72

9. $5.47 + $2.99 + $6.49 = ______

10. $27.89 + $13.25 = ______

11. $0.19 + $0.45 + $0.22 = ______

12. $13.76 + $4.18 + $0.12 = ______

Problem Solving and Test Prep

13. Don bought a toy dinosaur for $2.89. He bought a football for $9.99. How much did Don spend in all?

____

14. Wendy bought a sticker for $0.30, a card for $2.95, and a purple pencil for $0.70. How much did Wendy spend in all?

____

15. Brad bought knee pads for $6.98. He bought elbow pads for $5.98. How much did Brad spend in all?

A $11.99 **C** $12.96

B $12.98 **D** $13.00

16. Daisy bought a puzzle book for $3.95, a gift bag for $2, and a ribbon for $0.75. How much did Daisy spend in all?

A $6.99 **C** $6.96

B $5.69 **D** $6.70

Name________________________________ Lesson 22.5

Subtract Money Amounts

Subtracting money amounts is like subtracting whole numbers. You can also estimate to see if your answer is reasonable.

Estimate. Then find the difference.

- Estimate to the nearest dollar.

$$\begin{array}{r} \$41.15 \\ -\ \$26.84 \\ \hline \end{array} \rightarrow \begin{array}{r} \$41.00 \\ -\ \$27.00 \\ \hline \$14.00 \end{array}$$

- Line up the decimal points and subtract. Write your answer in dollars and cents.

$$\begin{array}{r} \$41.15 \\ -\ \$26.84 \\ \hline \end{array}$$

Step 1	Step 2	Step 3	Step 4
Subtract the hundredths.	Regroup the ones as 10 tenths and subtract. Bring down the decimal point.	Regroup the tens as 10 ones and subtract.	Subtract the tens. Bring down the dollar sign.
$41.15 − $26.84 1	0 11 $41.15 − $26.84 .31	10 3 0 11 $41.15 − $26.84 4.31	10 3 0 11 $41.15 − $26.84 $14.31

- Since $14.31 is close to $14.00, your answer is reasonable.

So, the difference is $14.31.

Estimate. Then find the difference.

1. $0.76 − $0.28 ______ ______
2. $28.42 − $5.25 ______ ______
3. $18.67 − $3.44 ______ ______
4. $67.99 − $15.42 ______ ______
5. $0.45 − $0.38 ______ ______

Name ______________________________

Subtract Money Amounts

Estimate. Then find the difference.

1. $\begin{array}{r} \$0.99 \\ -\$0.51 \\ \hline \end{array}$

2. $\begin{array}{r} \$28.99 \\ -\$12.21 \\ \hline \end{array}$

3. $\begin{array}{r} \$15.02 \\ -\ \$3.98 \\ \hline \end{array}$

4. $\begin{array}{r} \$0.85 \\ -\$0.27 \\ \hline \end{array}$

5. $\begin{array}{r} \$11.99 \\ -\ \$9.21 \\ \hline \end{array}$

6. $\begin{array}{r} \$48.55 \\ -\$25.26 \\ \hline \end{array}$

7. $\begin{array}{r} \$98.02 \\ -\$1.89 \\ \hline \end{array}$

8. $\begin{array}{r} \$0.75 \\ -\$0.36 \\ \hline \end{array}$

9. $\begin{array}{r} \$15.47 \\ -\$6.49 \\ \hline \end{array}$

10. $0.62 − $0.25 = ______

11. $69.25 − $27.49 = ______

Problem Solving and Test Prep

12. Jorge wants to buy a ball. A soccer ball costs $8.86. A football costs $12.67. How much more does the football cost than the soccer ball costs?

____

13. Sara buys a sun hat for $7.89. She pays with a $10 bill. How much is Sara's change?

____

14. Lily is saving for a flute that costs $25.99. She has saved $18.02 so far. How much more money does Lily need to buy the flute?

A $7.97 **C** $8.97
B $7.98 **D** $8.01

15. Ken has saved $9.89. He wants to buy a CD that costs $19.78. How much more money does Ken need to buy the CD?

A $9.79 **C** $9.99
B $9.98 **D** $9.89

Name________________________ **Lesson 22.6**

Multiply and Divide Money Amounts

Multiply and divide money amounts like you do whole numbers.
You can also estimate to see if your answer is reasonable.

Multiply.

Step 1	Step 2	Step 3
Estimate.	Multiply like you would whole numbers. Regroup if needed.	Write the answer using the dollar sign and decimal point.
$\$6.13 \rightarrow \$6.00$ $\times\ 6 \quad \times\ 6$ $\$36.00$	$\overset{1}{6}13 \times 6 = 3678$	$\$\overset{1}{6}.13 \times 6 = \$36.78$

So, $\$6.13 \times 6 = \$36.78$

Since $36.78 is close to your estimate of $36.00, your answer is reasonable.

Divide.

Step 1	Step 2	Step 3
Estimate.	Divide like you would whole numbers.	Write the answer using the dollar sign and decimal point.
$\$9.36 \div 3$ ↓ $\$9.00 \div 3 = \$3.00$	$3\overline{)936} = 312$ -9 03 -3 06 -6 0	$3\overline{)\$9.36} = \$3.12$ -9 03 -3 06 -6 0

So, $\$9.36 \div 3 = \$3.12$

Since $3.12 is close to your estimate of $3.00, your answer is reasonable.

Find the product.

1. $\$5.17 \times 8$ **2.** $\$4.68 \times 5$ **3.** $\$8.24 \times 3$ **4.** $\$2.03 \times 4$ **5.** $\$1.92 \times 9$

Divide.

6. $\$7.24 \div 2$ **7.** $\$6.18 \div 3$ **8.** $\$5.55 \div 5$ **9.** $\$7.16 \div 4$ **10.** $\$2.98 \div 2$

______ ______ ______ ______ ______

Name ____________________

Lesson 22.6

Multiply and Divide Money Amounts

Find the product.

1. $\begin{array}{r} \$7.31 \\ \times \quad 6 \\ \hline \end{array}$

2. $\begin{array}{r} \$3.58 \\ \times \quad 3 \\ \hline \end{array}$

3. $\begin{array}{r} \$5.25 \\ \times \quad 5 \\ \hline \end{array}$

4. $\begin{array}{r} \$9.19 \\ \times \quad 4 \\ \hline \end{array}$

5. $\begin{array}{r} \$6.23 \\ \times \quad 7 \\ \hline \end{array}$

6. $\begin{array}{r} \$8.09 \\ \times \quad 8 \\ \hline \end{array}$

7. $\begin{array}{r} \$4.14 \\ \times \quad 9 \\ \hline \end{array}$

8. $\begin{array}{r} \$8.98 \\ \times \quad 2 \\ \hline \end{array}$

9. $\begin{array}{r} \$1.75 \\ \times \quad 6 \\ \hline \end{array}$

10. $\begin{array}{r} \$7.04 \\ \times \quad 5 \\ \hline \end{array}$

Divide.

11. $5.79 ÷ 3 = ______
12. $9.50 ÷ 5 = ______
13. $7.60 ÷ 4 = ______
14. $4.97 ÷ 7 = ______
15. $6.56 ÷ 8 = ______
16. $3.84 ÷ 6 = ______

Problem Solving and Test Prep

17. Farmer Carson's tractor uses 5 gallons of diesel fuel each hour when he's harvesting his crops. Diesel fuel costs $2.53 per gallon. How much does it cost to run Farmer Carson's tractor for 1 hour?

____

18. Jay, Erik, John, and Bill have $13.00 in all. They want to buy 1 pizza for $10 and 4 drinks for $1 each. Do they have enough money?

____

19. Freddy wants to buy toy cars. He has $5.00. Toy cars cost $2.50 each. How many toy cars can Freddy buy?

A 1
B 5
C 2
D 3

20. Sabine has $5.25. She buys 3 juice pops. She doesn't get any change. How much does 1 juice pop cost?

A $1.50
B $1.75
C $1.85
D $1.25

Name____

Tell Time

You can tell time on a digital clock and an analog clock.

On a **digital** clock, the first one or two numbers show the **hour**. The two numbers to the right of the hour show the **minutes.** Write the time as 2:12.

2:12 is read as two twelve or twelve minutes after two.

On an **analog clock**, the small hand points to the hour.

The large hand points to the **minutes.**

Write the time. Then write two ways you can read the time.

The minute hand on the clock is pointing at 6.

A **half hour** has 30 minutes.

The hour hand is between 4 and 5.

Write 4:30.

Read 4:30 as four thirty, half past four, or thirty minutes after four..

Write the time. Then write two ways you can read the time.

1. 

2. 

3. 

______________ ______________ ______________

______________ ______________ ______________

______________ ______________ ______________

Name____________________________________ Lesson 22.7

Tell Time

Write the time. Then write two ways you can read the time.

1. 

2. 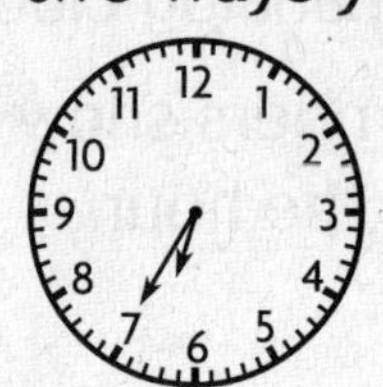

3. 

4. 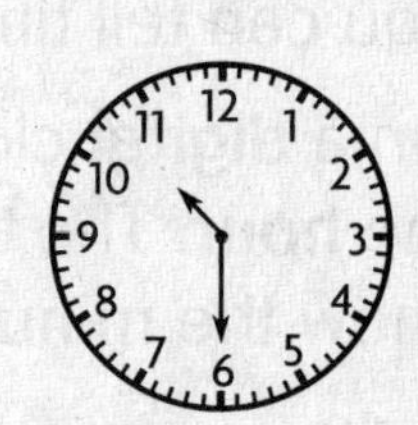

______ ______ ______ ______

______ ______ ______ ______

For 5–12, write the letter of the clock that shows the time.

a. 

b. 

c. 

d. 

5. eight twenty-two ____

6. 11:40 ____

7. fifteen minutes before three ____

8. two forty-five ____

9. twelve ten ____

10. twenty minutes before twelve ____

11. twenty-two minutes after eight ____

12. 12:10 ____

Problem Solving and Test Prep

13. Tom said he'd meet his sister, Karen, at eleven. Tom looks at the clock. He sees that it's eleven minutes before eight. Does Tom have time to walk 40 feet to meet Karen? Explain.

____

____

14. What does Mary's digital watch look like when it's six minutes before two?

____

15. Burt woke up at a quarter to seven. Which is one way to write this time?

A 7:15 **C** 6:45

B 7:45 **D** 6:15

16. Elena ate dinner at 20 minutes before six. What is one way to write this time?

A 5:40 **C** 6:40

B 6:20 **D** 5:20

Name____________________________ **Lesson 22.8**

A.M. and P.M.

Twelve o'clock during the day is called **noon**. Many people eat lunch at noon. Write **P.M.** for times in the afternoon and evening. When the clock shows 12:00 again, it is **midnight**. Write **A.M.** for times after midnight and before noon.

Write the time for the following activity. Use A.M. or P.M.

put on pajamas

You put your pajamas on at night.
The time on the clock is 8:12. But is it A.M. or P.M.?

In the A.M. hours, you wake up, get dressed, and go to school.

In the P.M. hours, you finish school, go home, do homework, and go to sleep.

Night is P.M.
So, the time is 8:12 P.M.

Write the time for each activity. Use A.M. or P.M.

1. school gets out

____

2. eat dinner

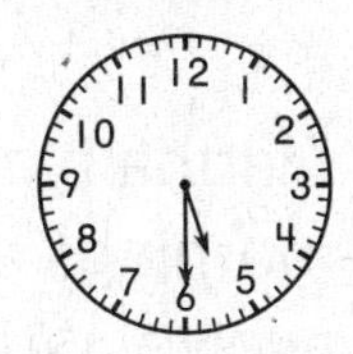

____

3. get up in the morning

____

4. school begins

____

5. time for lunch

____

6. recess

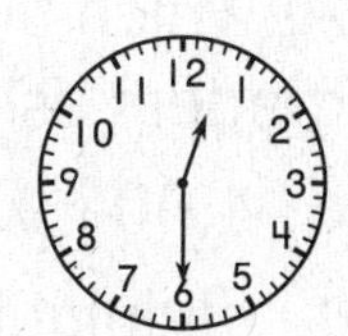

____

7. walk the dog

____

8. Sleep

____

Name________________________________

A.M. and P.M.

Write the time for each activity. Use A.M. or P.M.

1. 

play basketball

____

2. 

eat lunch

____

3. 

go to the library

____

4. 

eat breakfast

____

Write the time by using numbers. Use A.M. or P.M.

5. eight twenty in the morning

____

6. five minutes after three in the afternoon

____

7. fifteen minutes before eleven at night

____

8. six forty-five in the morning

____

Problem Solving and Test Prep

9. Martha plays soccer every Saturday morning at 10 o'clock. Write this time using numbers. Use A.M. or P.M.

____

10. Martha plays soccer on Sunday mornings at twelve minutes to twelve. Write the time using numbers. Use A.M. or P.M.

____

11. At which time are most third graders awake?
   - **A** 4:00 P.M.
   - **B** midnight
   - **C** 3:00 A.M.
   - **D** 11:00 P.M.

12. At which time are most third graders asleep?
   - **A** 4:00 P.M.
   - **B** midnight
   - **C** 3:00 P.M.
   - **D** 11:00 A.M.

Spiral Review

For 1–5, write the fraction for the given decimal.

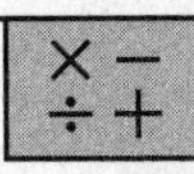

1. $0.3 =$

2. $0.8 =$

3. $0.4 =$

4. $0.10 =$

5. $0.49 =$

For 10–12, list the possible outcomes for each.

10. 

Kendra flips the coin.

____

11. 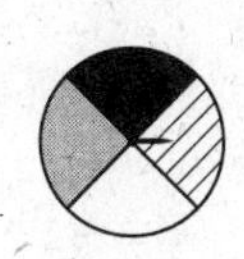

Jonah spins the spinner.

____

12. 

Jenna pulls a marble from the bag.

____

For 6–9, draw lines to match the name with its description.

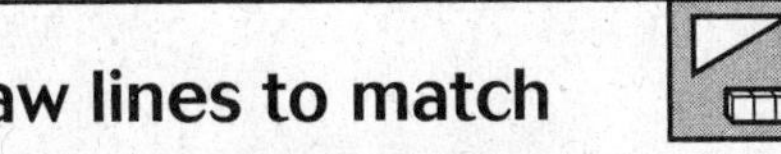

Name	Description
6. sphere	solid figure that is not a sphere and has no faces
7. cube	solid figure with six rectangular faces
8. cone	solid figure with no faces
9. rectangular prism	solid figure with six square faces

For 13–15, find the missing factor.

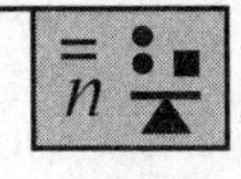

13. $(6 \times \square) \times 3 = 18$

_____

14. $4 \times (4 \times \square) = 32$

_____

15. $(5 \times \square) \times 2 = 90$

_____

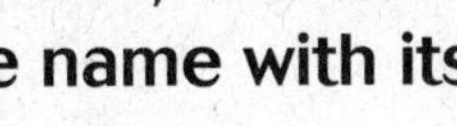

Name________________________________ Lesson 22.9

Model Elapsed Time

The time it takes for something to happen is the elapsed time. You start brushing your teeth at 7:00. You finish at 7:03. The elapsed time is 3 minutes.

Use a clock to find the elapsed time.

Start: 4:20 A.M.

End: 5:00 A.M.

- Begin with the start time. Show 4:20 on your clock. The hour hand should point between 4 and 5. The minute hand should also point at 4.
- Now look at the end time. The hours change, so you will move the hour hand of your clock.
- Count 1 hour. Move the hour hand until it points at 5.
- The minutes also change. So you will move the minute hand of your clock.
- Count 40 minutes. Move the minute hand until it points at 12.
- Check if A.M. changed to P.M. A.M. did not change to P.M.

So, the elapsed time is 40 minutes.

Use a clock to find the elapsed time.

1. Start: 3:15 A.M.
End: 4:30 A.M.
Elapsed time:

____

2. Start: noon
End: 2:25 P.M.
Elapsed time:

____

3. Start: 11:44 A.M.
End: 12:03 P.M.
Elapsed time:

____

4. Start: 6:32 P.M.
End: 7:56 P.M.
Elapsed time:

____

5. Start: 5:18 P.M.
End: 6:07 P.M.
Elapsed time:

____

6. Start: 10:47 P.M.
End: 1:15 A.M.
Elapsed time:

____

Name ____________________

Model Elapsed Time

Use a clock to find the elapsed time.

1. Start: 3:15 P.M.
   End: 5:25 P.M.
   ____
2. Start: 12:11 P.M.
   End: 6:19 P.M.
   ____
3. Start: 9:55 A.M.
   End: 12:05 P.M.
   ____

Write the elapsed time in hours and minutes.

4. 67 minutes ____________
5. 190 minutes ____________
6. 210 minutes ____________
7. 131 minutes ____________

Tell what time it will be.

8. 25 minutes after 1:15 P.M. ____________
9. 90 minutes after 11:15 P.M. ____________
10. 2 hours after 10:30 A.M. ____________
11. 57 minutes after 6:30 P.M. ____________
12. 74 minutes after 2 A.M. ____________
13. 2 hours 30 minutes after 10:30 A.M. ____________

Problem Solving and Test Prep

14. Joyce left the house thirty-five minutes after 8:20 A.M. What time did Joyce leave the house?
    ____
15. Evelyn took a walk at 5:05 P.M. and arrived home 40 minutes later. What time did Evelyn arrive home?
    ____
16. Which time is 1 hour and 20 minutes after 8:30 P.M.?
    A 9:50 P.M.
    B 8:50 P.M.
    C 9:30 P.M.
    D 9:50 A.M.
17. Which time is 3 hours and 15 minutes after 7:45 A.M.?
    A 11:00 P.M.
    B 10:45 A.M.
    C 11:00 A.M.
    D 10:15 A.M.

Name______________________________

Use a Calendar

A calendar shows the days, weeks, and months of the year. There are 7 days in a week and 12 months in a year.

You can use a calendar to find the elapsed time of any event.

Use the calendars below to solve.

Today is February 11. Nikki began reading her book January 30. She reads her book every day. Counting today, for how many days has Nikki been reading her book?

January						
Sun	Mon	Tue	Wed	Thu	Fri	Sat
		1	2	3	4	5
6	7	8	9	10	11	12
13	14	15	16	17	18	19
20	21	22	23	24	25	26
27	28	29	30	31		

February						
Sun	Mon	Tue	Wed	Thu	Fri	Sat
					1	2
3	4	5	6	7	8	9
10	11	12	13	14	15	16
17	18	19	20	21	22	23
24	25	26	27	28		

- January is the month before February.
- Nikki began reading on January 30 and is still reading on February 11.
- Count the days.
- There are 13 days.

So, Nikki has been reading her book for 13 days.

For 1-2, use the calendar.

1. Today is February 2. Rebecca's birthday is on February 8. Including today, how many days is it until Rebecca's birthday?

____

2. Ben's parents are going out of town at 3:00 p.m. on February 21. They will return at 2:00 p.m. on February 24. How long will Ben's parents be gone?

____

Name____________________

Use a Calendar

For 1-4 use the calendars.

May						
Sun	Mon	Tue	Wed	Thu	Fri	Sat
				1	2	3
4	5	6	7	8	9	10
11	12	13	14	15	16	17
18	19	20	21	22	23	24
25	26	27	28	29	30	31

June						
Sun	Mon	Tue	Wed	Thu	Fri	Sat
1	2	3	4	5	6	7
8	9	10	11	12	13	14
15	16	17	18	19	20	21
22	23	24	25	26	27	28
29	30					

July						
Sun	Mon	Tue	Wed	Thu	Fri	Sat
		1	2	3	4	5
6	7	8	9	10	11	12
13	14	15	16	17	18	19
20	21	22	23	24	25	26
27	28	29	30	31		

1. Barbara planted grass seed on June 2. The grass started to grow on June 18. How many days did it take for the grass to start growing?

____

2. The school T-shirt sale runs for two weeks. The sale starts on May 6. On what date does the school t-shirt sale end?

____

3. Today is July 5, and it's 9:00 A.M. Fran's family is going on a trip beginning July 8 at noon. For how long must Fran wait for her trip to start?

____

4. On May 15, Ms. Paolo said that there would be a field trip in 3 weeks. On what date will there be a field trip?

____

Problem Solving and Test Prep

5. Jamal's family went to his cousin's house. The family left home at 8:00 a.m. on July 2. They returned home at 7:00 p.m. on July 5. For how long was Jamal's family away from their home?

____

6. Today is the second Saturday in June. How many weeks and days is it until the Fourth of July? Use the calendar above to solve this problem.

____

7. Madeline's birthday is January 3. Jeremy's birthday is January 18. How many days after Madeline's birthday is Jeremy's birthday?

   A 13 days
   B 2 weeks
   C 14 days
   D 15 days

8. It is 8:00 a.m. on Tuesday. Alexa's aunt will visit at 3 p.m. on the upcoming Sunday. How long is it until Alexa's aunt visits?

   A 7 days 5 hrs.
   B 5 days 6 hrs.
   C 5 days 7 hrs.
   D 4 days 7 hrs.

Name____ **Lesson 22.11**

Sequence Events

The order that things happen is called the **sequence**.
A sequence tells what happened first, second, third, and so on.

You can use a time line to sequence events. The events on the left of the time line happened before, or earlier, than the events on the right.

Use the time line below to answer the following question.

Which movie was shown earlier, *The Little Mermaid* or *Cinderella*?

Years in Which Movies Were First Shown

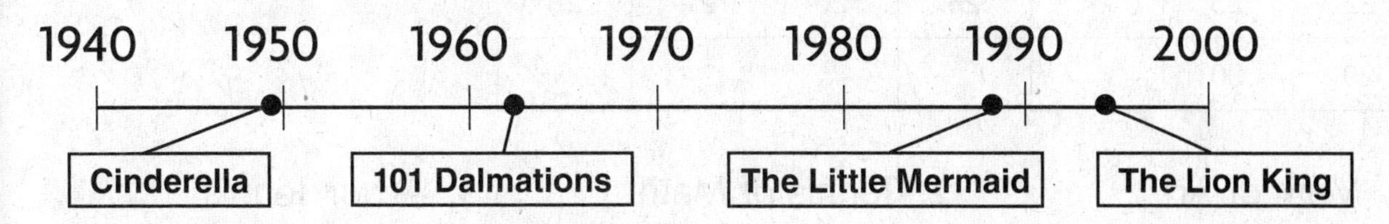

You can find the answer in two ways.

ONE WAY

Remember that the events on the left happened earlier than events on the right.

So, the movie that is farther to the left happened first.

Cinderella is farther to the left than *The Little Mermaid.*

So, *Cinderella* was shown earlier.

ANOTHER WAY

You can read the timeline and compare the dates or times of each event.

Cinderella was first shown in 1950.

The Little Mermaid was first shown in between 1980 and 1990.

1950 is earlier than 1989.

So, *Cinderella* was shown earlier.

For 1–4, use the timeline above.

Which movie was shown earlier?

1. *101 Dalmatians* or *The Lion King*

____

2. *The Little Mermaid* or *101 Dalmatians*

____

Which movie was shown later?

3. *Cinderella* or *101 Dalmatians*

____

4. *The Little Mermaid* or *Cinderella*

____

MR 1.1 Analyze problems by identifying relationships, distinguishing relevant from irrelevant information, sequencing and prioritizing information, and observing patterns.

Name________________________________

Sequence Events

Use the calendar and the key. Which activity is earlier?

June						
Sun	Mon	Tue	Wed	Thu	Fri	Sat
1	2	3	4	5 Tennis	6	7
8	9	10	11	12	13	14
15	16	17	18	19	20	21
22	23	24	25	26	27	28
29	Math 30 Art					

Art	1:00 PM
Math	9:00 AM
Tennis	10:00 AM

1. Math or Art ____________

2. Tennis or Math ____________

3. Art or Tennis ____________

Problem Solving and Test Prep

4. Liz read her book at 10:00 A.M., played ball at 8:00 P.M., and went to bed at 10:00 P.M. Which activity is earliest?

____

5. Austin went shopping May 9, Muriel went shopping April 2, and Cindi went shopping July 20. Who went shopping first?

____

6. Use the July calendar. Which event happens first?

A dance **C** swimming
B fireworks **D** Ted's visit

7. Use the July calendar. Which event comes last?

A dance **C** swimming
B fireworks **D** Ted's visit

July						
Sun	Mon	Tue	Wed	Thu	Fri	Sat
		Ted's visit 1	2	3	4 Fireworks	5
6	7	8	9	10	11	12
13	14	15 Swimming	16	17	18 Dance	19
20	21	22	23	24	25	26
27	28	29	30	31		

Name ________________________________ **Lesson 23.1**

Length

When you measure the distance between two points, you are measuring length. You can use customary units to measure length.

Which unit would you use to measure the length of a shoe?

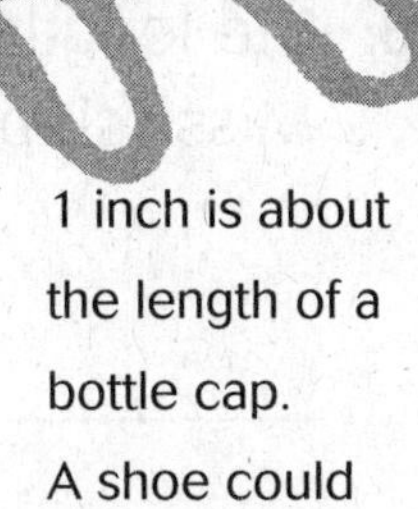

1 inch is about the length of a bottle cap. A shoe could be measured in inches.

1 foot is about the length of a folder. A shoe could be measured in feet.

1 yard is about the length of a water ski. A shoe could not be measured in yards.

1 mile is about 4 times around an outdoor sports track. A shoe could not be measured in miles.

A shoe could be measured in inches or feet.

Choose the unit you would use to measure each.
Write *inch, foot, yard,* or *mile*.

1. 

   ____

2. 

   ____

3. 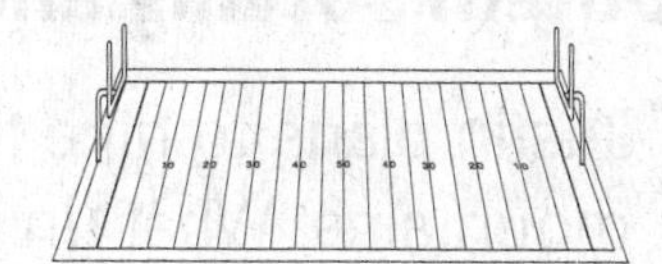

   ____

4. the width of a door

   ____

5. the distance from Texas to Colorado

   ____

6. the length of a pizza box

   ____

7. the length of a bar of soap

   ____

8. the length of a house

   ____

9. the distance from Earth to the Moon

   ____

Name________________________________ Lesson 23.1

Length

Choose the unit you would use to measure each. Write *inch, foot, yard* or *mile*.

1. 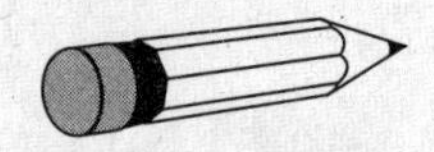

____

2. 

____

3. 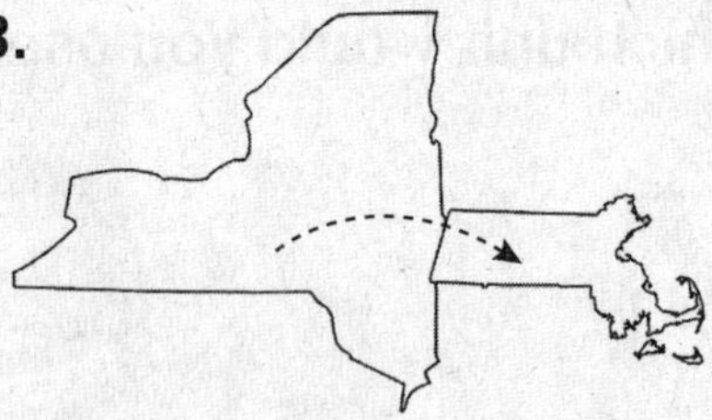

____

4. the length of a cereal box

____

5. the length of a spoon

____

6. the length of the Mississippi River

____

7. the length of a tea kettle

____

8. distance between 2 state capitals

____

9. the length of an automobile

____

Problem Solving and Test Prep

10. Justin plans to hike through the mountains. What unit best describes the distance Justin will hike?

____

11. Alex saw an adult shark at the aquarium. What unit best describes the length of the shark?

____

12. Lilly wants to measure the length of a bike. About how long might the bike be?

A 5 inches
B 5 feet
C 5 yards
D 5 miles

13. Tyler wants to measure the length of a book. About how long might the book be?

A 9 inches
B 9 feet
C 9 yards
D 9 miles

Name____

Lesson 23.2

Estimate and Measure Inches

You can use an inch ruler to measure the length of an object to the nearest inch and half inch.

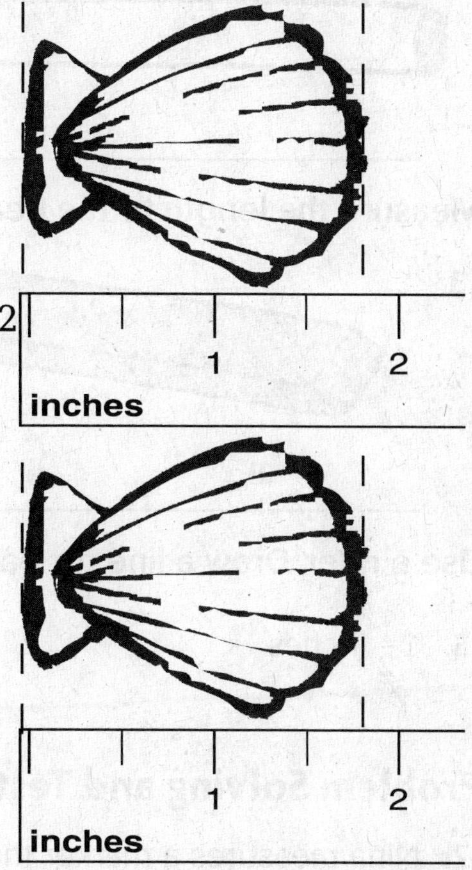

Measure the length of the seashell to the nearest inch.

The length of the seashell is between 1 inch and 2 inches.

The length of the seashell is closer to 2 inches than to 1 inch.

To the nearest inch, the length of the seashell is 2 inches.

Measure the length of the seashell to the nearest half-inch.

The length of the seashell is between $1\frac{1}{2}$ inches and 2 inches.

The length of the seashell is closer to 2 inches than to $1\frac{1}{2}$ inches.

To the nearest half-inch, the length of the seashell is 2 inches.

Measure the length to the nearest inch.

1. ____________

2. ____________

Measure the length to the nearest half-inch.

3. 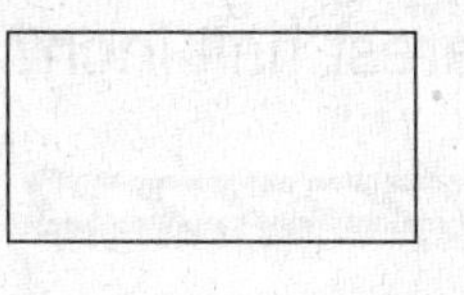 ____________

4. 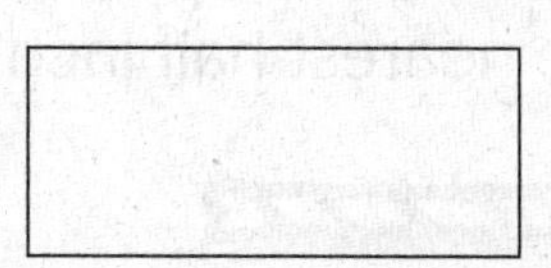 ____________

Name____

Lesson 23.2

Estimate and Measure Inches

Measure the length to the nearest inch.

1.

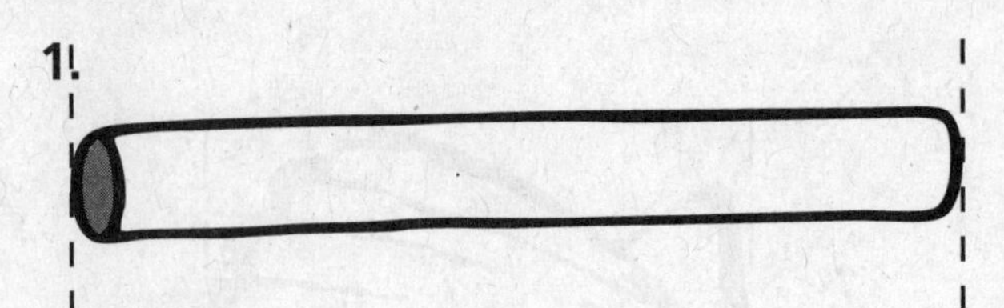

____

2.

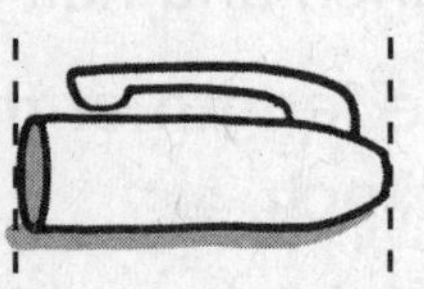

____

Measure the length to the nearest half inch.

3.

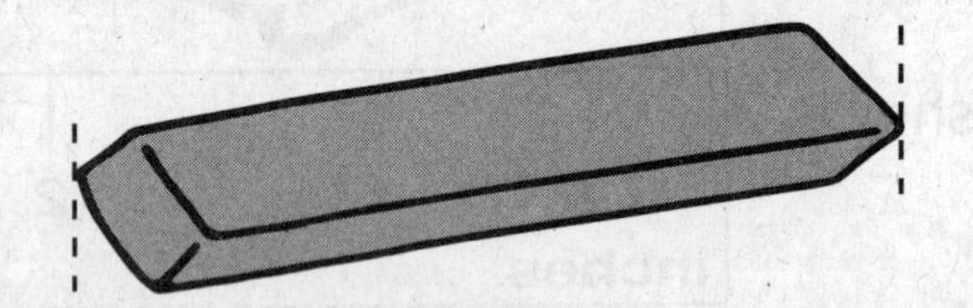

____

4.

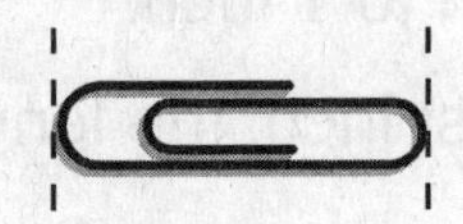

____

Use a ruler. Draw a line for each length.

5. $1\frac{1}{2}$ inches

6. 2 inches

Problem Solving and Test Prep

7. Nina measures a marker that is $2\frac{1}{2}$ inches long. Between which two inch-marks does the end of the marker lie?

____

8. What is the length of the card below to the nearest half inch?

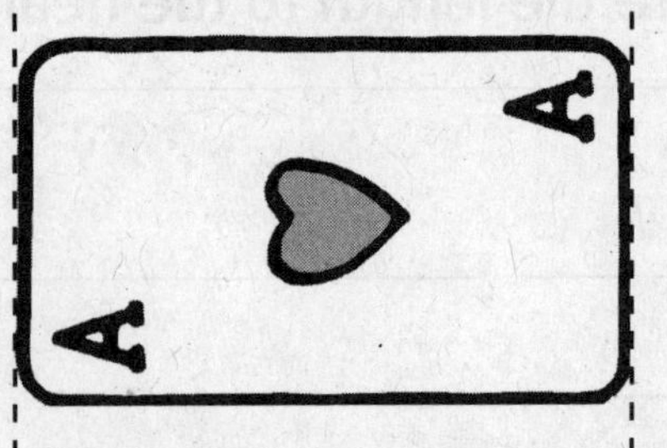

____

9. Which is the length of the string below to the nearest half inch?

A 1 inch
B $1\frac{1}{2}$ inches
C 2 inches
D $2\frac{1}{2}$ inches

10. Which is the length of the string below to the nearest half inch?

A 2 inches
B $2\frac{1}{2}$ inches
C 3 inches
D $3\frac{1}{2}$ inches

Name ______________________ **Week 34**

Spiral Review

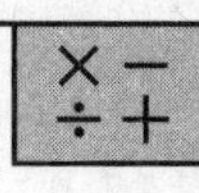

For 1–5, find each sum. Use subtraction to check.

1. $\begin{array}{r} 916 \\ +450 \\ \hline \end{array}$

2. $\begin{array}{r} 507 \\ +589 \\ \hline \end{array}$

3. $\begin{array}{r} 8,954 \\ +\ \ 647 \\ \hline \end{array}$

4. $\begin{array}{r} 6,784 \\ +2,069 \\ \hline \end{array}$

5. $\begin{array}{r} 3,109 \\ +5,317 \\ \hline \end{array}$

For 9–10, the line plot shows the results of spinning a spinner with sections of unequal size 15 times.

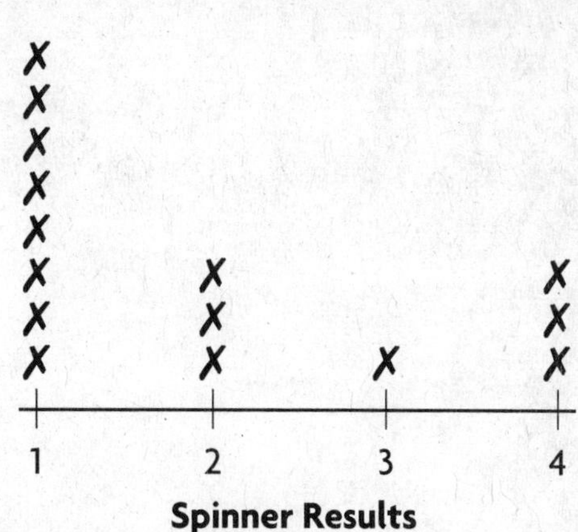

9. Which number probably had the largest section of the spinner?

____

10. Which numbers are equally likely to be spun?

____

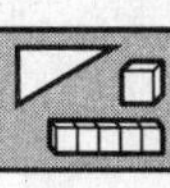

For 6–8, Find the volume of the solid figures.

Figure | Volume

6. 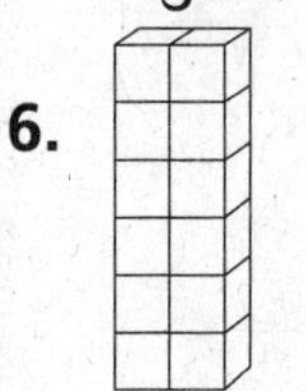 _____ cubic units

7. 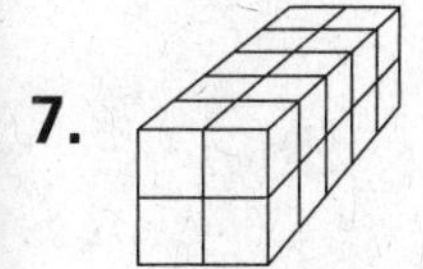 _____ cubic units

8. 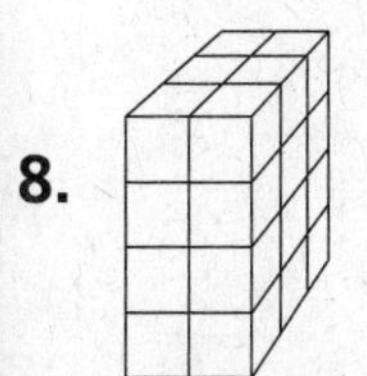 _____ cubic units

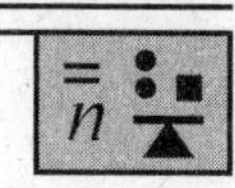

For 11–14, fill in the blank.

11. 24 months = _____ years

12. 60 minutes = _____ hour

13. 21 days = _____ weeks

14. 3 hours = _____ minutes

Name

Lesson 23.3

Estimate and Measure Feet and Yards

To estimate the length of an object, compare the object with a ruler or a yardstick.

Which is the better unit of measure to use to find the length of a bathtub: 2 feet or 2 yards?

Think of feet and yards in terms of a 12-inch ruler.

1 foot = 12 inches or one 12-inch ruler 2 feet = 24 inches or two 12-inch rulers	1 yard = three feet or three 12-inch rulers 2 yards = 6 feet or six 12-inch rulers

A bathtub is longer than 2 regular rulers or 2 feet, so 2 yards is the better unit to measure the length of a bathtub.

Choose the better unit of measure.

1. the width of a car

8 feet or 8 yards

2. the length of a bed

6 feet or 6 yards

3. the length of a shoe

1 foot or 1 yard

4. the length of a dog

4 feet or 4 yards

5. the length of a game board

3 feet or 3 yards

6. the length of a limousine

5 feet or 5 yards

MG 1.1 Choose the appropriate tools and units (metric and U.S.) and estimate and measure the length, liquid volume, and weight/mass of given objects.

Name____________________

Estimate and Measure Feet and Yards

Choose the better unit of measure.

1. the length of a bed

8 feet or 8 yards

2. the length of a puppy

1 foot or 1 yard

3. the length of a soccer field

100 feet or 100 yards

4. the length of a pickup truck

5 feet or 5 yards

5. the length of a sofa

6 feet or 6 yards

6. the length of a tennis court

80 feet or 80 yards

Find two objects in the classroom to match each length.

Draw them and label their length.

7. about 1 yard

8. about 1 foot

9. Jamie plans to knit a sweater. She needs 12 feet of yarn. She has 3 yards of yarn. Does Jamie have enough yarn to knit the sweater? Explain.

____

____

____

Name____ **Lesson 23.4**

Capacity

Capacity is the amount a container will hold. **Cup (c), pint (pt), quart (qt),** and **gallon (gal)** are customary units used to measure capacity.

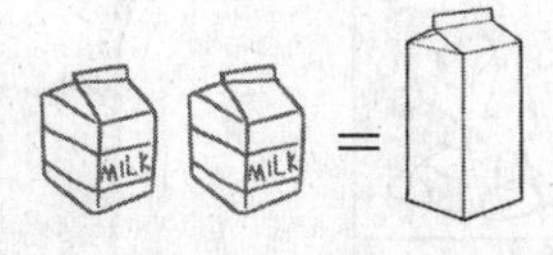

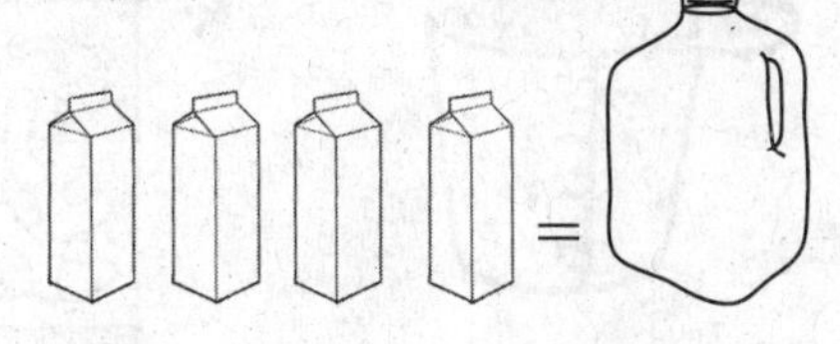

2 cups = 1 pint 2 pints = 1 quart 4 quarts = 1 gallon

Which unit would you use to measure the capacity of a mug of cocoa?

The size of the mug shows that it has the capacity to hold either a cup, or a pint, of cocoa.

Choose the unit you would use to measure each. Write *cup, pint, quart,* or *gallon*.

1. 

____

2. 

____

3. 

____

4. 

____

5. 

____

6. 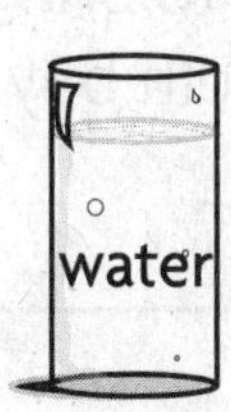

____

Name____

Capacity

Choose the unit you would use to measure each. Write *cup, pint, quart,* or *gallon*.

1. 

____

2. 

____

3. 

____

4. 

____

5. 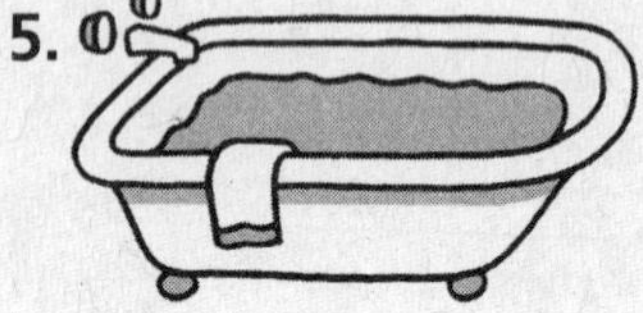

____

6. 

____

7. 

____

8. 

____

9. 

____

10. 

____

11. 

____

12. 

____

13. Beth needs to bring a gallon of juice to a party. She bought 2 quarts of juice. Did Beth buy enough juice? Explain.

____

____

Name____

Weight

Weight is a measure of how heavy an object is. The customary units for measuring weight are **ounces (oz)** and **pounds (lb)**.

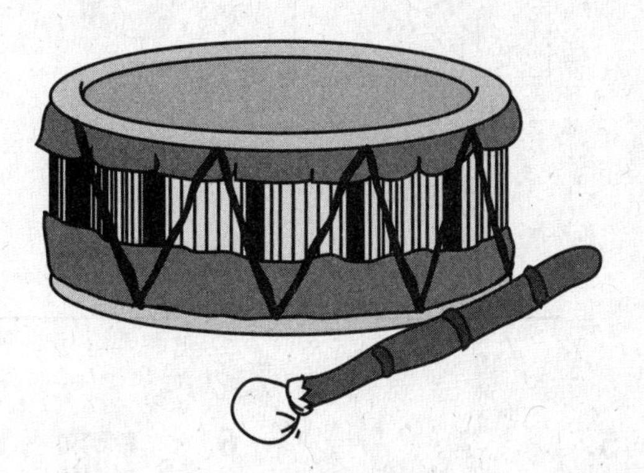

There are 16 ounces in a one pound.

An ounce weighs about as much as 9 pennies.

Would you use ounces or pounds to measure the weight of the drum?

A drum usually weighs a great deal more than 16 portions of 9 pennies, or 1 pound.

So, **pounds** would be used to measure the weight of the drum.

Choose the unit you would use to measure each. Write *ounce* or *pound*.

1.  ____________

2.  ____________

3.  ____________

4.  ____________

5.  ____________

6. 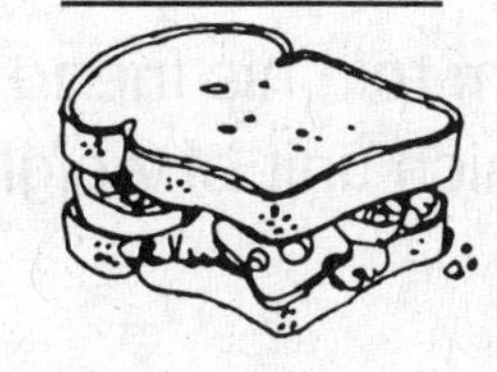 ____________

Name____ **Lesson 23.5**

Weight

Choose the unit you would use to weigh each. Write *ounce* or *pound*.

1. 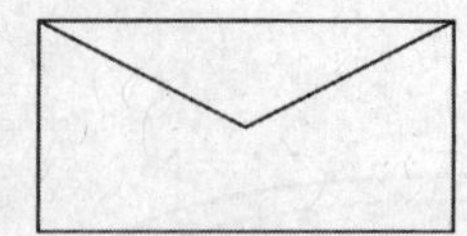
____

2. 
____

3. 
____

4. 
____

5. 
____

6. 
____

7. 

____

8. 
____

Find two objects in the classroom to match each weight.
Draw them and label their weight.

9. about 5 pounds

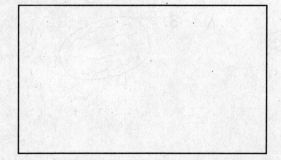

10. about 4 ounces

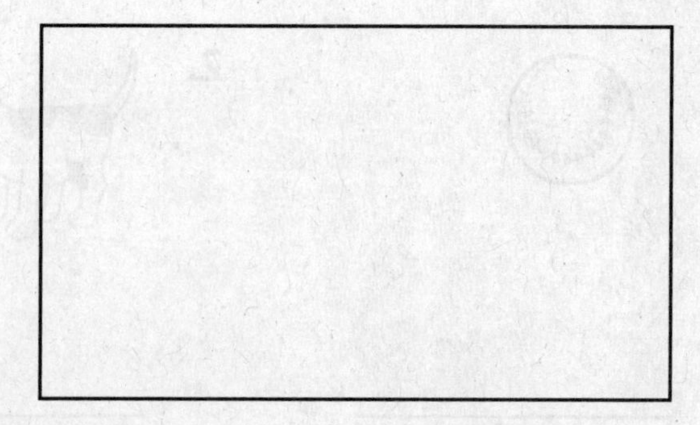

11. Sam told his friend that his puppy weighs 48. He did not give the unit. Which unit of weight should Sam have said after 48, ounces or pounds?

____

Name________________________ Lesson 23.6

Estimate or Measure

In some problems you will need to *measure* to solve. In other problems, you can *estimate* to solve. Look at the problem below.
Choose estimate or measure.

Morgan needs 36 inches of wire for each side of a rabbit pen. Should she estimate or measure the wire?

Estimate, when an exact measurement is not needed.	Measure, when an exact measurement is needed.	Morgan needs exactly 36 inches of wire for each side of the rabbit pen.

So, Morgan needs to measure the wire to build the rabbit pen.

Choose *estimate* or *measure*.

1. Alice wants to make a frame for her school picture. The size of the picture is 5 inches by 7 inches. Should Alice estimate, or measure the size for her picture frame?

   ____

2. Andrew needs 4 feet of cardboard to construct his model airplane. Should Andrew estimate, or measure, the length of the cardboard?

   ____

3. Tara is baking brownies. She wants to add walnuts to the recipe. Should Tara estimate, or measure, the number of walnuts?

   ____

4. Greg boiled water in a pot to cook spaghetti. Should Greg have estimated, or measured, the water in the pot?

   ____

MG 1.1 Choose the appropriate tools and units (metric and U.S.) and estimate and measure the length, liquid volume, and weight/mass of given objects.

Name ______________________________

Estimate or Measure

Choose *estimate* or *measure*.

1. Samantha needs four tablespoons of lemonade mix to make 1 quart of lemonade. Should Samantha estimate or measure the lemonade mix?

____

2. Anna fills up the bathtub to take a bath. Should Anna estimate or measure the water?

____

3. Charlie has to be 52 inches tall to enter into the fun house. Should the fun house estimate or measure Charlie's height?

____

4. Andrew needs a long piece of string for his kite. Should Andrew estimate or measure the piece of string?

____

Problem Solving and Test Prep

5. Nicole plans to make a dress and needs to buy 4 yards of fabric. Should Nicole estimate or measure the length of the fabric?

____

6. Wayne makes a collage with different-sized pieces of paper. Should Wayne estimate or measure the sizes of the pieces of paper?

____

7. Which amount should you measure?

   A distance from class to gym
   B amount of water in a fish tank
   C distance a person can throw a baseball
   D length of a person's foot when buying new shoes

8. Which amount should you estimate?

   A amount of flour in a recipe
   B amount of fence needed for a yard
   C amount of dirt placed in a flower pot
   D number of cups of water in a quart

Name____________________________________ Lesson 23.7

Algebra: Rules for Changing Units

You can use a rule to change units. Look at the problem below.

Table of Measures
Length
12 inches = 1 foot
3 feet = 1 yard
Capacity
2 cups = 1 pint
4 cups = 1 quart
2 pints = 1 quart
8 pints = 1 gallon
4 quarts = 1 gallon
Weight
16 ounces = 1 pound

Find ☐ yards = 9 feet

The table shows 3 feet = 1 yard.

The smaller unit, 9 feet, is being changed into a larger unit, yards.

To change a smaller unit to a larger unit, divide 9 by the number of feet in 1 yard.

$9 \div 3 = 3$

So, 3 yards = 9 feet.

You can also make a table to help you change units.

Find ☐ pints = 12 cups

2 cups = 1 pint.

The smaller unit, 12 cups, is being changed into pints.

To change a smaller unit to a larger unit, *divide*.

Divide the number of cups by 2 to get the number of pints.

cups	2	4	6	8	12	16
pints	1	2	3	4	6	8

So, 6 pints = 12 cups.

Complete using the Table of Measures.

1. 32 ounces = ______ pounds

2. ______ pints = 4 quarts

3. 12 quarts = ______ gallons

4. ______ inches = 2 feet

5. 6 feet = ______ yards

6. 8 cups = ______ pints

Name ______________________________ Lesson 23.7

Algebra: Rules for Changing Units

USE DATA For 1–6, use the table below.

Table of Measures		
Length	**Capacity**	**Weight**
12 inches = 1 foot	2 cups = 1 pint	16 ounces = 1 pound
3 feet = 1 yard	4 cups = 1 quart	
	2 pints = 1 quart	
	8 pints = 1 gallon	
	4 quarts = 1 gallon	

1. 9 feet = __________ yards

2. __________ cups = 2 quarts

3. 2 pounds = __________ ounces

4. 16 pints = __________ gallons

5. 36 inches = __________ feet

6. 10 pints = __________ quarts

Problem Solving and Test Prep

7. John made applesauce. He added 3 pounds of apples into a pot of boiling water. How many ounces of apples did John add to the pot of boiling water?

____

8. Jorge makes a ceramic punch bowl. His bowl can hold 6 pints of water. How many quarts of water can Jorge's punch bowl hold?

____

9. Which rule can you use to find the number of cups in a 3 pints?

A Multiply the number of cups by 2.

B Multiply the number of pints by 2.

C Divide the number of cups by 2.

D Divide the number of pints by 2.

10. How many quarts are in 2 gallons?

A 4

B 8

C 12

D 16

Name ________________________________ **Week 35**

Spiral Review

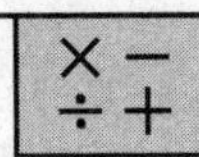

For 1–5, find the product.

1. $\begin{array}{r} 19 \\ \times 7 \\ \hline \end{array}$

2. $\begin{array}{r} 34 \\ \times 8 \\ \hline \end{array}$

3. $\begin{array}{r} 65 \\ \times 4 \\ \hline \end{array}$

4. $\begin{array}{r} 510 \\ \times 6 \\ \hline \end{array}$

5. $\begin{array}{r} 528 \\ \times 5 \\ \hline \end{array}$

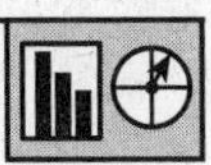

For 9–10, answer the questions using the line plot.

Number Cube Rolls	1	2	3	4	5	6
	X X	X X X X X	X X X	X	X X	

Number Cube Rolls

9. The line plot shows what happened when a number cube was rolled. What number would you expect to be rolled next?

______

10. Which number was rolled the most often?

____

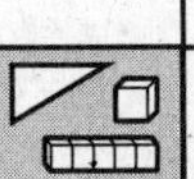

For 6–8, draw lines to match the object with the best unit of measure.

6. 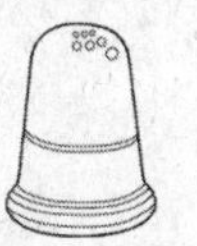 milliliter

7. 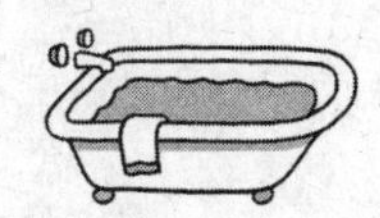 liter

8.  kiloliter

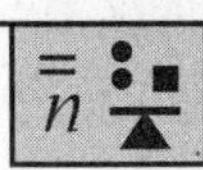

For 11–15, fill in the blank.

11. 2 feet = ______ inches

12. 4 yards = ______ feet

13. 2 yards = ______ inches

14. 48 inches = ______ feet

15. 15 feet = ______ yards

Name____________________ Lesson 23.8

Problem Solving Workshop Skill: Choose a Unit

Mary enjoys drinking cocoa in the morning. Does Mary's mug hold about 2 cups or about 2 quarts?

1. What are you asked to find?

____

____

2. Think about how customary units of capacity are related. Which is the smaller customary unit of capacity, cup or quart?

____

3. How many cups are in 1 quart? ____________

4. How can you use the choose a unit strategy to solve the problem?

____

____

5. Does Mary's mug hold about 2 cups or about 2 quarts?

____

6. How can you check your answer?

____

____

Choose a unit to solve.

7. Karen plans to make a picture frame for a collage she had made on a piece of construction paper. Will the picture frame be 10 inches long or 10 feet long?

____

8. Matt plans to buy a bag of sand to add to the sandbox in his backyard. Should Matt buy 25 ounces of sand or 25 pounds of sand?

____

MG 1.1 Choose the appropriate tools and units (metric and U.S.) and estimate and measure the length, liquid volume, and weight/mass of given objects.

Name______________________________

Problem Solving Workshop Skill: Choose a Unit

Problem Solving Skill Practice

Solve by choosing the better unit of measure.

1. Mr. Brill wants to measure the distance from each goal line to the half-field line of a soccer field. Which customary unit of length will Mr. Brill use?

____

2. Allison makes juice for herself and her 3 friends. Which customary unit does Allison use to measure the amount of juice she makes?

____

3. George measures how much water his kitchen sink holds. Which customary unit of capacity does George use?

____

4. Julie measures the length of her sister's hair. Which customary unit of length does Julie use?

____

Mixed Applications

5. **Pose a Problem** Marco measures how much water his coffee mug holds. Which customary unit of capacity does Marco use?

____

6. Gracie bought 6 cans of cat food, and 3 cat toys. Each can of cat food cost $2. How much did Gracie spend on cat food?

____

7. Patrick rode his bike 10 miles, then 4 more miles, and then ate 2 sandwiches. How far did Patrick ride his bike in all?

____

8. There were 26 students at the park on Monday. About 15 of these students were girls. About how many students at the park on Monday were boys?

____

Name________________________________

Fahrenheit Thermometer

Temperature is the measure of how hot or cold something is. The customary unit of temperature is **degrees Fahrenheit (°F).**

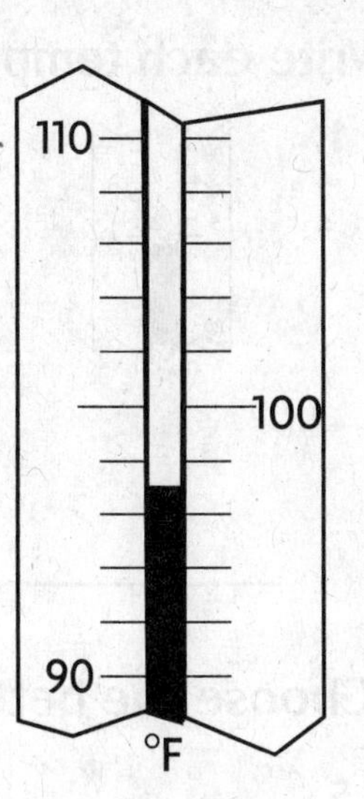

In °F, what is the temperature on the thermometer at the right?

From one mark to the next represents 2 degrees.

The top of the bar is between two marks, rather than directly on a mark.

The top of the bar is between 96°F and 98°F.
So, the temperature on this thermometer is 97°F.

Which would be the better temperature for ice skating, 90°F or 30°F?

Ice skating is a cold weather activity, not a warm weather activity.

30°F is a much cooler temperature than 90°F.

So, the better temperature for ice skating is 30°F.

Write each temperature in °F.

1. 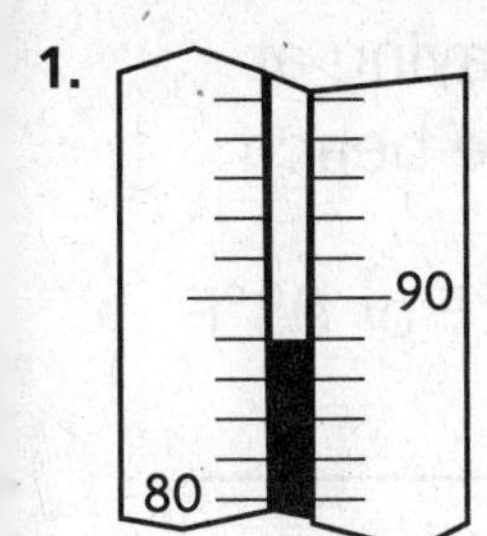

____

2. 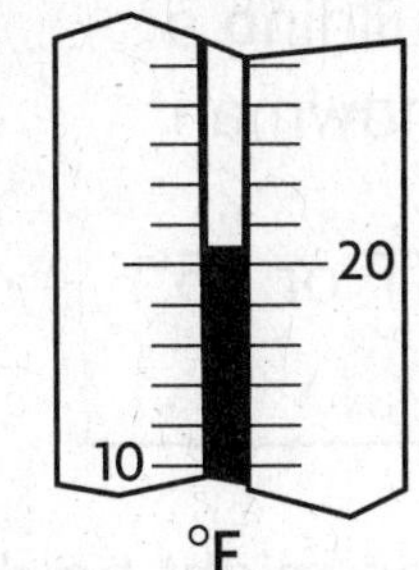

____

3. 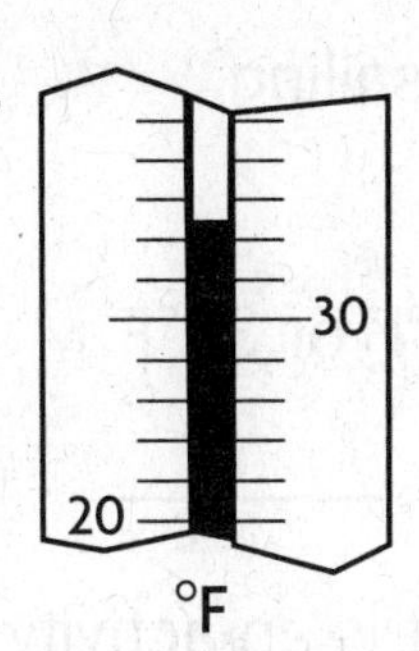

____

4. 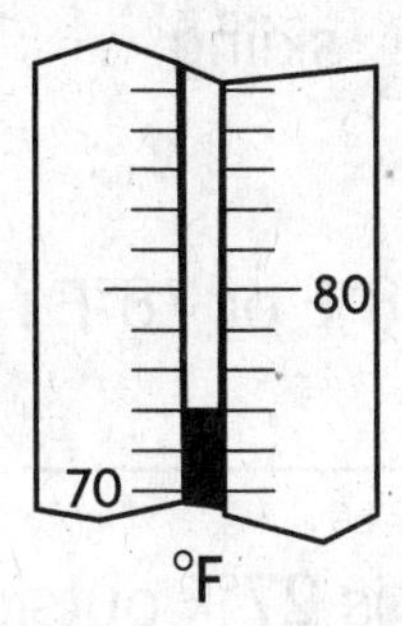

____

Choose the better temperature for each activity.

5. 

bike riding

71°F or 12°F

____

6. 

playing golf

38°F or 69°F

____

7. 

playing hockey

30°F or 77°F

____

8. 

having a picnic

80°F or 42°F

____

Name ____

Fahrenheit Temperature

Write each temperature in °F.

1. 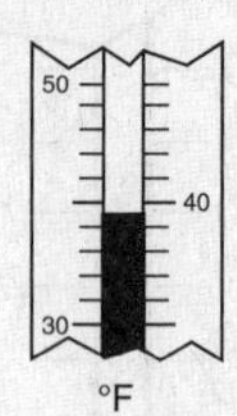

____

2. 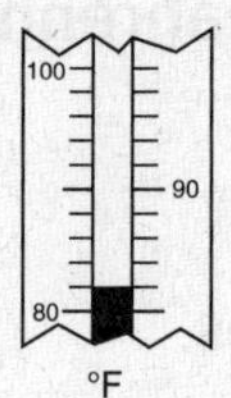

____

3. 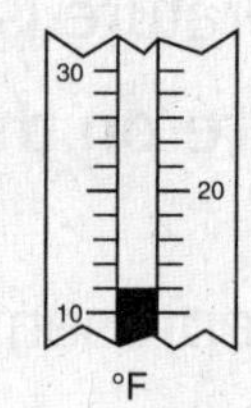

____

4. 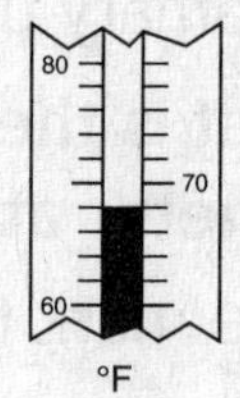

____

Choose the better temperature for each activity.

5. 

skiing

28°F or 78°F

____

6. 

sailing

82°F or 32°F

____

7. 

building a snowman

65°F or 25°F

____

8. 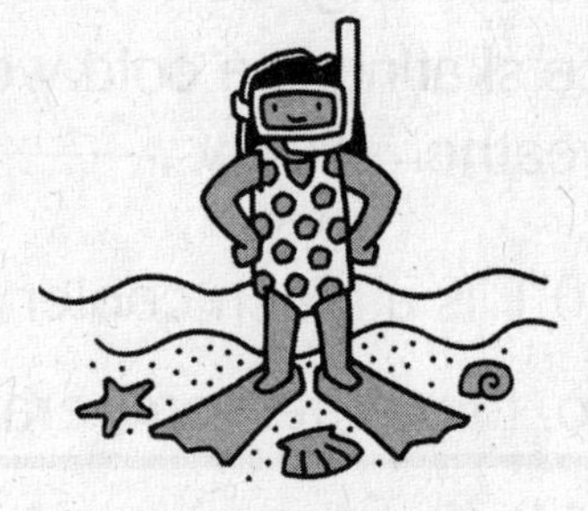

playing at the beach

53°F or 93°F

____

9. It is 27°F outside. What is an activity Jeanne might be doing outside? What clothes do you think Jeanne might wear?

____

____

____

Name____

Length

You can use centimeters, decimeters, meters, and kilometers to measure length and distance.

Use **centimeters (cm)** and **decimeters (dm)** to measure the length of *smaller* objects.	Use **meters (m)** to measure the length of *very large objects* and *short distances.*	Use **kilometers (km)** to measure the length of *long distances*.

Choose the unit you would use to measure the length of the highway below. Write *cm*, *m*, or *km*.

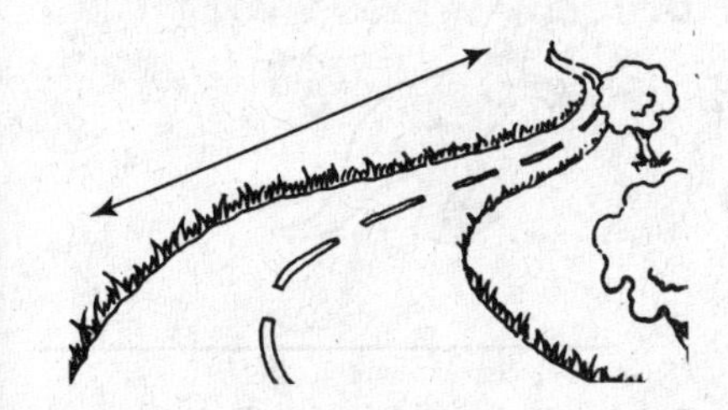

The length of the highway is a long distance.

So, use km to measure the length of the highway.

Choose the unit you would use to measure each. Write *cm*, *m*, or *km*.

5.

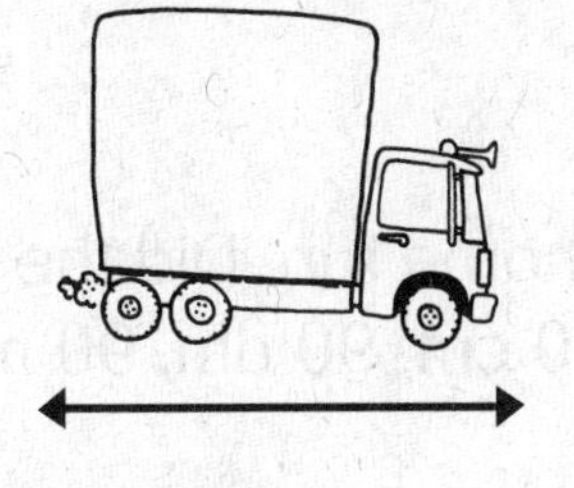

____

6.

____

7.

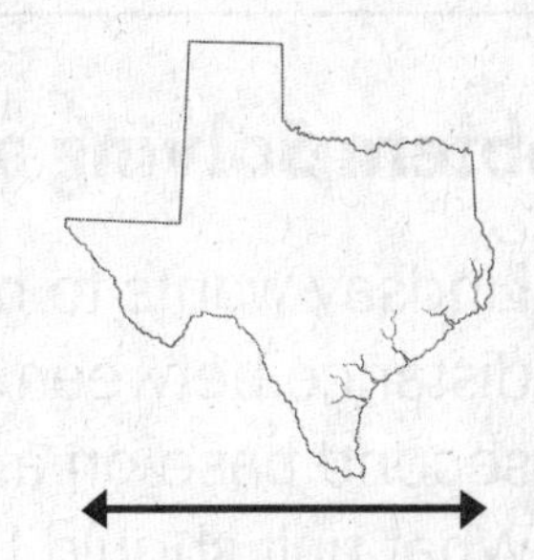

____

8.

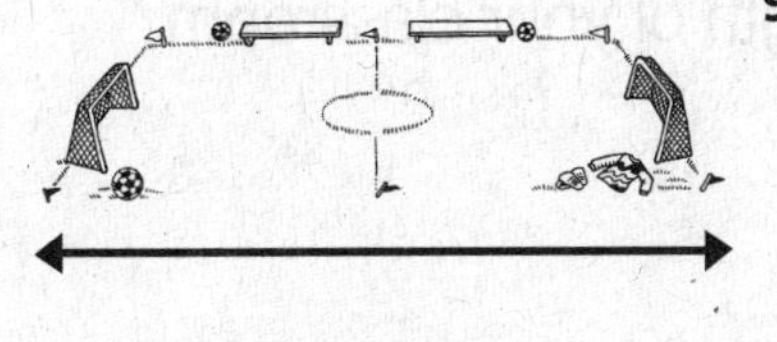

____

9.

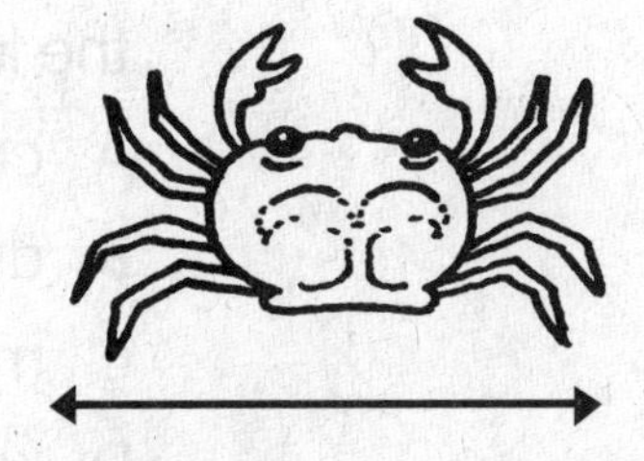

____

10.

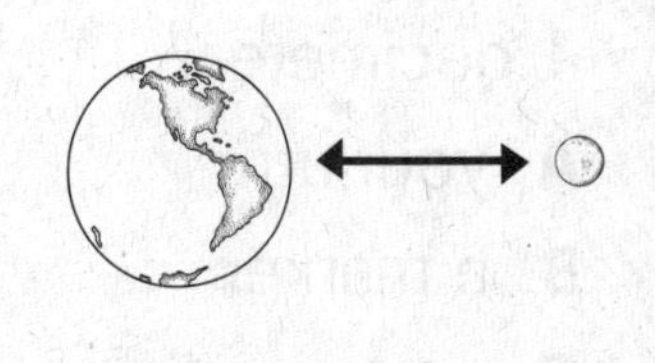

____

Name______________________________

Length

Choose the unit you would use to measure each.
Write *cm, m,* or *km*.

1.  ______________

2. 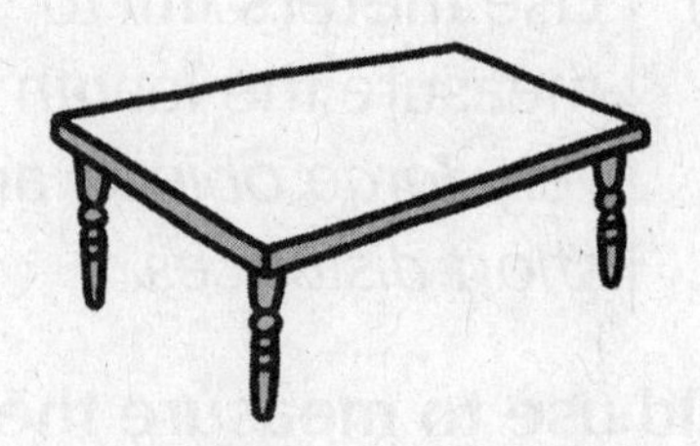 ______________

3. 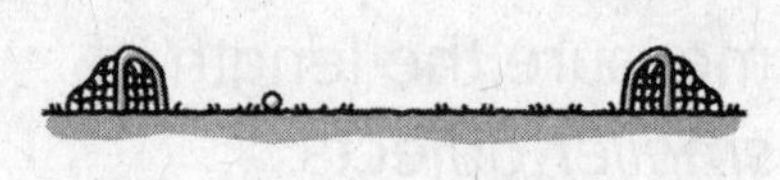 ______________

4.  ______________

5.  ______________

6.  ______________

7. distance between two towns ______________

8. width of a book ______________

9. height of a building ______________

10. length of a fire truck ______________

11. distance to the moon ______________

12. length of your hand ______________

Problem Solving and Test Prep

13. Lindsay wants to measure the distance between first and second base on a baseball field. What unit should Lindsay use?

____

14. Pedro hit a home run. Did the ball travel 90 cm, 90 dm, 90 m, or 90 km?

____

15. Which has a length of about 1 decimeter?

A your arm
B a marker
C a paper clip
D your big toe

16. Which unit would you use to measure the length of your classroom?

A cm
B dm
C m
D km

Centimeters and Decimeters

You can use a centimeter ruler to measure the length of an object to the nearest centimeter and decimeter.

10 centimeters = 1 decimeter

Estimate the length in centimeters. Then use a centimeter ruler to measure the length of the crayon below to the nearest centimeter.

Estimate: about 4 cm

Now measure the length of the crayon with a centimeter ruler.

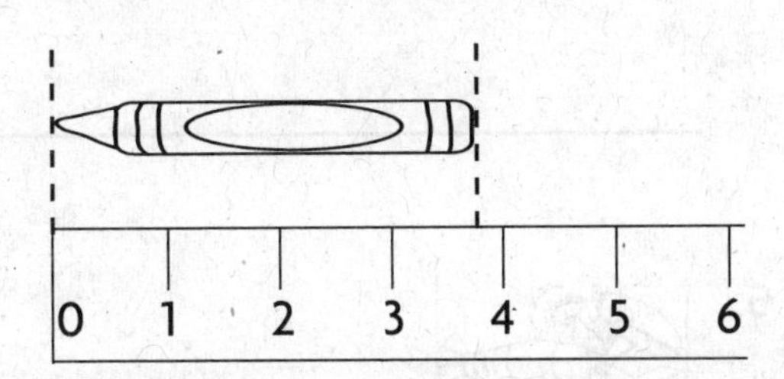

Line up one end of the crayon with the 0 on the ruler.

The end of the crayon is closer to 4 centimeters than to 3 centimeters.

So, the length of the crayon to the nearest centimeter is 4 cm.

Estimate the length in centimeters. Then use a centimeter ruler to measure to the nearest centimeter.

1. 

____

2. 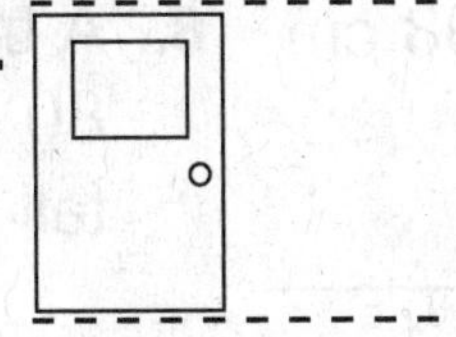

____

3. 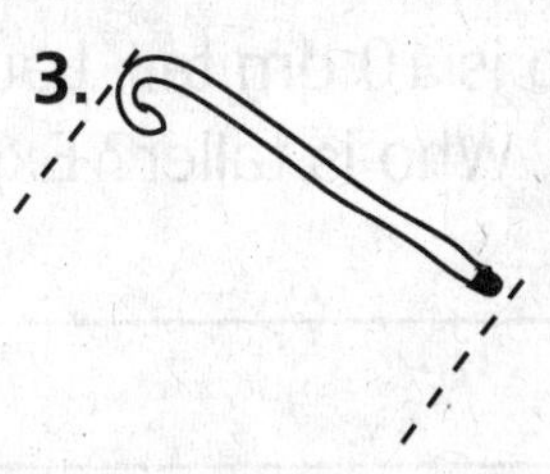

____

4. 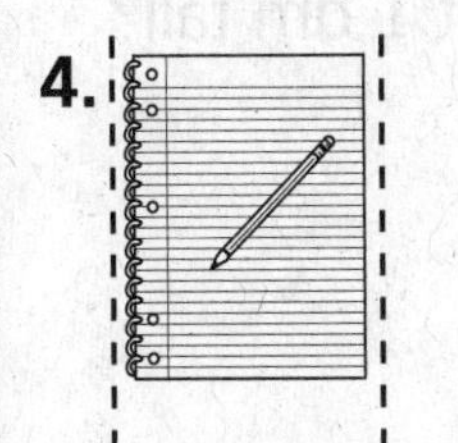

____

5. 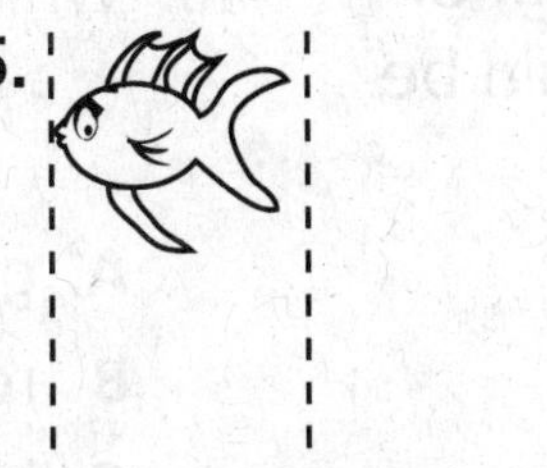

____

6. 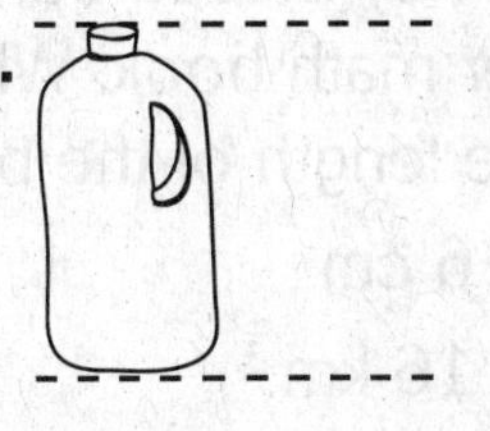

____

Name ____________________

Centimeters and Decimeters

Estimate the length in centimeters. Then use a centimeter ruler to measure to the nearest centimeter.

1.

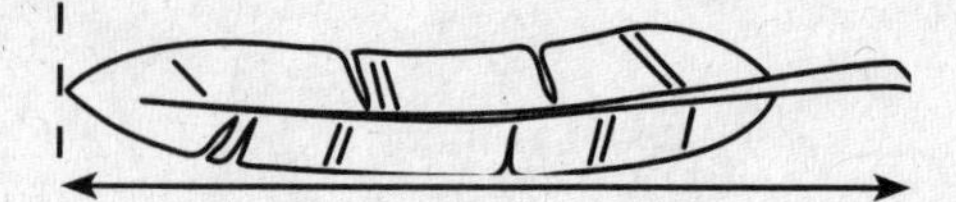

____

2.

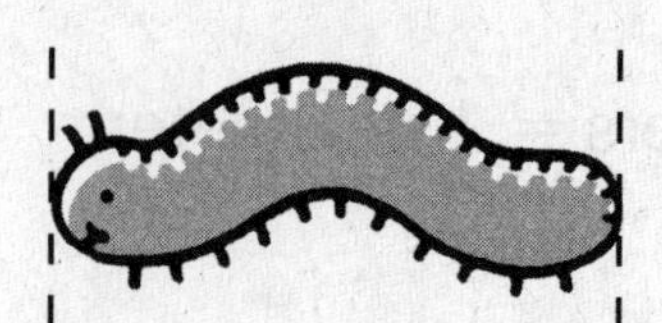

____

3.

____

4.

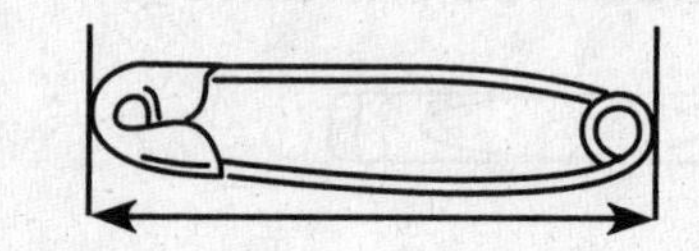

____

Choose the better estimate.

5.

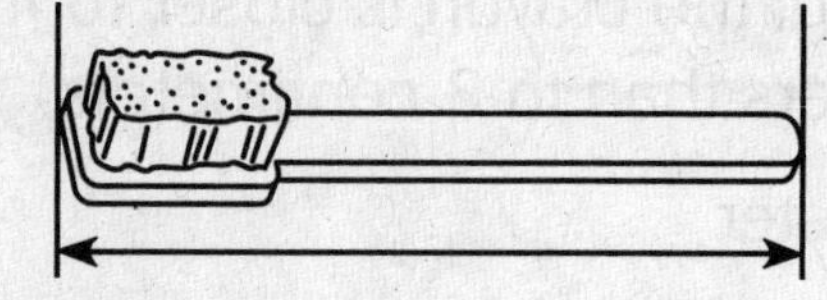

5 cm or 5 m?

6.

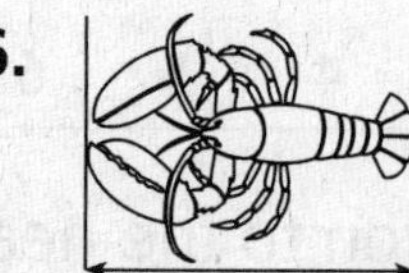

2 dm or 2 cm?

Problem Solving and Test Prep

7. Leo is 10 dm tall. Lauren is 98 cm tall. Who is taller? Explain.

____

____

8. A tree in Miguel's front yard is 80 dm tall. How many centimeters tall is the tree?

____

9. Shirley measured the length of her math book. Which could be the length of the book?

A 6 cm

B 16 km

C 26 cm

D 46 dm

10. Which object is about 1 dm tall?

A giraffe

B rooster

C lamppost

D soup can

Name____________________

Meters and Kilometers

A **meter (m)** is used to measure *long lengths* or *short distances.*

A **kilometer (km)** is used to measure *long distances*

Choose the unit you would use to meaure the length of the soccer field below. Write *m* or *km*.

A soccer field is a short distance.

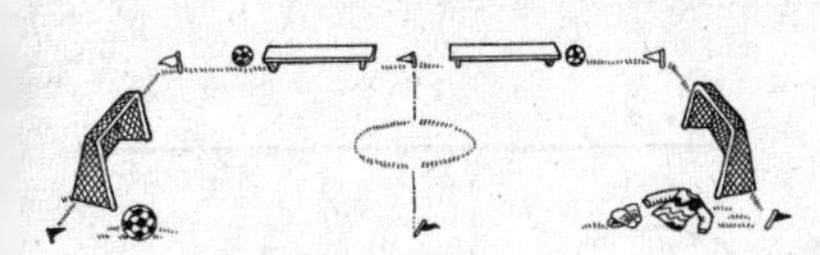

So, use *m* to measure the length of the soccer field.

Choose the unit you would use to measure each. Write *m* or *km*.

1. 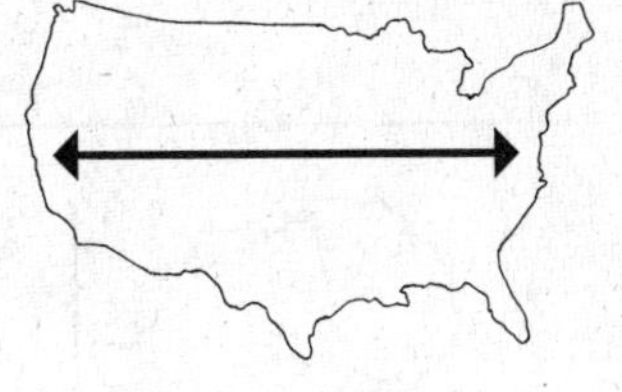
____

2. 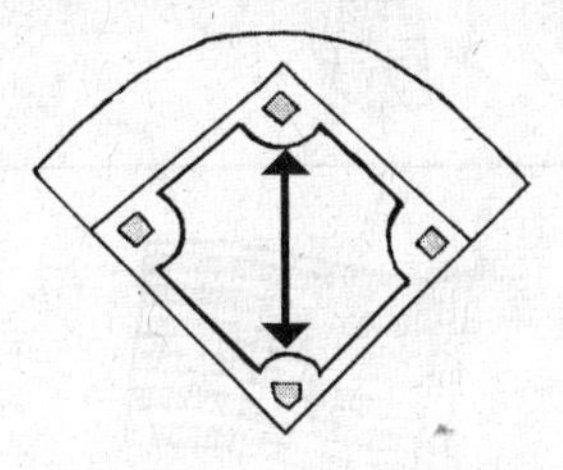
____

3. 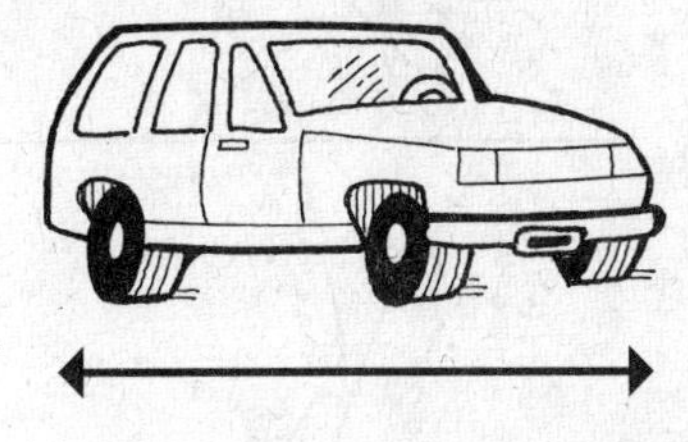
____

4. length of a river
____

5. height of a house
____

6. height of a grizzly bear
____

7. 
____

8. 

____

9. 
____

10. 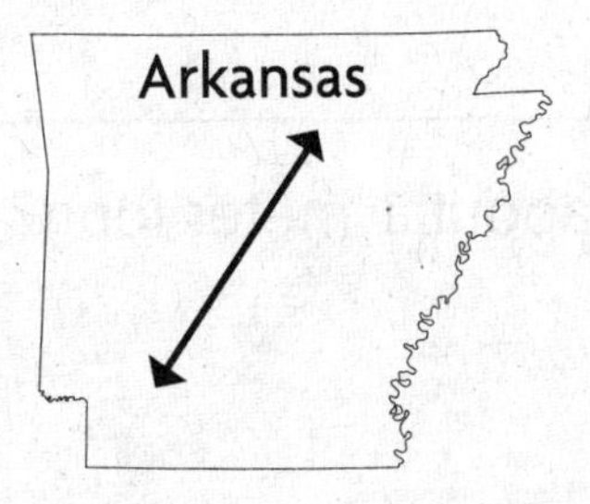

____

11. 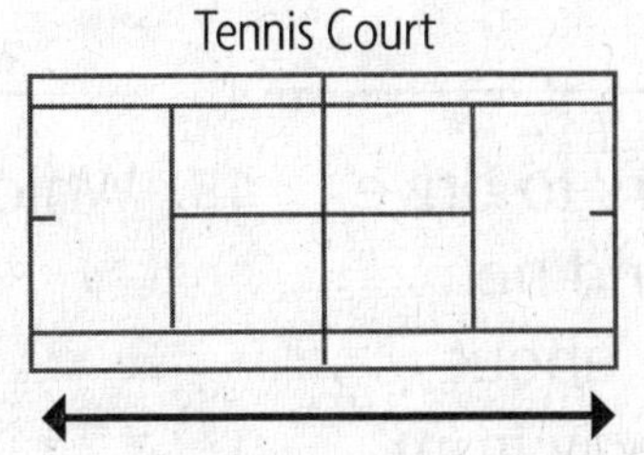

____

12. 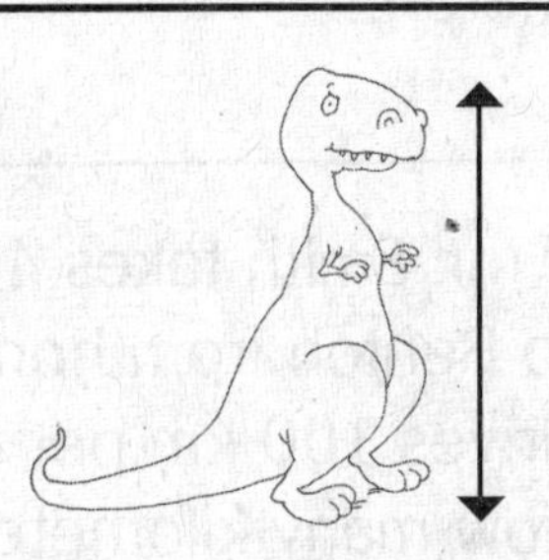
____

MG 1.1 Choose the appropriate tools and units (metric and U.S.) and estimate and measure tht length, liquid volume, and weight/mass of given objects.

Name____

Meters and Kilometers

Choose the unit you would use to measure each. Write *m* or *km*.

1. 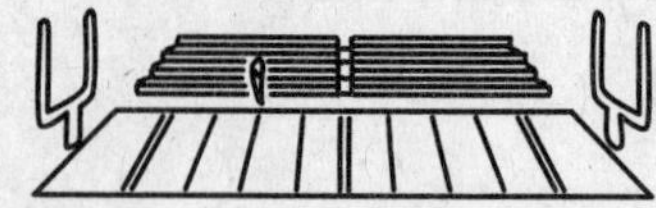

____

2. 

____

3. 

____

4. 

____

5. 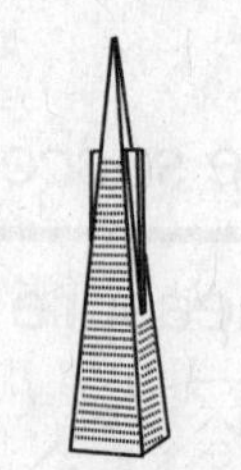

____

6. 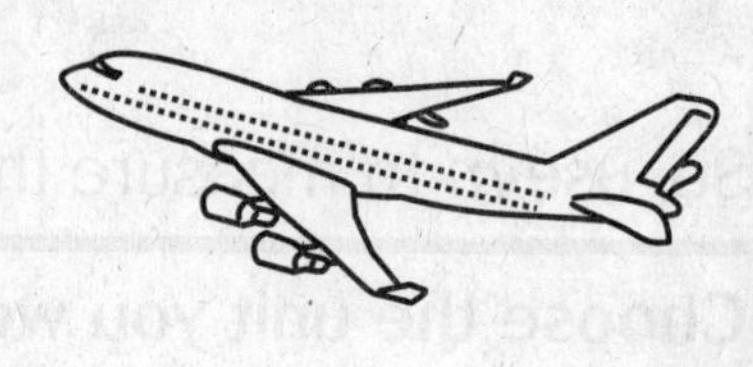

____

7. 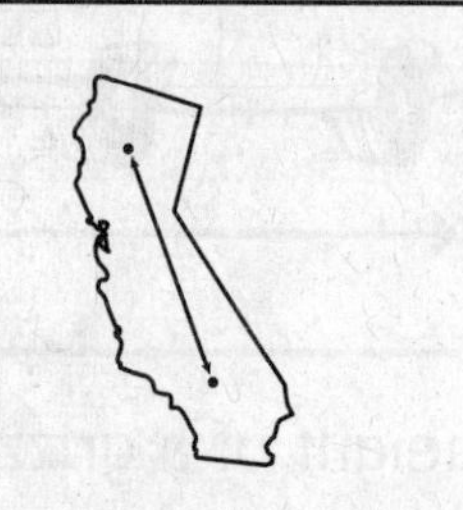

____

8. 

____

9. 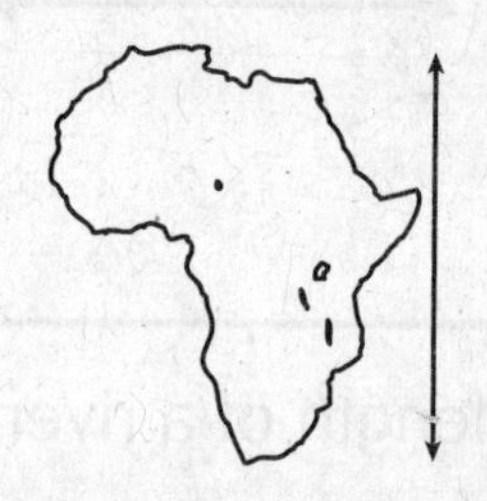

____

Problem Solving and Test Prep

10. The world's tallest mountain is Mount Everest in the Himalayas in Asia. It is about 8,708 meters tall. Is Mount Everest taller or shorter than 9 kilometers? By how many meters?

____

11. The tallest mountain in North America is Mount McKinley in Alaska. It is about 96 meters taller than 6 kilometers. About how many meters tall is Mount McKinley?

____

12. If Mr. Smith takes 4 hours to drive to Benton from home, and he drives 100 km per hour, about how many kilometers away from Mr. Smith's home is Benton?

A about 4 km **C** about 400 km

B about 40 km **D** about 4,000 km

13. Which is about 1 meter long?

A shoe **C** river

B pencil **D** umbrella

Name____ **Week 36**

Spiral Review

For 1–5, find the difference. Write the answer in simplest form

1. $\frac{7}{8} - \frac{1}{8} =$

2. $\frac{3}{4} - \frac{1}{4} =$

3. $\frac{8}{9} - \frac{3}{9} =$

4. $\frac{7}{10} - \frac{3}{10} =$

5. $\frac{11}{12} - \frac{5}{12} =$

For 11–14, tell whether each event is *likely, unlikely, certain,* or *impossible* for this spinner with sections of equal size for one spin.

11. spinner landing on a black section ________

12. spinner landing on a gray section ________

13. spinner landing on a number ________

14. spinner landing on a color ________

For 6–10, fill in the blank.

6. 1 L = ________ mL

7. 1 cm = ________ mm

8. 1 m = ________ cm

9. 1,000 g = ________ kg

10. 1 m = ________ mm

For 15–17, write a number sentence to solve.

15. Mr. Coe signed 10 forms today. If he signs the same number of forms every day, how many forms will Mr Coe sign in 7 days?

____

16. Ping cleaned out the pantry. She threw away 2 items from each shelf. There are 9 shelves. How many items did Ping throw away?

____

17. Zoe cut four tarts into 12 pieces each. How many pieces of tart did Zoe create?

____

Name____ Lesson 24.4

Capacity

Metric units used to measure capacity are **milliliter (mL)** and **liter (L)**.

1 mL is the amount of liquid in about 20 small drops. Milliliters are used to measure the capacity of small containers such as a spoon.

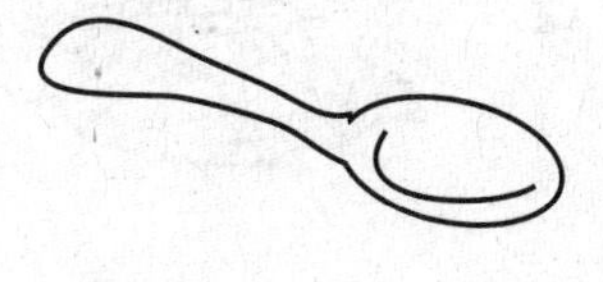

1 L is the amount of liquid in a store bought bottle of water. Liters are used to measure the capacities of large containers.

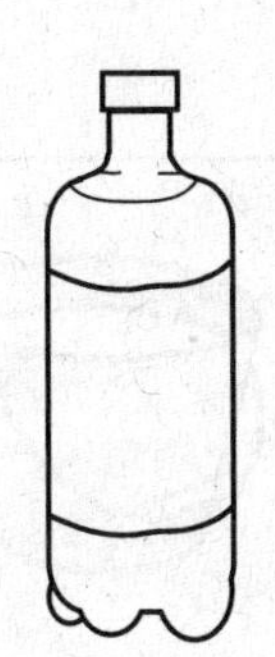

Choose the unit you would use to measure the capacity of the fish tank below. Write *mL* or *L*.

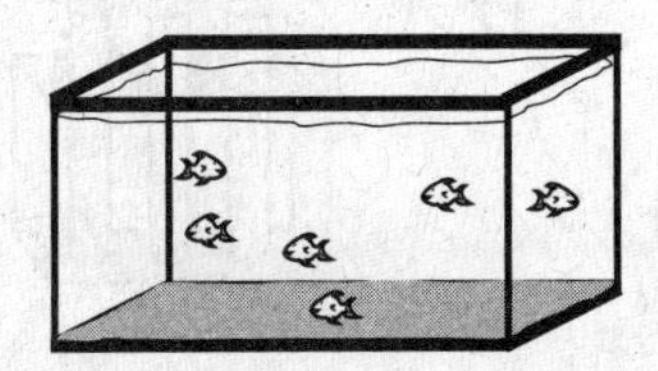

The fish tank is a large container.

So use *L* to measure the capacity of the fish tank.

Choose the unit you would use to measure the capacity of each container. Write *mL* or *L*.

1.  ____________

2.  ____________

3. 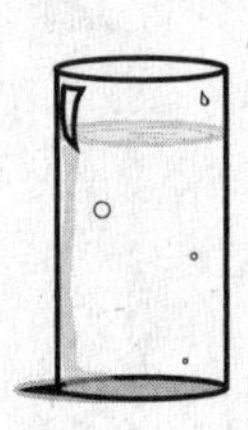 ____________

4.  ____________

4.  ____________

6.  ____________

7.  ____________

8.  ____________

Name ____________________

Capacity

Choose the unit you would use to measure the capacity of each. Write *mL* or *L*.

1.  ____________

2.  ____________

3. 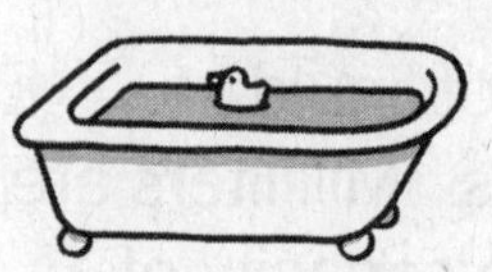 ____________

4.  ____________

5.  ____________

6.  ____________

7.  ____________

8.  ____________

9.  ____________

10.  ____________

11.  ____________

12. 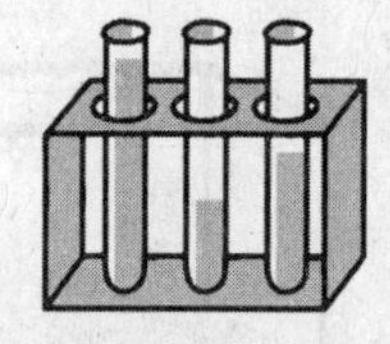 ____________

13. In the space at the right, draw and label a picture of a container that has a capacity less than 1 liter.

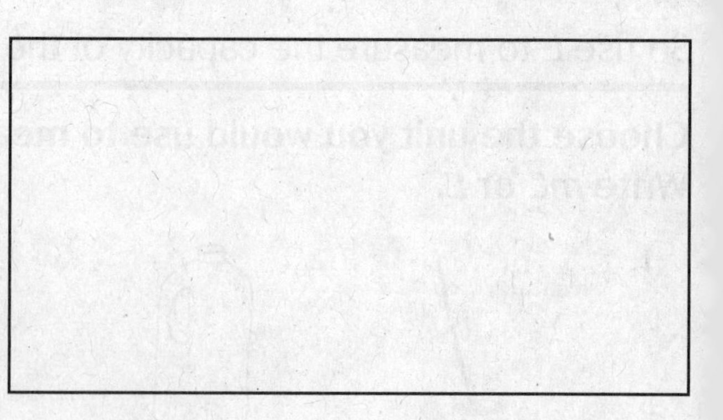

Find each missing number.

14. ______ mL = 3 L

15. ______ L = 6,000 mL

16. 9,000 mL = ______ L

17. 10 L = ______ mL

18. 20,000 mL = ______ L

19. ______ L = 13,000 mL

Mass

The **gram (g)** and the **kilogram (kg)** are units of **mass**. You can use units of mass to measure how heavy an object is.

Grams (g) are used to measure the mass of *light* objects such as strawberries.

1,000 grams = 1 kilogram

Choose the unit you would use to find the mass of the chair below. Write *g* or *kg*.

The chair is heavier even than a thick book.

So, use *kg* to measure the chair.

Choose the unit you would use to find the mass of each. Write *g* or *kg*.

1. 

____

2. 
____

3. 
____

4. 
____

5. 
____

6. 
____

7. 
____

8. 
____

9. 
____

10. 
____

11. 
____

12. 
____

13. 
____

14. 
____

15. 
____

16. 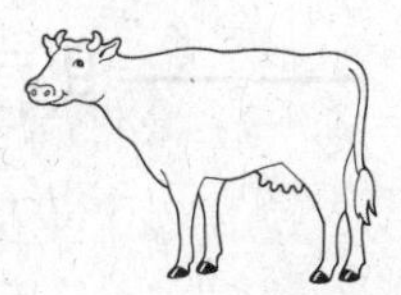
____

MG 1.1 Choose the appropriate tools and units (metric and U.S.) and estimate and measure the length, liquid volume, and weight/mass of given objects.

Name____

Mass

Choose the unit you would use to find the mass of each. Write *gram* or *kilogram*.

1. 
____

2. 
____

3. 
____

4. 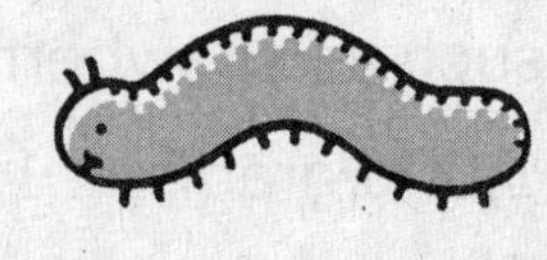
____

5. 
____

6. 
____

7. 
____

8. 
____

9. 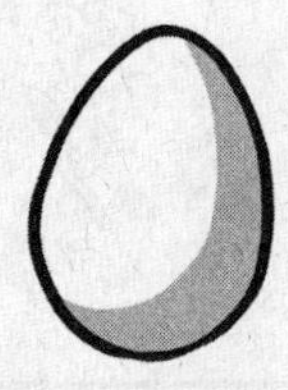
____

10. 
____

11. 

____

12. 
____

13. In the space at the right, draw and label an object that has a mass greater than 1 kilogram.

Find each missing number.

14. ________ g = 6 kg

15. 12,000 g = ______ kg

16. 20 kg = ________ g

Name____ **Lesson 24.6**

Algebra: Rules for Changing Units

You can use a table of measures to help change units.

Metric Table of Measures
Length
10 centimeters (cm) = 1 decimeter (dm)
100 centimeters (cm) = 1 meter (m)
10 decimeters (dm) = 1 meter (m)
1,000 meters (m) = 1 kilometer (km)
Capacity
1,000 milliliters (mL) = 1 liter (L)
Mass
1,000 grams (g) = 1 kilogram (kg)

To change a smaller unit to a larger unit, divide.
To change a larger unit to a smaller unit, multiply.

Copy and complete. Use the Table of Measures.
Change kilograms to grams.
larger unit: ________

☐ g = 9 kg

Look at the table. It take 1,000 grams to equal 1 kilogram.
So, kilogram is the larger unit.

To change a larger unit to a smaller unit, multiply.
Multiply: 9 kg × 1,000.
9 × 1,000 = 9,000
So, 9,000 g = 9 kg.

Copy and complete. Use the Table of Measures.

1. Change decimeters to meters.
   larger unit: ________
   ☐ dm = 5 m

2. Change milliliters to liters.
   larger unit: ________
   3,000 mL = ☐ L

3. Change kilograms to grams.
   larger unit: ________
   ☐ g = 5 kg

4. Change centimeters to meters.
   larger unit: ________
   1,000 cm = ☐ m

5. Change liters to milliliters.
   larger unit: ________
   ☐ mL = 2 L

6. Change kilometers to meters.
   larger unit: ________
   ☐ m = 7 km

Name________________________________ Lesson 24.6

Algebra: Rules for Changing Units

Complete each problem. Use the table of measures.

Metric Table of Measures
Length
10 centimeters = 1 decimeter
100 centimeters = 1 meter
1,000 meters = 1 kilometer
Capacity
1,000 milliliters = 1 liter
Mass
1,000 grams = 1 kilogram

1. Change centimeters to meters.

Larger unit: ________

600 cm = ________ m

2. Change liters to milliliters.

Larger unit: ________

________ mL = 5 L

3. Change decimeters to meters.

Larger unit: ________

________ m = 100 dm

4. Change kilograms to grams.

Larger unit: ________

12 kg = ________ g

5. Change grams to kilograms.

Larger unit: ________

________ kg = 7,000 g

6. Change milliliters to liters.

Larger unit: ________

8 L = ________ mL

7. Change kilometers to meters.

Larger unit: ________

3 km = ________ m

8. Change meters to centimeters.

Larger unit: ________

10 m = ________ cm

Problem Solving and Test Prep

9. A male African elephant can grow to a height of 396 dm. A male Asian elephant can grow to a height of 2,900 cm. How many decimeters taller can a male African elephant grow than a male Asian elephant can grow?

____

10. A mother African elephant averages about 250 dm in height. A newborn baby African elephant averages 89 cm in height. How many centimeters shorter is a newborn baby African elephant than its mother?

____

11. Which expression can you use to find the number of milliliters in 10 liters?

A $1,000 \div 10$ **C** 10×100

B $10,000 \div 10$ **D** $10 \times 1,000$

12. Which is true?

A 3,000 cm = 30 dm

B 4 L = 4,000 mL

C 5 kg = 500 g

D 600 m = 6 km

Name____________________ Lesson 24.7

Problem Solving Workshop Strategy: Compare Strategies

Mr. Tanner's family room is 6 meters wide. How many decimeters wide is the room?

Read to Understand

1. What are you asked to find? ____________________

Plan

2. Can making a table help you solve the problem? Explain. ____________________

3. Can Acting It Out help you solve the problem? Explain. ____________________

Solve

4. Write the answer to the problem in a complete sentence. ____________________

Check

5. Write a multiplication sentence showing your work. ____________________

Solve by making a table or acting it out.

6. Tyler draws a line that is 20 centimeters long. How many decimeters long is the line? ____________________

7. Paris used 5 L of orange juice in a fruit punch recipe. How many milliliters of orange juice is this? ____________________

Name ______________________________ Lesson 24.7

Problem Solving Workshop Strategy: Compare Strategies

Problem Solving Strategy Practice

Make a table or act it out to solve.

1. Belinda found a horseshoe crab on the beach. The horseshoe crab measured 40 cm in length. How many decimeters long was the horseshoe crab?

____

2. Silas made a sand castle with a moat, or ditch, around it. He poured 3 L of seawater into the moat. How many milliliters of seawater did Silas pour into the moat around his sand castle?

____

Mixed Strategy Practice

3. Lucia can carry 4,000 mL of seawater in her pail. How many liters of seawater can Belinda carry in her pail?

____

4. Together, Belinda and Silas collected 40 seashells. Belinda collected 10 more seashells than Silas. How many seashells did they each collect?

____

USE DATA For 5–6, use the graph.

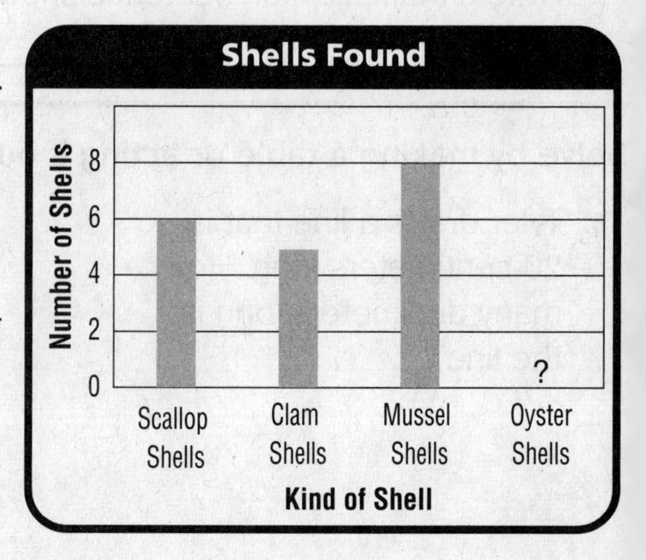

5. If you were to also count the number of oyster shells found, the total number of shells found was 23. How many oyster shells were found?

____

6. Lucia wanted to save two types of shells listed in the graph. What combinations of shells could Lucia choose to save?

____